INTRODUCTION TO
MANUFACTURING
PROCESSES

INTRODUCTION TO
MANUFACTURING PROCESSES

John A. Schey

Professor
Department of Mechanical Engineering
University of Waterloo, Ontario

McGraw-Hill Book Company

New York   St. Louis   San Francisco
Auckland   Bogotá   Düsseldorf
Johannesburg   London   Madrid
Mexico   Montreal   New Delhi
Panama   Paris   São Paulo
Singapore   Sydney   Tokyo   Toronto

INTRODUCTION TO
MANUFACTURING PROCESSES

4 5 6 7 8 9 0   DODO   8 3 2 1 0

This book was set in Helvetica by Progressive Typog-
raphers. The editors were B. J. Clark and Frances A. Neal;
the designer was Joseph Gillians; the cover was
designed by John Hite; the production supervisor was
Robert C. Pedersen. The drawings were done by
ANCO/Boston.
R. R. Donnelley & Sons Company was printer and binder.

Drawing of Nasmyth's shaper on dedication page is
reproduced by permission. British Crown Copyright,
Science Museum, London.

Library of Congress Cataloging in Publication Data

Schey, John A
    Introduction to manufacturing processes.

    Includes index.
    1. Manufacturing processes. I. Title.
TS183.S33   670   76-29004
ISBN 0-07-055274-6

*To Gitta,
John, and
my parents*

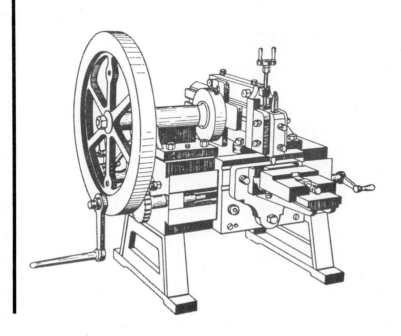

CONTENTS

PREFACE

Manufacturing is the lifeline of all industrialized societies. Without it, few nations could afford many of the amenities that improve the quality of life for their citizens yet all too often are taken for granted.

Despite its obvious importance to society in general and engineering in particular, manufacturing has been neglected in most engineering schools, particularly in North America. The hopeful assumption of the educational theory prevailing in the 1960s was that the young engineer, well versed in the fundamentals, would readily apply technical knowledge to the real problems he encountered in his professional career. Consequently, the more applied—and often purely descriptive—courses, including those on manufacturing, were eliminated from most curricula. It soon became clear, however, that the transition from fundamentals to practice was by no means as natural and easy as had been hoped for, and the practicing engineer often came to question the relevance of the tidy, well-defined solutions characteristic of engineering science courses to the rather messy, open-ended problems of real life. Hence, the integration of basics into courses of applied orientation is again becoming a required part of engineering education.

The task is not easy in any specific field, but it becomes especially demanding in manufacturing. Many processes—developed over the centuries through the perseverance of gifted "natural" engineers—defy exact solutions, and when a solution is found, it may be of little help in dealing with practical problems. This does not mean, however, that fundamentals cannot be applied; on the contrary, they can aid not only in understanding existing processes but also in developing new ones.

Manufacturing is really nothing else but the art and science of transforming materials into usable—and salable—end products. Starting from the premise that this transformation is best achieved with the co-operation, rather than against the objection, of the material, the treatment in this book builds on the interactions between material properties and process conditions. Processes and equipment are described in only as much detail as is essential for an understanding of the more fundamental arguments that finally evolve into a judgment of the advantages and limitations of various processes. Manufacturing technology is inextricably interwoven with concerns for cost and productivity; therefore, an attempt is made to develop also an understanding of the competitive nature of processes, and thus to help in arriving at a reasoned choice of a technologically sound and economically attractive manufacturing process.

There are a great number of books available that discuss in detail the "how" of manufacturing; this book aims at the "why" and "under what conditions." It builds, therefore, on advances made in recent years in various scientific and engineering disciplines, but always with an eye on applicability and relevance to manufacturing processes. Many of the principles discussed are too complex to be expressed mathematically. Others are amenable to quantitative analysis but, within the confines of the present treatment, only the most useful and relevant methods of calculation can be given, and then without proof.

The material is kept to the minimum and can be fitted into an intensive one-term course if some of the more descriptive passages are treated as reading assignments. At the same time, expansion to a full year needs only the added emphasis of a field close to the instructor's interests. It is assumed that the student will have had introductory courses in engineering materials and in mechanics, although most concepts are defined, however briefly, to aid recapitulation and self-study.

Most readers will perhaps agree that the inch system is dead, but one must also admit that it is still a rather vigorous corpse, particularly in North America. To make the transition less painful, conventional USCS and SI units are used side-by-side, except in those tables that would become too awkward to handle.

I would hope that practicing engineers too will find some food for thought in this book, if not in their own specialty, in the ever-broadening fields of alternative and competitive processes and materials.

At this point I would like to express my gratitude to numerous colleagues, among them W. Rostoker (University of Illinois at Chicago Circle), K. J. Schneider (California State Polytechnic University), and Z. Eliezer (University of Texas at Austin) who read the entire text, K. G. Adams, I. Bernhardt, H. W. Kerr, H. R. Martin, P. Niessen, K. F. O'Driscoll, A. Plumtree, D. M. R. Taplin, B. M. E. van der Hoff (University of Waterloo), H. W. Antes (Drexel University), G. F. Bolling (Ford Motor Company), and M. Field (Metcut Research Associates) who read various sections and offered helpful criticism. I am also indebted to G. E. Roberts for his assistance, and to many students, most notably S. M. Woodall, D. L. Agarwalla and J. V. Reid, for their help with problem solutions.   My wife, Gitta, has worked with me for many months, and without her help this book would have remained but an unrealized plan.

John A. Schey

LIST OF SYMBOLS

A	contact area, cross-sectional area (instantaneous)	E_1	specific cutting energy for 1-mm undeformed chip thickness
A_f	final cross-sectional area (at fracture)	F	shear force
		F	frictional force
A_m	average cross-sectional area of barreled cylinder	F_s	shear force in shear plane
		G	shear modulus
A_0	original (starting) cross-sectional area	H_d	drop height in hammer
		I	current
A_t	total cross-sectional area	K	strength coefficient in cold working
A_1	cross-sectional area after deformation	
B	constant $= 2V/f(d_0 - d_1)$	L	length of contact zone between tool and workpiece
C	strength coefficient in hot working	M	mass
C	Taylor constant (cutting speed in fpm for 1 min tool life)	M	atomic weight
		M_T	total torque
C_E	composition of a eutectic	M_s	temperature at which martensite transformation begins
C_L	composition of liquid	
C_0	alloy composition	N	rotational frequency (rpm)
C_S	composition of solid crystals	N_t	number of pieces cut with one tool
C_f	cost associated with floor-to-floor time	P	applied force
		P_R	resultant force in cutting
C_0	cost of starting material per unit weight	P_a	force in axial upsetting of cylinder
		P_b	bending force
C_{pr}	production cost of one piece	P_c	cutting force
C_s	value of scrap per unit weight	P_d	deep-drawing force
C_t	cost of one tool	P_{dr}	drawing force
C_{tp}	tool cost for one workpiece	P_e	extrusion force
E	Young's modulus	P_n	normal force on cutting-tool face
E_c	specific cutting energy	P_r	rolling force
E_h	energy of hammer blow	P_s	shearing force (maximum)

P_t　thrust force on cutting tool
Q　pressure-multiplying factor
Q_a　Q for axial upsetting of a cylinder
Q_c　Q for impression and closed-die forging
Q_{dr}　Q for drawing of wire
Q_e　Q for extrusion (ram pressure)
Q_{fe}　multiplying factor for energy requirement in forging
Q_i　Q for inhomogeneous deformation
Q_p　Q for plane-strain upsetting
R　resistance
R_a　average surface roughness
R_b　radius of bending die
R_d　draw-die radius
R_e　extrusion ratio, reduction ratio
R_f　final radius of bent part
R_m　rate of charge for machine tool
R_{min}　minimum bending radius
R_0　rate of pay of machine tool operator
R_p　punch radius
R_t　maximum surface roughness
S　solubility
T　temperature
T_L　liquidus temperature
T_S　solidus temperature
T_g　glass-transition temperature
T_i　temperature of invariant reaction
T_m　melting point (K)
T_1　intermediate temperature
V　volume
V_t　rate of material removal
W　weight
W_e　weight per unit area removed by an electric current
W_f　final weight
W_0　original weight
Z_f　multiplying factor for feed
Z_v　multiplying factor for cutting speed
a　side dimension of HCP prism
c　height of HCP prism
d　average grain size
d　instantaneous diameter of workpiece
d_m　average diameter of barreled cylinder
d_0　original diameter of workpiece
d_p　punch diameter

e_c　compressive engineering strain
e_f　tensile engineering strain at fracture
e_t　tensile engineering strain
e_u　engineering tensile strain at necking
f　feed (in cutting)
g　gravitational acceleration
h　instantaneous height
h_c　undeformed chip thickness
h_m　mean or average height
h_0　original height or thickness
h_1　height after deformation
h_2　deformed chip thickness, height after deformation
i　ISO tolerance unit
j　current density
k_y　Hall-Petch slope
l　instantaneous length
l_f　final length at fracture
l_0　original length
m　strain-rate sensitivity exponent
n　strain-hardening exponent
n　Taylor exponent
p　interface pressure
p_a　average interface pressure in axial upsetting
p_e　average extrusion (ram) pressure
p_g　partial gas pressure
p_i　indentation pressure
p_m　mean indicated pressure
p_p　average interface pressure in plane-strain upsetting
q　reduction of area in tension test
r　r value (a measure of anisotropy)
r_0　r value in rolling direction
r_{45}　r value at 45° to rolling direction
r_{90}　r value transverse to rolling direction
r_m　mean r value
t_c　net cutting time
t_{ch}　tool-changing time
t_f　floor-to-floor time
t_1　loading and unloading time
t_r　reference tool life
t_s　solidification time
v　velocity of deforming tool
v_s　standard (reference) velocity of cutting

w	instantaneous width	ϵ	natural (logarithmic) strain
w_b	opening of bending die	ϵ_m	mean (average) strain
z	valence	ϵ_u	uniform strain
θ	cutting tool relief angle	$\dot{\epsilon}$	strain rate (instantaneous)
θ	wetting angle	$\dot{\epsilon}_m$	mean (average) strain rate
Φ	inhomogeneity factor in drawing	η	dynamic viscosity
α	solid solution species	η	efficiency
α	die half-angle in extrusion or wire drawing	μ	coefficient of friction
		σ	normal stress; true stress
α_b	angle of bending	σ	strength
α_{max}	angle of acceptance in rolling	σ_{eng}	engineering stress $= P/A_0$
β	constant in Hall-Petch equation	σ_f	flow stress
β	solid solution species	σ_{fm}	mean flow stress
γ	shear strain	σ_0	basic yield stress
γ	interfacial energy	$\sigma_{0.2}$	yield stress for 0.2% plastic strain
γ_{SL}	interfacial energy between solid and liquid	τ	shear stress
		τ_f	flow strength in shear
γ_{SV}	interfacial energy between solid and vapor	τ_i	interface shear strength
		ϕ	shear angle in cutting
γ_c	cutting tool rake angle	ψ	friction angle in cutting
$\dot{\gamma}$	shear strain rate		

INTRODUCTION TO
MANUFACTURING
PROCESSES

1

INTRODUCTION

Manufacturing is a human activity that pervades all phases of life. It has been defined as "the making of goods and articles by hand or, especially, by machinery, often on a large scale and with division of labor." It has been practiced for several thousand years, beginning with the production of stone, ceramic, and metallic articles. The Romans already had factories for the mass production of glassware, and many activities, including mining, metallurgy, and the textile industry, have long employed the principle of division of labor. Nevertheless, much of manufacturing remained an essentially individual activity, practiced by the craftsman and his apprentices, until the advent of the Industrial Revolution at the end of the 18th century. Since then, development has been marked by the rapid succession of milestones in various fields (Table 1.1). The sum of these advances has created an abundance of goods, and the impact of the growth of manufacturing has been felt by everyone.

Manufacturing has often been cast as the villain on the stage of human development. Indeed, the Industrial Revolution began with little concern for the very people who made it possible. Yet the factory was the alternative willingly chosen by the masses seeking to escape their rural existence; the idyllic, pastoral qualities seem to have existed mostly in the imagination of poets, while the reality was burdened with famine and disease. Modern demographic studies show that the misery of rural life prompted people to crowd into cities even before the Industrial Revolution. The growth of industry—which parallels the growth of manufacturing—has since led to undeniable advances, not only in providing an abundance of material possessions, but also in creating the economic basis for genuine improvements in the quality of life.

TABLE 1.1 HISTORICAL DEVELOPMENT OF MANUFACTURING UNIT PROCESSES

Year	Casting	Deformation	Joining	Machining	Ceramics	Plastics
4000 B.C.	Stone, clay molds	Bending, forging (Au, Ag, Cu)	Riveting	Stone, emery, corundum, garnet, flint	Earthenware	Wood, natural fibers
2500	Lost wax (bronze)	Shearing, sheet forming	Soldering, brazing	Drilling, sawing	Glass beads, potter's wheel	
1000		Hot forging (iron), wire drawing (?)	Forge welding, gluing	Iron saws	Glass pressing, glazing	
0 A.D.		Coining (brass), forging (steel)		Turning (wood), filing	Glass blowing	
1000		Wire drawing			Stoneware, porcelain (China)	
1400	Sand casting, cast iron	Water hammer		Sandpaper	Majolica, crystal glass	
1600	Permanent mold	Tinplate can, rolling (Pb)		Wheel lathe (wood)		
1800	Flasks	Deep drawing, rolling (steel), extrusion (Pb)		Boring, turning, screw cutting	Plate glass; porcelain (Germany)	
1850	Centrifugal, molding machine	Steam hammer, tinplate rolling		Shaping, milling, copying lathe	Window glass from slit cylinder	Vulcanization

Year	Casting	Forming	Joining	Machining / Material removal	Glass	Polymers
1875		Rail rolling, continuous rolling		Turret lathe, universal mill, vitrified wheel		Celluloid, rubber extrusion, molding
1900		Tube rolling, extrusion (Cu)	Oxyacetylene, arc welding, electrical resistance welding	Geared lathe, automatic screw machine, hobbing, high-speed steel, synthetic SiC, Al_2O_3		
1920	Die casting	W wire (from powder)	Coated electrode			Bakelite, PVC casting, cold molding, injection molding
1940	Lost wax for engineering parts, resin-bonded sand	Extrusion (steel)	Submerged arc		Automatic bottle making	Acrylics, PMMA, P.E., nylon, synthetic rubber, transfer molding, foaming
1950	Ceramic mold, nodular iron, semiconductors	Cold extrusion (steel)	TIG welding, MIG welding, electroslag	EDM		ABS, silicones, fluorocarbons, polyurethane
1960			Plasma arc	Manufactured diamond	Float-glass	Acetals, polycarbonate, polypropylene

1.1   THE ECONOMIC ROLE OF MANUFACTURING

In the final analysis, there are only two substantial, basic sources of wealth: the material resources of this planet, and the skills and energies its people expend in utilizing these resources. Even though agriculture and mining are of prime importance, they represent only 5–8 percent of the gross national product (GNP) in industrially developed states, while manufacturing claims the largest single share of it. In fact, one could argue that the mark of an industrially developed state is the proportionately large contribution of manufacturing to the national wealth. A review of typical data for industrialized and developing nations is instructive. While there are obvious differences in productivities and production profiles and thus the magnitudes reported are not necessarily directly comparable, the overall conclusion is clear: for nations otherwise similarly endowed with natural resources and human talent, there are large differences in the standard of living, at least as expressed in per capita GDP (gross domestic product = GNP less income from foreign countries). These differences can be, very approximately, related to the contribution of manufacturing to overall economic activity (Fig. 1.1).

FIG. 1.1   Per capita Gross Domestic Product and the growth of manufacturing. (Data extracted from *Encyclopaedia Britannica*, 15th ed., 1974, and *Encyclopaedia Britannica 1975 Book of the Year*, 1975, for countries with populations over 20 million.)

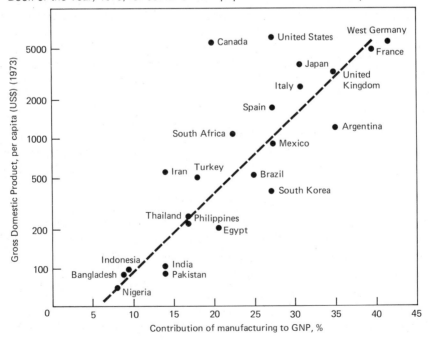

However, it should be noted that, for the economic well-being of a nation, a large manufacturing component is a necessary but not sufficient condition. It is essential that manufacturing should also be competitive, not only locally but—with the shrinking of our world—also on a global basis. Manufacturing productivity is a key issue of economic development, and nations falling behind in this respect find their living standards gradually eroding.

Exceptional natural resources may, for a short time, boost living standards but, judging from experience to date, only manufacturing can create a permanent basis of economic well-being. Manufacturing includes, of course, the production of nondurables and semidurables. In the narrower sense adopted here, we will limit ourselves to the manufacture of "hardware," articles of production and consumption in the durable and semidurable categories.

1.2 MANUFACTURING AS A TECHNICAL ACTIVITY

Without manufacturing, there is little need for engineers and technologists or, indeed, for many of the people who are engaged in supporting activities. Manufacturing is a central function of most technically educated people, although they often fail to recognize it themselves. Every technical decision carries a manufacturing implication: it usually narrows the choices available in both materials and processes, and it has a marked effect on costs. To give but a few examples:

1 In deciding on the operating temperature of a gas turbine the primary concern is efficiency, but this decision also defines what materials will survive in service, and this in turn may severely limit the choices of manufacturing methods.

2 Decisions regarding the safety, pollution level, and gasoline consumption of a passenger automobile also affect its strength, weight, and structure, thus also the materials and manufacturing techniques that are needed.

3 A management-oriented example: a decision to enter a market at a profitable output level hinges on a consideration of available, possible, and economical manufacturing techniques, not only within the company but also at both local and worldwide competitors.

4 In a very general way, decisions must also recognize that at the end of their useful life, manufactured products may be disposed of (often creating problems) or recycled (thus saving raw materials and energy). The redirection of manufacturing processes in such a way as to facilitate recycling instead of disposal is an increasingly evident trend and is going to shape manufacturing activities of the future.

Figure 1.2 is a simplified sketch to show some of the many interactions among various technical activities. It will be noted that many technical activities provide essential inputs to manufacturing processes; at the same time, manufacturing creates many of the machines that are needed for the conversion of energy and raw materials, and for construction, transportation, and communication activities. These industries, together with the individual consumer, dictate the range of products that manufacturing has to provide.

The manufacturing process itself is a series of complex interactions between materials, machines, energy, and people. It begins with the creation of individual parts that will finally make up a finished product. The processes involved in making individual parts are called unit processes, and these form the subject of this book.

While it will be discussed only very briefly, it is obvious and indicated in Fig. 1.2 that the finished parts will have to be made into an end product through assembly operations. In both unit and assembly operations, automation and computer control play increasingly important roles. Technical improvements are meaningful only if costs are also controlled along with quality, and this requires efficient organization of all phases of manufacture. Organizational and management aspects can be neglected or ignored only at the expense of the competitiveness of the entire

FIG. 1.2 Major interactions between manufacturing and other industrial activities.

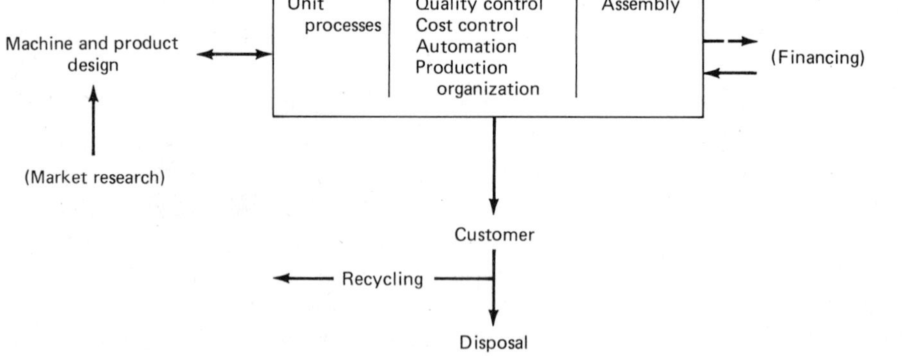

manufacturing activity. It would be futile, however, to attempt to increase productivity merely by organizing a given process to its ultimate degree of efficiency, because a more original mind may have already invented a new process that, by the time it is fully developed, will make the old one noncompetitive and outdated.

The multiple interactions indicated above require that the many engineers and technologists (including manufacturing, materials, mechanical, industrial, and system specialists) who make up a manufacturing team should know enough about manufacturing technologies to arrive at technically correct and economically justifiable process choices. This book addresses itself to active and potential members of such teams.

1.3 INTERACTION BETWEEN DESIGN AND MANUFACTURING

Outside the manufacturing group, but in closest collaboration with it, is the designer, who strives to create a product that will:

1 Fulfill a function; that is, operate satisfactorily over the expected life of the product. This function must be fulfilled at a reasonable cost, therefore neither excess performance (tolerances, compositional specifications, etc.) nor excess life are needed (although the latter aspect is undergoing some modifications on account of energy considerations). In general, it is found that a product satisfying the minimum requirements can be produced at some minimum cost. Quality and performance can often be increased and thereby the selling price quite substantially raised with relatively little increase in the cost of manufacturing. Further improvements are likely to lead to much higher manufacturing cost and only marginally increase customer appeal; thus the selling price cannot be raised proportionately. Consequently, there is always a point beyond which quality and performance should not be improved. This point is determined through the cooperation of design and manufacturing teams.

2 Present a desirable appearance and therefore appeal to the customer. This is usually assured by the industrial designer whose work must be closely coordinated with manufacturing, since a seemingly minor change in shape or surface appearance may often present (or remove) enormous manufacturing problems.

3 Serve the customer; that is, the man-machine relationship must be observed.

4 Be easily maintained over its intended life. At the end of its life it must be readily disposed of or recycled. This means that not only must ease of assembly be considered, but ease of disassembly or recycling as well.

The process of design involves many steps and no generalizations can be made about their manufacturing consequences. The most persistent and ever-present problem is, however, the choice of tolerances and surface finishes. All too often, this choice is more an afterthought than the crucial decision that exercises an overwhelming influence over process selection and cost. Because of its importance, this topic will be discussed in Chap. 11.

1.4 MATERIALS IN MANUFACTURING

Manufacturing is essentially the art of transforming raw materials or, very often, semifinished products into goods and articles (whether they are means of production or articles of consumption).

The designer chooses materials to satisfy criteria which can be, very broadly, grouped into three sets (Table 1.2).

All the performance criteria must be satisfied, however, at a cost that is commensurate with the willingness of the potential customer to pay for the finished article. Since raw material costs are usually a smaller part of the total cost, all decisions regarding materials of construction must be made with an eye on the feasibility of manufacture.

Voluminous handbooks exist which show various properties of materials, usually classified according to composition. Since there are thousands of potential materials to be considered, such classifications are of

TABLE 1.2 PERFORMANCE CRITERIA OF MATERIALS

Physical and chemical properties	Mechanical properties	Shape and size
Electrical and thermal (conductivity, insulation)	Strength (static, dynamic, fatigue)	Geometry
	Ductility	Dimensions
Chemical reactivity (corrosion)	Toughness (crack resistance)	Tolerances
		Surface finish
Optical (color, transparency, fluorescence)	Wear resistance	Surface appearance
Magnetic		Weight
Change of state (softening, melting)		

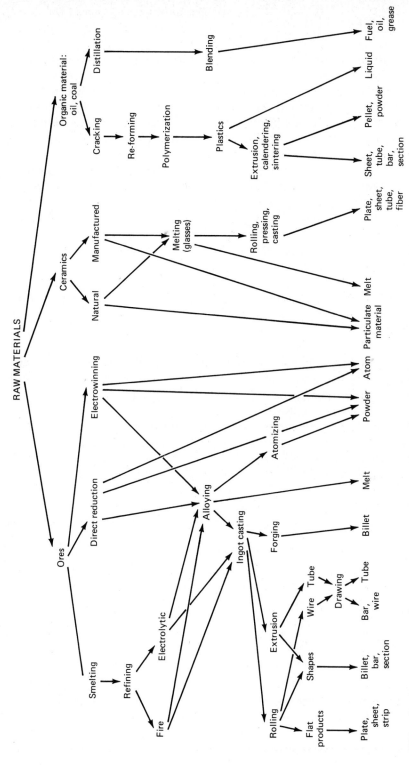

FIG. 1.3 Alternative processing routes for the starting materials of manufacturing unit processes.

little value unless some more generally applicable guidelines can be brought to bear on the problem. Such guidelines allow the designer to consider the broadest possible group or groups of materials without prematurely restricting the choice to one or two materials and thus immediately limiting the possibilities of manufacture. A premature decision may set a cost that will make the final product uneconomical or noncompetitive, no matter how ingenious the design may be.

Because manufacturing processes aim at creating a usable end product or component, the starting material is often the result of prior operations. While many possible routes of primary processing are available, even the few indicated in Fig. 1.3 show that the same starting material may often be obtained through a number of alternative routes—some of them much shorter than others—from raw material to semifabricated product. It would, however, be too hasty to conclude that the more complex processes are necessarily more expensive. Very often, economy is a matter of scale; thus, it is still possible to buy steel strip at a lower price than powder, partly because of the vast quantities produced in strip form.

Metals are the most generally employed engineering materials, and the growth of their production (especially that of steel) has often been taken as an indicator of industrial development. With the increasing so-

TABLE 1.3 SELECTED PRODUCTION DATA AND PROPERTIES FOR MATERIALS OF MANUFACTURING

Material	World production,* 10^6 metric tons	Melting point, °C	Density, kg/m³	Elastic modulus, 10^3 MPa	Resistivity at 20°C, 10^{-8} Ω·m	Thermal conductivity at 20°C, J/m·s·°C
Iron	634 (steel)	1536	7900	210	9.7	75
Aluminum	11.0	660	2700	70	2.7	240
Copper	7.0	1083	8900	122	1.7	400
Zinc	5.2	419	7100	90	5.9	115
Lead	3.6	327	11300	16	21.0	35
Nickel	0.6	1455	8900	210	6.8	90
Magnesium	0.26	649	1700	44	4.0	160
Tin	0.2	232	5800	42	11.0	65
Titanium	0.06	1670	4500	106	6.8	25
Plastics	43.3†		900–2200	3–10	(10^{12})	0.2

* 1972 data, compiled from *Metal Statistics 1974*, American Metal Market, Fairchild Publications, Inc., New York, 1974.

† 1973 data, from *Encyclopaedia Britannica Book of the Year, 1975*, Chicago, 1975.

phistication of many products and the increasing use of plastics, these re-lationships are not valid any more, particularly in the industrialized states. Nevertheless, metals remain indispensable. Steel still represents an overwhelming portion of total metal production (Table 1.3), but other metals offer unique properties and some of them (notably Mg and Ti) would become much more important if they could be extracted with a smaller energy outlay. Polymers (plastics) play an increasingly important role and, while some readjustment is necessary during raw material (oil) shortages, their position is not likely to suffer greatly. Although they are not shown in Table 1.3, the various natural and manufactured ceramics represent a vast source of engineering materials, many of them used also in the manufacturing industries.

1.5 SCOPE AND PURPOSE OF BOOK

The number of manufacturing processes in existence defies enumeration, let alone description, in a single volume. There are a number of encyclo-pedic books in which details on individual processes can be found.

The first step in dealing with manufacturing processes—for both those who practice them and those who make use of them—is an under-standing of basic principles underlying the myriad individual techniques. These principles are discussed with special reference to metals in Chap. 2.

The application of principles to the major classes of unit processes is shown in Fig. 1.4 and explained in Chaps. 3 through 9. The reader inter-ested in further details or theories of processes will find ample material in the readings suggested at the end of each chapter. These readings were chosen to alleviate the dilemma of depth vs. breadth of treatment. A fully quantitative treatment would require excessive length and could easily obscure the larger issues; a fully qualitative treatment would give little guidance in process selection. The compromise adopted here attempts to retain all the sound scientific principles that must be brought to bear on the subject; in particular, it attempts to emphasize the mutual constraints exerted by materials and processes on each other (this point is elaborated in Chap. 11). As much quantitative information is given as is essential for a well-informed process choice, without proofs or derivations that could—and do—fill up a library of specialized books.

Unit processes on their own do not constitute manufacturing: thus some broader aspects of manufacturing are touched upon in Chap. 10.

The purpose of this treatment is to give a broad overview of manufac-turing processes. It presumes a course in (or equivalent familiarity with) the properties of materials. It should serve as an introduction to the sub-ject not only for the manufacturing specialist but also for mechanical,

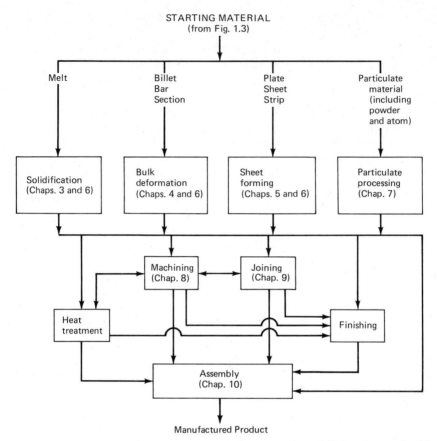

FIG. 1.4   Major classes of manufacturing unit processes and their coverage in this volume.

materials, industrial, and systems engineers and technologists whose activities are largely directed at some facet of manufacturing.   Learning is, indeed, a lifelong process, and suggestions for further reading will be found below, in a list representing only a small fraction of the vast literature and, partly for space limitations, only that in English.

Further Reading

HISTORY:

AITCHISON, L.: *A History of Metals,* Macdonald and Evans, London, 1960.

DERRY, T. K., and T. I. WILLIAMS: *A Short History of Technology,* Oxford University Press, London, 1961.

GENERAL MANUFACTURING TEXTBOOKS:

BEGEMAN, M. L., and B. H. AMSTEAD: *Manufacturing Processes,* Wiley, New York, 1969.

CLARK, D. S.: *Engineering Materials and Processes,* 3d ed., International Textbook Co., Scranton, Pa., 1959.

DE GARMO, E. P.: *Materials and Processes in Manufacturing,* 3d ed., Macmillan, Toronto, 1963.

DOYLE, L. E., and others: *Manufacturing Processes and Materials for Engineers,* 2d ed., Prentice-Hall, Englewood Cliffs, N.J., 1961.

LINDBERG, R. A.: *Materials and Manufacturing Technology,* Allyn and Bacon, Boston, 1968.

MOORE, H. D., and D. R. KIBBEY: *Manufacturing: Materials and Processes,* Grid, Inc., Columbus, Ohio, 1975.

PATTON, W. J.: *Modern Manufacturing Processes and Engineering,* Prentice-Hall, Englewood Cliffs, N.J., 1970.

RHINE, C. R.: *Machine Tools and Processes for Engineers,* McGraw-Hill, New York, 1971.

GENERAL TEXTS WITH MORE MATHEMATICAL TREATMENT:

ALEXANDER, J. M., and R. C. BREWER: *Manufacturing Properties of Materials,* Van Nostrand, London, 1963.

DATSKO, J.: *Material Properties and Manufacturing Processes,* Wiley, New York, 1966.

RADFORD, J. D., and D. B. RICHARDSON: *Production Engineering Technology,* 2d ed., Macmillan, London, 1974.

THOMAS, G. G.: *Production Technology,* Oxford University Press, London, 1970.

GENERAL COVERAGE WITH EMPHASIS ON DESIGN:

NIEBEL, B. W., and A. B. DRAPER: *Product Design and Process Engineering,* McGraw-Hill, New York, 1974.

TRUCKS, H. E.: *Designing for Economical Production,* Society of Manufacturing Engineers, Dearborn, Mich., 1974.

GENERAL COVERAGE WITH DETAILS OF PROCESSES:

BAUMEISTER, T. (ed.): *Marks' Standard Handbook for Mechanical Engineers,* 7th ed., McGraw-Hill, New York, 1967.

BOLZ, R. W. (ed.): *ASME Handbook,* vol. 4, *Metals Engineering—Processes,* McGraw-Hill, New York, 1958.

BOLZ, R. W.: *Production Processes: the Productivity Handbook,* Conquest Publications, Novelty, Ohio, 1974.

DALLAS, D. B. (ed.): *Tool and Manufacturing Engineers Handbook,* 3d ed., Society of Manufacturing Engineers, McGraw-Hill, New York, 1976.

HORTON, H. L., (ed.): *Machinery's Handbook,* 20th ed., Industrial Press, New York, 1975.

STANDEN, A., (ed.): *Kirk-Othmer Encyclopedia of Chemical Technology,* 2d ed., Interscience, New York, 1963.

CONFERENCE PROCEEDINGS (UP-TO-DATE COVERAGE):

COLWELL, L. V., et al. (eds.): *International Conference: Manufacturing Technology,* American Society of Tool and Manufacturing Engineers, Dearborn, Mich., 1967.

SHAW, M. C., et al. (eds.): *International Research in Production Engineering,* American Society of Mechanical Engineers, New York, 1963.

TOBIAS, S. A., and F. KOENIGSBERGER, (eds.): *Advances in Machine Tool Design and Research,* proceedings of annual conferences since 1960.

Proceedings of the North-American Metalworking Research Conference, Mc-Master University, 1973: University of Wisconsin–Madison, 1974; Carnegie-Mellon University, 1975; Battelle-Columbus, 1976.

SELECTED JOURNALS WITH GENERAL COVERAGE:
American Machinist
International Journal of Production Research
International Journal of Machine Tool Design and Research
Machinery
Machinery and Production Engineering
Manufacturing Engineering
Manufacturing Engineering Transactions
Metal Progress
Production
Production Engineer (contains abstracts of papers in all fields)

2

MANUFACTURING PROPERTIES OF METALS AND ALLOYS

There are several thousand engineering materials (metals, alloys, polymers, and ceramics) in everyday engineering use. Their behavior during processing in the liquid or solid state shows an enormous variability, and a catalog of their manufacturing properties would be a bewildering encyclopedia. There are, however, some basic principles one can identify which govern the behavior of a large majority of materials. By necessity, such rules are generalizations, and exceptions to them will always be found. Nevertheless, these rules can give a broad first indication of what materials and processes may possibly fulfill the requirements of the final product, and thus allow a more intelligent design of products and processes. These principles will be first identified for metals and their alloys, and extended to ceramics and polymers in a later treatment of their processing technologies (Chaps. 6 and 7). The discussion that follows will appear to be a review of known material properties to many readers but, hopefully, the bias toward the manufacturing viewpoint will be evident.

2.1 PURE METALS

While absolutely pure metals are seldom used, they provide a convenient frame of reference for a discussion of some basic properties.

2.1.1 Melting and Solidification

A very important property of pure metals is that upon heating they melt at a characteristic temperature, the *melting point.* For convenience, this is expressed in conventional reference units such as degrees Fahrenheit or Celsius, although it must be recognized that for the metal itself the only reference points (unless there is a phase transformation in the solid state) are absolute zero and the melting point. When the latter is expressed in absolute degrees Kelvin, this T_m gives the end point of the *homologous temperature scale,* which allows normalization of metal behavior (Fig. 2.1).

In the molten state, the atoms of metals occupy statistically random positions, although some temporary short-range order—a remnant of their *crystalline structure*—exists even then. On cooling to the melting point, the atoms occupy sites dictated by the crystalline structure. First, stable

FIG. 2.1 Characteristics of metal behavior related to the homologous temperature, with pure copper and lead as examples.

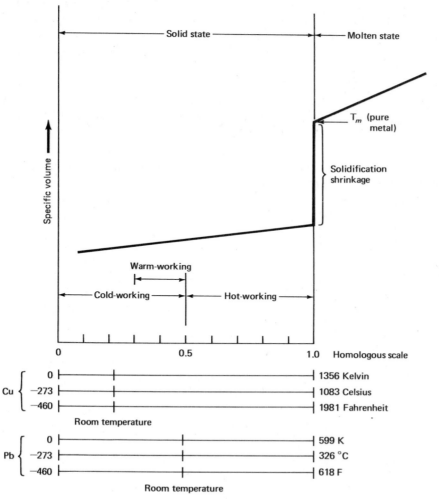

nuclei form, either from the pure melt (*homogeneous nucleation*) after cooling below the melting point (*undercooling*) or, more likely, by growing onto preexisting crystalline nuclei of different composition (*heterogeneous nucleation*), including the mold walls. The number of nuclei increases with undercooling and the presence of heterogeneous nuclei; therefore, the melt that is in direct contact with the surface of a mold and is thus cooled at the fastest rate solidifies with a fine, *equiaxed* crystalline *structure* (Fig. 2.2). Further solidification usually proceeds by the growth of favorably oriented nuclei, in the direction of heat extraction, since the latent heat of fusion must be continuously removed. This leads to the often-observed *columnar structure,* which may be very difficult to suppress except by *inoculation* of the melt with a large number of nuclei. The center of a solidified casting may well exhibit an equiaxed structure because heat extraction there is slower and almost omnidirectional. Directional growth may be arrested and a largely equiaxed structure secured by mechanical vibration that breaks the growing crystals. Since in pure metals there are no compositional differences, the only features visible under the microscope are *grain boundaries* (Fig. 2.2*b*).

Since solidification means rearrangement of the atoms into an ordered, crystalline structure, the volume shrinks (with a few exceptions such as Bi) and, unless further liquid is supplied, this *contraction* (Fig. 2.1) leads to the formation of a *shrinkage cavity* (*pipe,* Fig. 2.2). Shrinkage ef-

FIG. 2.2 Solidification of a pure metal showing (*a*) an advancing solidification front and (*b*) the resulting microstructure (across the columnar grains).

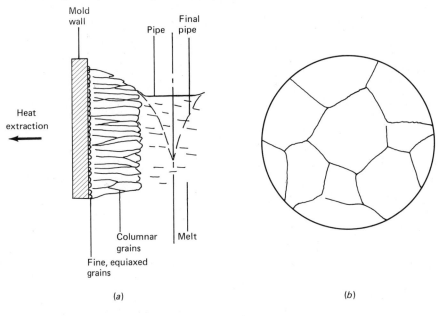

(*a*)

(*b*)

fects can be minimized and a fine-grained structure obtained by: (1) applying high *pressures* during solidification; (2) actually deforming the material while it solidifies (*melt-forging*); or (3) allowing the melt to cool
under continuous *agitation* prior to injecting it into the mold.   Once
solidified, the metal shrinks at a lower rate because of the decreasing thermal excursions of atoms.

2.1.2   Deformation

A solidified metal of crystalline structure is capable of changing shape
without fracture; that is, it is deformable.   When observed under an optical
microscope, *deformation* seems to take place by the slip of adjacent
crystal zones (Fig. 2.3*a*).   At high magnifications each *slip zone* appears
composed of many small steps, indicating displacement along preferred
slip planes (Fig. 2.3*b*).   On the atomic scale, slip takes place not by the
massive movement of entire adjacent crystal zones, but by the movement
of line defects (*dislocations*) in the preferred slip planes.   In the simplest
view, a dislocation could be regarded as an extra line or plane of atoms inserted into the structure (*edge dislocation,* Fig. 2.3*c*); thus it is only necessary to move this extra line of atoms along the slip plane instead of moving
hundreds of thousands of atoms at the same time over the entire slip surface.   Many of the deformation characteristics of metals can be inter-
preted by considering the ease with which these dislocations can move
and by considering obstacles that may impede or arrest their movements.
It is important to note that *shear stresses* must reach a critical value on the
slip planes before deformation can commence.

One might expect that dislocation movement (slip) should be easier
on planes that give the smoothest movement, the least bumpy ride; indeed, one finds that slip takes place most readily in the most closely
packed planes in the closest-packed crystallographic directions.   The
largest group of metals belongs to the face-centered cubic (FCC) structure
(Fig. 2.4*a*) in which there are four equivalent *closely packed planes* (the
{111} octahedral planes) with three equivalent *slip directions* ⟨110⟩, giving
a total of 12 independent *slip systems* (i.e., combinations of slip planes and
directions).   Thus if slip is limited on one plane because dislocations are
arrested, there is always a likelihood that some other slip system will be
oriented in the direction of the maximum deforming shear stress.   Thus
we can conclude that FCC metals should be deformable, essentially at all
temperatures.   Indeed, this is characteristic of such metals as Pb, Al, Cu,
and Ni.   This also applies to iron above 906°C, the transformation temperature to the FCC form (the so-called γ-iron).

In the body-centered cubic (BCC) structure, one cannot readily identify obviously close-packed planes, but a clearly closest-packed direction
is found in the body diagonal (Fig. 2.4*b*).   Therefore, these crystals slip in

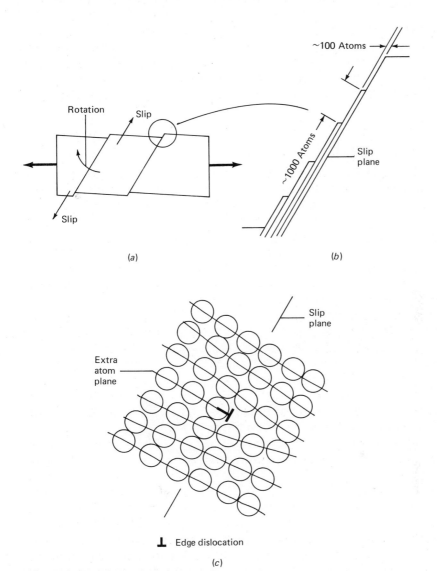

FIG. 2.3 Deformation of a single crystal in tension (*a*) on the macro scale, (*b*) at high magnification, and (*c*) on the atomic scale.

systems containing various planes that have the body diagonal $\langle 111 \rangle$ as a common slip direction, rather like a bunch of pencils that is deformed by sliding the pencils along their axes. This so-called *pencil-slip* allows extensive deformation, for example in α-iron (below 906°C) and in β-titanium (above 880°C).

The deformation of hexagonal close-packed (HCP) structures is governed by the ratio of height to side dimensions, the *c/a ratio* (Fig. 2.4*c*). In

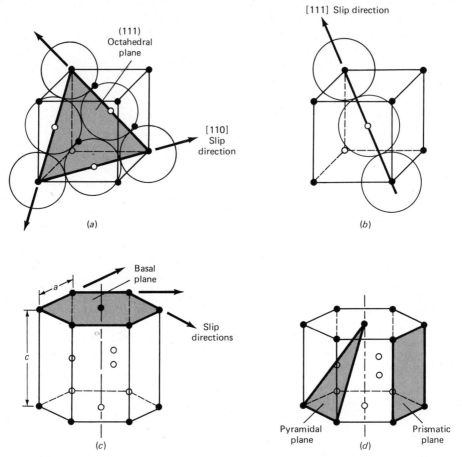

FIG. 2.4 Slip planes and directions in (a) FCC, (b) BCC, (c) hexagonal, large c/a ratio and (d) hexagonal, small c/a ratio structures.

an ideal HCP structure this ratio would be 1.6333. Some metals show a larger ratio; i.e., the basal planes are more widely separated. Slip then occurs only in these planes along the three equivalent closest-packed directions (*basal slip*, Fig. 2.4c, as in zinc c/a = 1.856). When the c/a ratio is less than the ideal, the atoms of the basal planes are effectively squashed into each other and slip is now prevented here; the material will choose slip planes either along the side of the prism or on a pyramidal surface (*prismatic* or *pyramidal slip*, Fig. 2.4d). The prime example of this behavior is α-titanium (c/a = 1.587 at temperatures below 880°C). A metal of close to the theoretical c/a ratio cannot slide readily along any of these planes and it is usually necessary to raise the temperature somewhat so that the increased freedom of atomic movement brings a number of slip systems into play. This is most clearly evidenced by magnesium ($c/a = 1.624$), which can take very little deformation at room temperature but deforms readily when heated to 220°C. Frequently, deformation is aided in HCP materials by *twinning* (which occurs when a part of the crystal flips over into a mirror-image position), which brings more slip planes into a favorable direction relative to the maximum shear stress.

2.1.3 Textures (Anisotropy)

Parts of crystals not only slide relative to each other during deformation, but the slip plane also rotates, into the direction of straining in tension and across the direction of straining in compression (Fig. 2.3*a*). This has important consequences in *polycrystalline* materials, particularly when only a limited number of slip systems is available. After deformation, there will be a noticeable alignment of crystallographic orientations in many of the crystals; thus the polycrystalline structure will show some of the *directional properties* typical of single crystals. The development of such *texture* is evident in variations of the elastic modulus, yield stress, elongation, and many other properties with the direction of testing, but a most convenient characterization is available through the measurement of strains in tensile tests on sheet-metal specimens.

In an *isotropic* (nontextured) *material*, deformations are identical in all directions. Thus, a sheet specimen would show the same width strain ϵ_w and thickness strain ϵ_t (and, by inference, also length strain ϵ_l, since the sum of strains is always zero when expressed as natural or logarithmic strain). Their ratio, or *r value*

$$r = \frac{\epsilon_w}{\epsilon_t} \qquad (2.1)$$

would be unity (Fig. 2.5*a*). It would not matter whether the specimen was cut in the rolling direction, across it, or at some intermediate angle (Fig. 2.5*b*); in an isotropic material

$$r_0 = r_{90} = r_{45} = 1$$

It is conceivable that the *r* values might vary in relation to the rolling direction, but would still average at unity

$$r_m = \frac{r_0 + r_{90} + 2r_{45}}{4}$$

This is denoted as *planar anisotropy* and would lead to such problems as earing in deep drawing (Sec. 5.5.2).

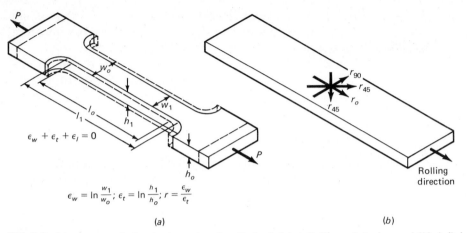

$$\epsilon_w + \epsilon_t + \epsilon_l = 0$$

$$\epsilon_w = \ln\frac{w_1}{w_o}; \; \epsilon_t = \ln\frac{h_1}{h_o}; \; r = \frac{\epsilon_w}{\epsilon_t}$$

(a)   (b)

FIG. 2.5 Measurement of anisotropy in a tensile test: (a) definition of strains and (b) definition of test directions.

Or, r values may be identical in all directions but deviate from unity

$$r_0 = r_{90} = r_{45} \gtrless 1$$

This is then described as *normal anisotropy,* because deformation of the test specimen in thickness (normal to the sheet surface) is greater or less than that in width.

This is most readily evident with hexagonal materials in which the limited number of slip systems leads to the development of a texture after relatively small (20–30 percent) deformations, with most of the basal planes aligned perpendicular to the application of the rolling force, that is, with basal planes almost parallel to the sheet surface. When a tensile test specimen cut from such a sheet is elongated in a tensile-testing machine, deformation will be highly anisotropic. In materials with a high *c/a* ratio sliding is limited to the basal planes (Fig. 2.4c), thus the thickness of the sheet is reduced while its width is hardly affected (just as a card pack can be elongated by sliding the cards over each other, Fig. 2.6a). The r value becomes very small, typically 0.2 in zinc. The deformation of a tensile specimen cut from a material with a low *c/a* ratio shows a dramatically different behavior. Since slip now takes place on prismatic and/or pyramidal planes (Fig. 2.4d), the sheet thickness is hardly reduced at all; instead, most of the deformation takes place by rearrangement of the hexagonal prisms, leading to a marked reduction in the width of the specimen (Fig. 2.6b). The r value could, theoretically, reach infinity, but in practice seldom exceeds 6, the value for titanium.

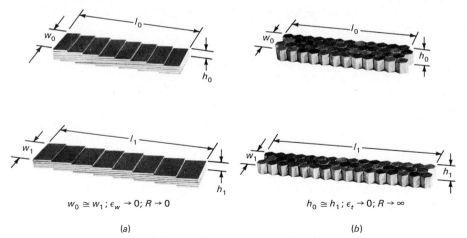

$$w_0 \cong w_1; \epsilon_w \to 0; R \to 0 \qquad\qquad h_0 \cong h_1; \epsilon_t \to 0; R \to \infty$$

(a)   (b)

FIG. 2.6 Variation of r value in the deformation of sheets with hexagonal crystal structure: (a) basal slip with high c/a ratio and (b) prismatic slip with low c/a ratio.

Metals with the FCC structure possess a great many equivalent slip systems; therefore only much later (typically after more than 50 percent reduction) do they develop a texture which will be less marked in its effect on subsequent deformation in a tensile test.   A completely randomly oriented polycrystalline FCC material behaves as though it were isotropic ($r = 1$).   The common slip direction in BCC materials can be exploited to give r values ranging from 1 to 2.   This is important for deformation processes such as deep drawing (Chap. 5).

Normal and planar anisotropy may, of course, appear simultaneously.

2.1.4   Effect of Grain Size

Grain boundaries may be sources of dislocations, but they also present barriers to dislocation propagation.   Therefore, it is generally found that, below $0.5T_m$, the strength of a material increases with decreasing *grain size* according to the Hall-Petch[1] relationship

$$\sigma = \sigma_0 + k_y d^{-1/2} \tag{2.2}$$

where d is the average grain size of the material, σ_0 a basic yield stress, and k_y the Hall-Petch slope.   In some metals an even better agreement is found between strength and the size of *subgrains* (relatively strain-free subunits

---

[1] E. O. Hall, *Proc. Phys. Soc. London, 64B:*747, 1951; N. J. Petch, *J. Iron Steel Inst. London, 173:*25, 1953.

within the larger crystal, arranged with a slight crystallographic misorientation relative to each other).

Since grain-size effects are found in all metals and alloys, control of grain size is one of the most powerful tools of manufacturing for improving both manufacturing and service properties of materials.

2.1.5 Strain Hardening

The effects of stress on engineering materials are most readily observed in the tensile test (Fig. 2.7).

Deformation is initially all elastic and, if the specimen were unloaded in this stress range, all deformation would be recovered. The engineering tensile strain e_t increases linearly with the imposed stress σ and is smaller for a material with a high elastic modulus (Young's modulus) E:

$$e_t = \frac{l - l_0}{l_0} = \frac{\sigma}{E} \qquad\qquad (2.3)$$

where l_0 and l are the gage lengths before and after deformation, respectively.

FIG. 2.7   Force-displacement (or engineering stress-strain) curve of a strain-hardening material.

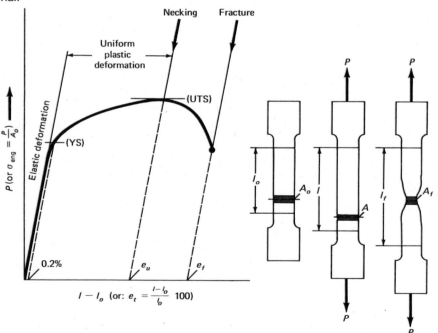

At some higher stress, plastic deformation begins. When the metal is deformed at a relatively low temperature, typically below one half of the absolute melting point ($T < 0.5T_\mathrm{m}$, Fig. 2.1), the movement of dislocations is limited mostly to their own slip plane. Obstacles such as grain boundaries, foreign atoms, and other dislocations impede or arrest their movement. This causes strain hardening: it takes a greater stress to make other dislocations move, and to generate new dislocations—in general, to maintain deformation.

For practical purposes, the stress sufficient to cause an easily measured strain is regarded as the yield stress (thus, at 0.2 percent permanent elongation, the stress is denoted $\sigma_{0.2}$). The cross section of the specimen reduces uniformly along its whole length, and strain hardening causes an increase in the force P required to maintain deformation, even though the cross-sectional area decreases from its original A_0 value to smaller A values. At a critical deformation level, strain hardening cannot counterbalance the loss of strength resulting from the ever-decreasing cross-sectional area, and a *neck* forms at the weakest point. Thereafter the imposed force P declines, while the deformation is concentrated in the already necked zone. Finally, fracture occurs at the minimum cross section A_f. The so-called *engineering* or *conventional stress-strain curve* is constructed purely as a matter of convenience by taking the instantaneous P force and dividing it by the original cross-sectional area A_0, entirely ignoring the uniform and localized deformation

$$\sigma_\mathrm{eng} = \frac{P}{A_0} \tag{2.4}$$

The maximum "stress" thus obtained is called the *ultimate tensile strength* (UTS) and, while it is the result of a technically unacceptable calculation, it is widely used in practice as a general guide value. As we will see, it also happens to give a fair estimate of the average flow stress required to maintain plastic deformation.

The technically correct way of calculating the *true stress* from the instantaneous force P by dividing by the instantaneous minimum cross-sectional area A,

$$\sigma = \frac{P}{A} \tag{2.5}$$

results in a *true stress-strain curve*. The strain is often calculated as *natural* or *logarithmic strain*

$$\epsilon = \ln\frac{l}{l_0} = \ln\frac{A_0}{A} \tag{2.6}$$

It is often found that past the initial yielding range, say $\epsilon > 0.05$, such a stress-strain curve (Fig. 2.8a) obeys a *power law,*

$$\sigma = K\epsilon^n \tag{2.7}$$

where K is the strength coefficient and n is the strain-hardening exponent. They are readily found by replotting the stress-strain curve on log-log paper: K is the stress at a strain of unity and n is the slope (measured on a linear scale) of the resulting straight line (Fig. 2.8b).

2.1.6 Ductility

The strain-hardening exponent (*n value*) is important in judging the *formability* of a material, because a large n value means greater resistance to necking by assuring that any incipient neck that may form will harden more rapidly than adjacent, nonnecked portions. It can be shown that, for a material that obeys the power law, the n value is numerically identical to the *uniform strain* ϵ_u, that is, the natural strain calculated for the onset of necking (Fig. 2.9a and b). A material with a high n value will be more suitable for forming operations involving tensile deformation, since the appearance of a neck on the deformed workpiece would be objectionable (Chap. 5).

The final engineering strain

$$e_f = \frac{l_f - l_0}{l_0} \tag{2.8}$$

FIG. 2.8   True stress-strain curve (*a*) of a material obeying the power law of strain hardening and (*b*) plot on log-log scale.

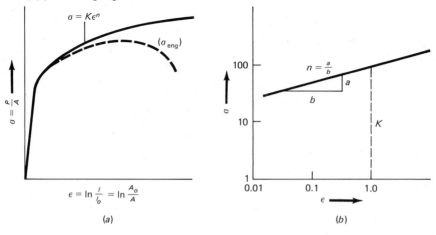

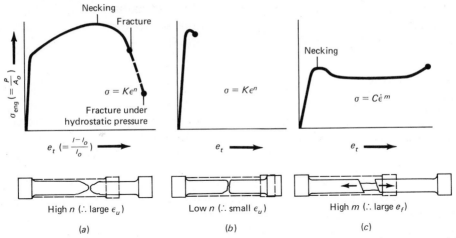

FIG. 2.9 Elongation of materials with (a) high and (b) low strain-hardening rates, and (c) high strain-rate sensitivity.

is commonly called *elongation.* As visible in Fig. 2.7, it includes both uniform elongation and localized elongation due to necking. It is therefore a composite value that is also sensitive to the *gage length* employed, since a shorter gage length will make the same material appear to have a larger elongation. For this reason, the gage length over which total elongation is measured must be stated, otherwise total elongation, a readily measurable quality-control indicator, would lose its meaning.

Another measure of the ductility of a material is the *reduction of area q* measured on the fractured test piece,

$$q = \frac{A_0 - A_f}{A_0} \tag{2.9}$$

As long as elongation is uniform, the *stress state* is *uniaxial.* When necking begins, deformation is localized and the smallest cross section is subjected to a complex stress state because contraction of the neck material is hindered by the neighboring thicker, nondeforming sections. This imposes radial tensile stresses on the neck material and thus, with the applied axial tension, creates a *triaxial tensile stress state* which endeavors to tear the material apart. At points of weakness (e.g., inclusions, voids) cavities begin to form which gradually grow and finally interconnect to result in *fracture.* Larger reduction in area, therefore, signifies greater ability to resist triaxial tensile stresses.

If the tensile test is conducted inside a chamber filled with pressurized fluid, a triaxial compressive stress (*hydrostatic pressure*) is superimposed on the test specimen. Yielding and necking of the material remain unchanged but void generation is suppressed and necking continues further before fracture (q increases, Fig. 2.9a). This will become of great importance in bulk deformation processes (Chap. 4).

2.1.7 Cold-Working

The ability of metals to harden during plastic deformation is often utilized in strengthening these metals by *cold-working,* for example, by rolling or drawing. Prior cold work causes the yield stress and UTS to rise in subsequent tensile testing (Fig. 2.10), while the increased dislocation density is reflected in much-reduced elongation. A composite curve that shows these changes (Fig. 2.11a) as a function of prior cold work reveals that the yield stress rises more rapidly and approaches the UTS, while ductility (expressed here by total elongation) drops fairly suddenly after a limited amount of prior cold work. The microstructure changes too; crystals (grains) become elongated in the direction of major deformation and they may develop directional properties (Sec. 2.1.3).

Many materials retain useful levels of ductility even in the strain-hardened condition; therefore cold-working is an important manufacturing approach to producing higher-strength materials.

2.1.8 Restoration Processes (Softening)

The effect of cold work can be mitigated or even eliminated by keeping the material at such higher temperatures that the increased thermal vibration of atoms allows greater dislocation mobility. At temperatures of typically $0.3–0.5T_m$, dislocations are mobile enough to form regular arrays and even annihilate each other on the same slip planes, giving rise to so-called *recovery*. This takes time (Fig. 2.11b); even though it does not change the grain structure, it restores some of the original softness (lower strength and greater ductility)—but, in most metals, not to the full extent.

Full *restoration* of the original soft condition is achieved by heating to temperatures above $0.5T_m$ when new and relatively low dislocation-density grains form (*nucleate*) and grow until all the cold-worked structure is *recrystallized.* Such *annealing* involves *diffusion* and is, therefore, greatly temperature- and also time-dependent (Fig. 2.11c). An equiaxed structure normally results, although a texture may be retained or developed (*annealing texture*). The temperature of $0.5T_m$ should be taken only as a very rough guide, since even minor amounts of alloying elements can substantially delay the formation of new grains and thus raise the recrystallization temperature.

In some metals restoration processes increase the ductility more than they decrease strength. Therefore, it is possible to control the final properties of the deformed product by first taking cold work to a fairly severe

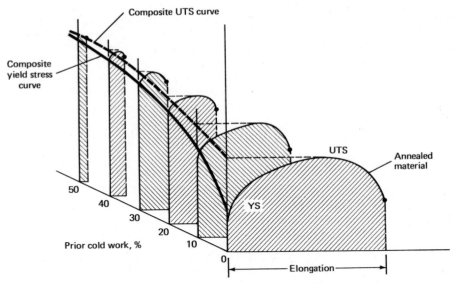

FIG. 2.10 Strain hardening due to cold working increases yield strength and UTS, and decreases elongation (and reduction of area).

FIG. 2.11 Changes in strength, ductility, and microstructure during (a) cold working, (b) recovery, and (c) recrystallization.

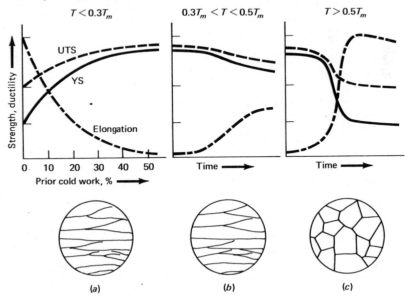

stage and following with a *recovery anneal* which restores much of the ductility without greatly lowering the strength (Fig. 2.12).

Cold-working followed by *recrystallization annealing* also affords the opportunity of controlling material properties by achieving an optimum grain size. The effect may be summarized in a three-dimensional diagram (Fig. 2.13). Recrystallization begins at lower temperatures with increasing cold work (note that the temperature increases toward the origin); grain size increases with increasing temperature, but heavier cold work leads to the formation of many more nuclei and therefore to a much smaller grain size. There is, of course, no recrystallization possible if cold work is zero, and the original grain size is retained. However, slightly increased dislocation densities resulting from very slight (say 2–4 percent) cold work encourage the formation of only a few nuclei, which can then grow to a large size. Such *critical cold work* is usually undesirable because of the poor mechanical properties of the coarse-grain structure (Sec. 2.1.4). On the other hand, very fine grain resulting from heavy cold work followed by annealing can give material of high strength without greatly diminishing the ductility of the material.

2.1.9 Hot-Working

We have noted that temperatures above $0.5T_m$ greatly facilitate the diffusion of atoms. This means that an arrested dislocation has the option of climbing, thereby moving into another, unobstructed atomic plane. If, therefore, deformation itself takes place at such elevated temperatures—between $0.5T_m$ and the solidus or, in general, at $0.7 < T_m < 0.8$ (*hot-working*)—many dislocations can immediately disappear, in fact, one

FIG. 2.12 Better combination of strength and ductility obtainable by partial annealing after cold-working. (From J. A. Schey, in *Techniques in Metals Research*, R. F. Bunshah (ed.), vol. 1, pt. 3, p. 1415, Interscience, 1968.)

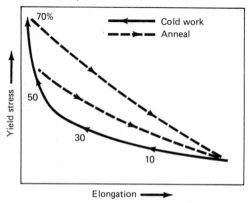

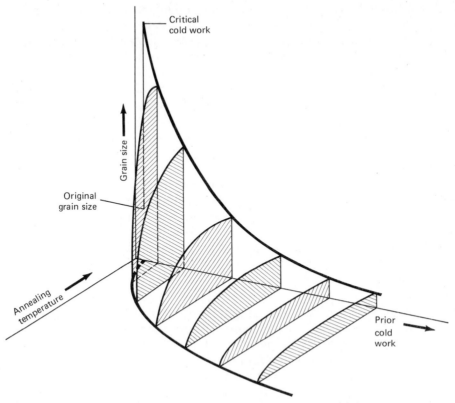

FIG. 2.13 The effect of prior cold work and annealing temperature on the grain size of the annealed material (for a constant annealing time).

finds that restoration processes work simultaneously with the slip processes. The resulting material will have a much lower dislocation density and, therefore, show less strain hardening than if it had been deformed by cold-working. The dislocation density may be kept low by either *dynamic* (simultaneous) *recovery* or *dynamic recrystallization* of the structure during working. The material cooled to room temperature often exhibits a recrystallized structure which, however, may result from recrystallization following deformation. The distinctive mark of hot-working is, therefore, not a recrystallized structure, but the simultaneous occurrence of dislocation propagation and restoration processes, with or without recrystallization during working. In general, the recrystallized structure becomes finer with lower deformation temperatures and faster cooling rates, and material of better properties is often obtained by controlling the finishing temperature.

Since all restoration processes require the movement of atoms, the time available for these processes is critical. This means that in hot-working there is substantial *strain-rate sensitivity*. *Strain rate* should not be confused with *deformation velocity;* in its simplest definition, strain rate is the instantaneous deformation (crosshead) velocity divided by the instantaneous length of the test piece during tensile deformation (Fig. 2.7),

$$\dot{\epsilon} = \frac{v}{l} \tag{2.10}$$

Thus, a tensile testing machine with a crosshead speed of 10 mm/s gives a strain rate of 1/s on a specimen of 10 mm length, but a strain rate of only 0.1/s on a specimen of 100 mm length.

For a constant temperature, the stress-strain curve may be fairly flat after initial yielding, indicating that strain hardening and restoration roughly balance each other (Fig. 2.14a), or the curve may be ascending or even show an inflection (*strain softening*). In the majority of instances, the strain-rate sensitivity can be expressed by the equation

$$\sigma = C\dot{\epsilon}^m \tag{2.11}$$

where *C* is a *strength coefficient,* and *m* is the *strain-rate sensitivity expo-nent.* If the stress-strain curves were all flat, a single set of constants would completely describe the behavior of the material. More frequently, different *C* and *m* values must be found for any given strain. When plotted on a log-log scale, the resulting straight lines give the value of *C* at a strain

FIG. 2.14 (a) Typical flow stress curves for hot working, and (b) plot of flow stress vs. strain rate on log-log scale.

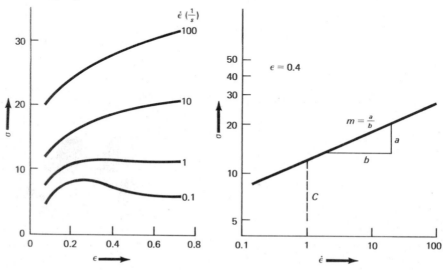

rate of unity and their slopes (again measured on a linear scale), define the strain-rate sensitivity exponent m (in the example of Fig. 2.14b, for a strain of $\epsilon = 0.4$).

For calculating purposes, experimentally determined C and m values (as, for example, in Tables 4.2 and 4.3) or flow stress curves must be used. It is worth noting, however, that time and temperature are equivalent in allowing restoration processes to exert their influence on flow stresses. Therefore, it is sometimes possible to express all hot-working flow stresses with a single curve that is a function of a velocity-(or strain-rate) modified temperature.

A high m value means, of course, that markedly higher forces are needed to deform the material at higher strain rates. It also means, however, greater resistance to fracture in tension. Since an incipient neck is the smallest cross section of the specimen, it is also the weakest. As deformation is momentarily concentrated in it, it is subjected to a higher strain rate than the adjacent nonnecked material; consequently, it presents greater resistance to deformation and does not deform further. Instead, the neck *spreads* until the entire length is deformed (Fig. 2.9c). Thus we find that next to n (the strain-hardening exponent), a high m value also indicates greater possible elongation. This will be important in stretching-type operations (Chaps. 5 and 6).

For a long while it was thought that strain-rate sensitivity in the cold-working of metals was small enough to be entirely ignored. It now appears that even small variations of m can substantially influence the onset of necking, therefore strain-rate sensitivity cannot be neglected even in cold-working. Typical values of the strain-rate sensitivity exponent are

Cold-working	$-0.05 < m < 0.05$
Hot-working	$0.05 < m < 0.3$
Superplasticity	$0.3 < m < m < 0.7$
Newtonian fluid	$m = 1$

Deformation at $0.3T_m$ to $0.5T_m$ is often denoted as *warm-working* (Fig. 2.1) and is characterized by reduced strain hardening, increased strain-rate sensitivity and somewhat lower flow stress relative to cold-working.

Superplasticity is a characteristic of some extremely fine-grain metals (most often alloys with a two-phase microduplex structure) in which high-temperature deformation takes place primarily by extensive grain-boundary sliding and accompanying diffusion. Deforming forces can be very low and, as long as strain rates are kept low enough to allow these unusual deformation mechanisms to prevail, the superplastic behavior is maintained and very large elongation values (up to several hundreds and even thousands of percent) are readily obtained. Thus, techniques developed for the forming of polymers (Chap. 6) can be applied also to metals.

After cooling from the superplastic temperature, many alloys develop substantial strength. Alternatively, the material may be made suitable for high-temperature service by destroying the superplastic properties through a heat treatment that coarsens the grain size.

2.2   ALLOYS

Most of the practical metals contain some intentional or unintentional additions; thus the behavior of *alloys* consisting of more than one element is of primary importance. The experimentally determined *equilibrium diagrams* (*phase diagrams*) are of tremendous value in judging the manufacturing properties of alloys.

2.2.1   Solid Solutions

Some atom species can readily substitute for each other in the crystal lattice, creating *substitutional solid solutions.* Occasionally, solid solubility extends over the entire concentration range; the alloys are single-phase at all compositions (Fig. 2.15a) and look like pure metals under the microscope, with only grain boundaries visible.

For a given composition C_0, solidification begins at a temperature given by the *liquidus*, T_L, and, under equilibrium conditions, ends when the

FIG. 2.15   Solid solution alloys: (*a*) equilibrium diagram, (*b*) solidification, (*c*) properties, and (*d*) inverse segregation.

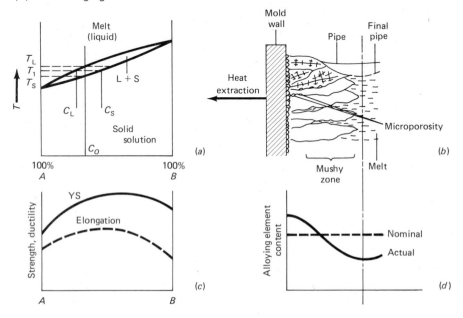

temperature of the *solidus, T_S,* is reached (Fig. 2.15a). At some intermediate temperature, T_1, the alloy is in a *mushy state:* solid crystals (of composition C_S) coexist with a liquid (of composition C_L). Their relative quantities are given by the *inverse lever arm rule.* At temperature T_1, the amount of solid is

$$S = \frac{C_0 - C_L}{C_S - C_L} \cdot 100\%$$

Obviously, the quantity of solid is vanishingly small (only nuclei are present) at the liquidus, provided there is no undercooling, and solid crystals grow gradually during cooling to the solidus.

The crystalline nature of metals manifests itself in *preferential growth* directions and this, coupled with factors relating to heat extraction and undercooling at the crystal surface, frequently results in *dendritic growth* (Fig. 2.15b) under practical cooling conditions. The *mushy zone* then extends deeply into the solidifying melt: In the example given in Fig. 2.15a, at temperature T_1, dendrite arms of composition C_S are surrounded by melt of composition C_L. On further cooling, more solid is formed and, if there were enough time available, its composition would move toward C_0 by diffusion of A atoms into the already solidified metal. Finally, all melt would disappear at T_S. The composition would by then be equalized everywhere and the structure would be undistinguishable from a pure metal under the microscope. In practice, however, diffusion is too slow and concentration gradients remain (Sec. 2.2.5). Furthermore, the intricate network of dendrite arms makes free movement of the liquid difficult, and spaces between arms may solidify without adequate supply of liquid to make up for solidification shrinkage. Consequently, *microporosity,* characterized by holes with ragged edges, is typical of solid solutions. The total shrinkage is similar to that of the pure metal, but the pipe is much smaller and a large proportion of the total shrinkage is in a *distributed* form.

The presence of slightly different size atoms in the lattice causes distortion of the slip planes and thus makes movement of dislocations somewhat more difficult. Consequently, the yield stress of solid solutions rises with increasing concentration of the solute element. The rate of strain hardening also increases and, since the strain-hardening exponent n signifies greater resistance to necking, solid solutions often actually exhibit improved ductility (Fig. 2.15c).

2.2.2 Yield-Point Phenomena

Another possibility for accommodating foreign atoms in a "solvent" lattice exists when the "solute" atoms are much smaller than the solvent atoms. They will then fit into the space existing between atoms in the basic lattice (*interstitial solid solution*). Very often they seek more comfortable sites where lattice defects have created voids in the structure. Most markedly, this is found with carbon and nitrogen in iron. These atoms are small enough to fit into the lattice; nevertheless, they tend to migrate to dislocations where distortion of the lattice provides more room (just below the extra row of atoms in Fig. 2.3c). In a sense, the solute atoms complete the lattice and immobilize (*pin*) the dislocations. Therefore a larger stress must be applied before the dislocation can break away from such a condensed atmosphere of carbon or nitrogen atoms.

This leads to the appearance of a *yield point* in the stress-strain curve of low-carbon steels; after the dislocations have broken away from the pinning carbon atoms, they multiply and move in large groups in the direction of maximum shear stress (very approximately, at 45° to the applied force). Such localized yielding creates a visible *strain band* or *Lüders line*. Successive generation of strain bands continues over the whole length of the specimen at a relatively low stress, giving the familiar *yield-point elongation* (Fig. 2.16a). On further straining the normal strain-hardening behavior is evident. If straining is interrupted and then immediately resumed, the original strain-hardening curve is rejoined. However, if sufficient time is allowed for the carbon atoms to seek out new dislocation sites (the carbon and nitrogen atmospheres condense), the yield-point phenomenon returns (broken line in Fig. 2.16a). This behavior is described as *strain-aging*. It leads to problems in stretching-type sheet-metalworking operations, because surface appearance is marred by the Lüders lines (Fig. 2.17).

Abnormal yielding, particularly stepwise yielding (Fig. 2.16b) is sometimes observed in other materials for causes related to negative strain-rate

FIG. 2.16 Examples of yield-point phenomena: (*a*) yield point and strain aging in mild steel; (*b*) serrated yielding in an aluminum alloy.

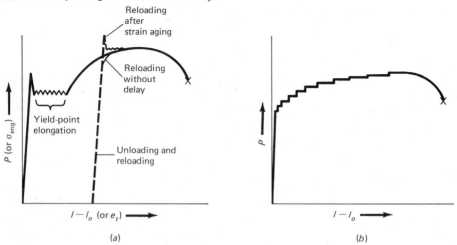

FIG. 2.17 Tensile specimen showing stretcher-strain marks; note reduced width at Lüders bands. (S. Kadela, University of Waterloo.)

sensitivity rather than dislocation pinning. Such *serrated yielding* is sometimes found in substitutional aluminum alloys and leads to the development of objectionable *stretcher-strain marks* when the sheet is deformed by stretching (Chap. 5).

2.2.3 Manufacturing Properties of Two-Phase Structures

In many alloys solidification leads to two-phase structures. Before discussing possible modes of solidification, it will be useful to inquire into the general principles that govern the manufacturing behavior of two-phase systems. First of all, it is necessary to recognize that the presence of two phases immediately implies the existence of an *interface* between them. Any interface, even a grain boundary in a pure metal, is a site of many unsatisfied, broken interatomic bonds that add up to an excess energy, the *interfacial energy* γ. The magnitude of interfacial energy is larger when the mismatch between adjacent atomic groupings is greater. Thus, the interfacial energy between the vapor and solid γ_{SV} of a substance, say metal, is much larger than the interfacial energy between its liquid and solid phases γ_{SL}. The relative magnitudes of interfacial energies between two dissimilar materials are readily judged by placing a liquid drop of one on top of a flat, solid surface of the other. The liquid drop sits in place (hence

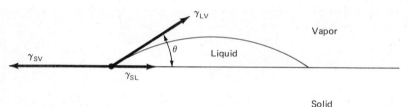

FIG. 2.18   Interfacial energies acting on sessile drop.

the name *sessile drop* technique), but is free to change its shape until the surface tensions establish a force equilibrium (Fig. 2.18):

$$\gamma_{SV} = \gamma_{SL} + \gamma_{LV} \cos \Theta \tag{2.12}$$

When $\Theta \leq 90°$ the surface is wetted and the drop spreads out; when $\Theta > 90°$ the surface is not wetted and the liquid forms, in the limit, a spherical droplet.

Wetting, then, is an indication of relative surface energies and, through these, a measure of the strength of the interface. Wetting is a sign of reasonable match between the contacting phases, and one can expect a wetted interface to resist stresses that might pull the phases apart. A nonwetted interface, on the other hand, can behave as a preexisting crack and thus impair mechanical properties. It should be remembered that surface tension is indeed a surface effect; therefore even the minutest amounts of a contaminant, segregated at the interface, can substantially reduce wetting.

With these preliminaries in mind, one would intuitively expect that the properties of a two-phase structure must depend on a number of factors, such as the properties, distribution, shape, and size of individual phases and the nature of the interfaces between them. Some basically different situations can be envisaged (Fig. 2.19):

1   Both phases are ductile and wet each other; they behave like a homogeneous body and properties can be estimated from the relative volumes of the two phases.

2   One of the phases is ductile, the other is brittle and wetted. In this case, the relative quantity, shape, and location of the brittle phase become dominant. If the brittle phase is the *matrix* (i.e., it surrounds the ductile phase), the resulting structure will be brittle. If the matrix is ductile, but coarse plates or needles of the brittle phase weave through it, the brittle phase causes *stress concentrations* on loading and the structure will be both weak and of low ductility. Finer plates confined within a grain and systematically aligned into a platelike *lamellar structure* will substantially strengthen the matrix but usually at the expense of ductility, since the brittle plates fracture during deformation and the stress raisers terminate

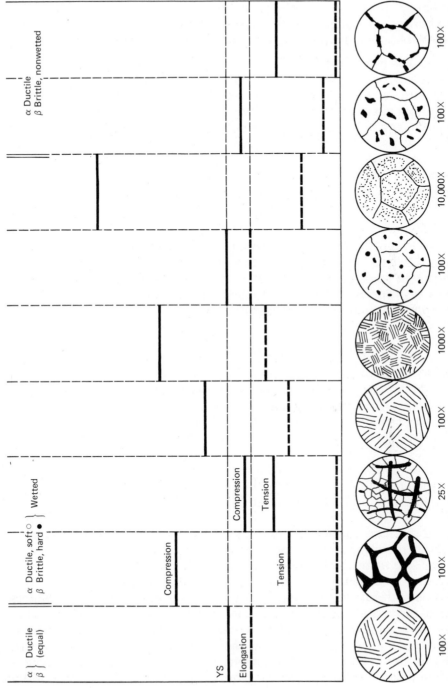

FIG. 2.19 Properties of two-phase structures as a function of the properties, wetting, shape, size, and distribution of the two phases.

plastic deformation in the ductile matrix. If the platelike structure is extremely fine and closely spaced so that the hard phase acts as a dislocation barrier, the composite structure may show high strength coupled with reasonable ductility, because rapid strain hardening delays necking.

3   The hard constituent is wetted by the soft phase and is in a roughly spheroidal form. It will not greatly affect the strength of the material if the hard particles are relatively coarse and widely spaced. Ductility is not impaired because dislocations circumnavigate such large blocks of harder material, and the *notch effect* is minimal because of the large notch radius. When the particle size becomes small enough to arrest or slow down dislocations, the structure will be strengthened and ductility reduced, depending on the total quantity and spacing of the hard particles. Great strengthening can be obtained by loading the ductile matrix with masses of particles (until the matrix becomes little more than a ductile cement), but then the ductility will greatly suffer.

4   The hard phase is not wetted by the matrix; the interface between them acts as a premade *crack*. Growth of these cracks and their coalescence into voids occurs readily; therefore nonwetted particles greatly impair the properties of materials, particularly if they are located on grain boundaries which tend to be less ductile in any event.

2.2.4   Alloys with Invariant Reactions

Alloys formed by elements of differing atomic radii often show *limited solid solubility* of each element in the other. When this limit is exceeded, the excess solute element population is rejected into a second phase.

Eutectic systems

In the example of Fig. 2.20a an alloy of composition C_0 begins to solidify by the rejection of α solid-solution nuclei which gradually grow until the temperature drops to the so-called *eutectic invariant temperature* T_i. Here the remaining liquid of C_E composition solidifies at a constant temperature by the simultaneous growth of α and β solid-solution phases. Since this requires grouping of the atoms, preferably by movement over short distances, the result is very often a lamellar structure. The solidified structure shows *proeutectic* (formed before the eutectic) α solid-solution crystals surrounded by a matrix of the *lamellar eutectic*. The mushiness typical of the solidification of solid solutions again results in some distributed shrinkage (microporosity). A melt of C_E eutectic composition solidifies, of course, at a constant temperature T_i; the structure is two-phased and contains—at reasonable cooling rates—fairly fine platelets of one phase in a matrix of the other phase. Because of the constant

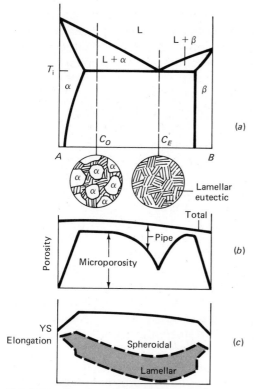

FIG. 2.20 Eutectic alloy system: (a) equilibrium diagram, (b) porosity, and (c) mechanical properties.

solidification temperature, the solidification front moves ahead almost like that of a pure metal and much of the shrinkage is accommodated in a central pipe (Fig. 2.20b).

The mechanical behavior of these materials can be readily deduced from the principles described earlier (Sec. 2.2.3). If the growth habit of the eutectic is lamellar, great benefits will derive from refining its structure. Further improvement may be made if the second phase can be encouraged to solidify in an essentially globular (spheroidal) form. Such *spheroidal eutectics* can show great ductility and desirable strength even when the second phase is weak or brittle (e.g., in nodular cast irons). The shaded area in Fig. 2.20c symbolizes this variation of properties with structure for the same composition.

In some systems, most notably in the iron-carbon system, a solid solution stable at a higher temperature decomposes into a two-phase structure on cooling below a critical temperature. Such *eutectoid transformation* can be often exploited for the control of manufacturing and service properties (see Sec. 2.2.8).

Peritectic systems

When the melting points of the two metals are greatly different, the invariant reaction is often of the *peritectic* type (Fig. 2.21). An alloy of composition C_0 first rejects α solid solution on cooling. At the invariant temperature T_i all the remaining liquid must transform into a β solid solution, which can be achieved only by the diffusion of B atoms into the already solidified α crystals. Thus a circumferential, peritectic diffusion reaction takes place and leads to a two-phase structure with properties that can be deduced from the properties of the two phases.

Intermetallic compounds, found in many systems with invariant reactions, have had great technical significance. These compounds form with distinct stoichiometric ratios and the unit cell must often accommodate very large numbers of atoms. As a group, intermetallics tend to be hard and brittle because no slip plane can be defined on which dislocations could propagate, except perhaps at high temperatures where dislocation mobility is much greater. When relatively small quantities of very fine intermetallic compound particles are uniformly distributed in a ductile, soft matrix, they become powerful strengthening constituents; if they are allowed to grow large, acquire unfavorable shape, or become poorly distributed, they can be detrimental. During working of an alloy they tend to align themselves in the direction of major deformation, and such *fibering* gives rise to markedly directional properties. In the direction of the major deformation they act as strengthening phases and impart great strength and reasonable ductility. Transverse to this major deformation direction, their alignment creates rows of relatively weak planes, therefore the so-

FIG. 2.21   Peritectic alloy system.

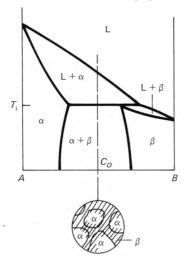

called *short-transverse properties* (both strength and ductility) of such alloys tend to be inferior.

2.2.5 Nonequilibrium Solidification

Until now we have assumed that cooling conditions during solidification allow the attainment of complete equilibrium. In reality, diffusion processes are too slow to establish equilibrium, and, particularly when the solidification temperature range is wide, *concentration gradients* due to lack of diffusion are found in solid solutions.

In a system like that shown in Fig. 2.22, solidification of a melt of composition C_O begins with the rejection of α solid-solution crystals of composition C_1. On further cooling, the crystal not only grows but its composition would also have to become enriched in the B element as dictated by the solidus. If time is insufficient to allow B atoms to diffuse into the already solidified core, the core remains leaner in the alloying element. The excess B atoms are retained in the melt, and solidification does not end when the T_S equilibrium solidus temperature is reached; instead, it

FIG. 2.22 Nonequilibrium solidification of a solid solution resulting in nonequilibrium eutectic on grain boundaries.

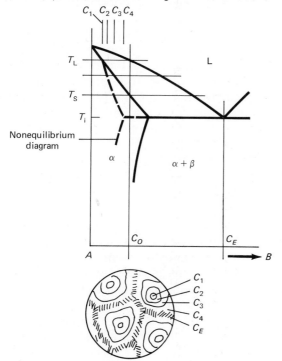

continues by the gradual deposition of richer and richer layers. The *nonequilibrium solidus* shown as a broken line in Fig. 2.22 represents the average composition of the solid. It is even possible that a liquid phase remains until the T_i eutectic temperature is reached, when the remaining liquid finally solidifies along the grain boundaries as a eutectic.

If the B element (which could also be an A_mB_n intermetallic compound) is brittle, the solidified structure will also be brittle, even though from the equilibrium diagram one would judge it to be a ductile solid solution. Such an alloy will also suffer from *hot shortness* when heated above the T_i eutectic temperature; even though it is nominally a solid solution and thus should be readily deformable, it will actually separate at the grain boundaries where the nonequilibrium, low-melting eutectic is present. Hot-short fracture is readily identified by its ragged appearance as it follows the grain boundaries. Sometimes the presence of an unsuspected contaminant which forms a low-melting eutectic may have a dramatic effect on hot workability. A striking example is sulfur in excess of 0.004 percent (i.e., 40 parts per million) in nickel, or high-nickel superalloys.

Because the centers (cores) of crystals grown during nonequilibrium solidification have a different composition, it is usual to refer to this phenomenon as *microsegregation* or *coring.* If necessary, it can be removed by heating the alloy just below its solidus temperature for prolonged periods; such *homogenization* eliminates most of the concentration gradients unless one of the elements has a very low diffusivity in the solvent metal. This is the case, for example, with phosphorus in iron, which remains at the original grain boundaries even after prolonged or repeated heating and even deformation (hot-working). Since the original grain structure becomes elongated in the direction of major deformation, deep etching of the cross section of a deformed steel structure (*macroetching*) reveals the concentration gradients in a readily visible form, and is thus a useful tool in judging the soundness of material flow (*flow lines* in Fig. 4.17).

The solidification of solid solutions (Fig. 2.15) under nonequilibrium conditions also creates large-scale concentration gradients (*macrosegregation*). The growth of dendrites is accompanied by shrinkage, and liquid of higher alloying-element content will be supplied to the spaces between dendrites as long as there is free access. When the center of the casting finally solidifies, supply of such liquid is no longer available; therefore the center tends to show a lower average alloying-element concentration than the surface zones (Fig. 2.15*d*). This is described as *inverse segregation,* in contrast to *normal* segregation that occurs when solidification is accompanied by volume increase (or gas evolution). In the latter case, the solidification zone moves out more smoothly (as in Fig. 2.2*a*), and thus the liquid enriched in alloying elements is pushed toward the center.

2.2.6　Limited Liquid Solubility

Occasionally, a metal or a compound is not dissolved even in the liquid metal, and such *limited liquid solubility (immiscibility)* can be exploited for practical purposes. Thus, lead is insoluble in copper and copper alloys, and improves their performance for bearing applications because the soft lead globules included in the copper, bronze, or brass matrix serve as an internal lubricant. Lead also makes many metals more machinable because it acts as an internal lubricant and also helps to break the chip (Sec. 8.2.3). It can, however, have detrimental effects on workability (Sec. 4.2.2): it makes α-brass hot-short, while it is harmless in β-brass, because it remains well distributed within the grains instead of separating out at the grain boundaries.

2.2.7　Precipitation Hardening

It is frequently found that at elevated temperatures the greater thermal excursions of the atoms in the lattice allow more solute element atoms to be accommodated. With decreasing temperature, the solubility decreases as indicated by the *solvus line* in Fig. 2.23. When the composition is between minimum and maximum solubility, say at C_0, and cooling is slow enough to allow the rejected solute atoms to regroup into the stable phase (in this case an intermetallic compound), the resultant two-phase structure exhibits the greater ductility of the now leaner α solid solution, little affected by the relatively few massive intermetallic particles. This *fully annealed condition* is, therefore, suitable for cold-working.

　　Since diffusion is time-dependent, its suppression is possible and this offers some intriguing possibilities. First the alloy must be heated up into a temperature range that assures complete dissolution of all second-phase particles (*solution treatment,* in the α temperature range). Rapid cooling (*quenching*) will retain the solute elements in a *metastable solid solution.* The excess solute atoms may subsequently be precipitated in a controlled fashion, in the form of the intermetallic compound, by heating the alloy at a temperature below the solvus (*artificial aging*) or sometimes even by holding at room temperature (*natural aging*). Higher temperatures and longer holding times allow more time for diffusion and therefore for coarsening of the precipitate particles; as explained in Sec. 2.2.3, this means that the strength gradually drops while ductility increases. By a proper choice of the precipitate size, a desirable compromise between strength and ductility can be reached.

　　The metastable solid solution obtained by quenching the solution-treated material is readily deformable, because a supersaturated solid solution behaves rather like other solid solutions: the strength is greater than in the annealed state but ductility is not unduly reduced because of the typically high strain-hardening exponent of solid solutions (as in Fig. 2.15c).

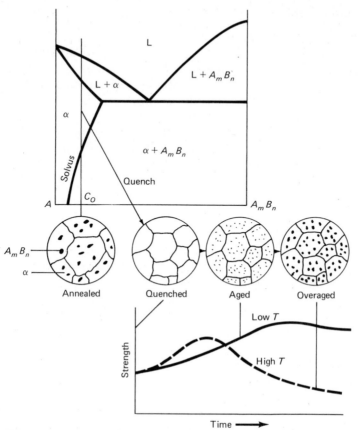

FIG. 2.23   Age-hardenable alloy system.

Cold-working of this structure introduces a high dislocation density which then induces precipitation of the second phase in an extremely finely distributed form on subsequent aging. Thus, exceptionally favorable combinations of strength and ductility may be obtained. Atomic mobility is greater during deformation; therefore aging may be initiated or enhanced by the deformation process itself. Greater strength, but with limited ductility, is attained by cold-working the fully heat-treated (precipitation-hardened) structure.

2.2.8   Eutectoid Transformation in Steel

Precipitation hardening is but one example of manipulating metastable phases. There are a number of other metastable situations that can be exploited, foremost of which is the eutectoid transformation occurring in steels.

The principles are most readily demonstrated on a steel of eutectoid composition (0.8% C). At high temperatures, carbon in the FCC γ-iron forms an interstitial solid solution (*austenite*). On cooling below the transformation temperature of 723°C, the austenite decomposes into a low-carbon, BCC α-iron solid solution (*ferrite*) and the compound Fe_3C (*cementite*). This involves diffusion and thus time, as shown by the familiar *time-temperature-transformation diagram* (Fig. 2.24).

Just below the transformation temperature few nuclei are formed, but they grow rather rapidly; consequently, on slow cooling (line 1) the structure will contain coarse cementite platelets in a ferrite matrix. Such coarse lamellar *pearlite* is relatively soft but not very ductile. On faster cooling (line 2), transformation is somewhat delayed and a metastable austenite exists until the time and temperature of transformation is reached. A great many nuclei form and the structure will consist of much finer but still lamellar pearlite. If the steel is cooled very rapidly and then held at an intermediate temperature, say around 350°C, transformation occurs along line 3 and results in *bainite,* in which the lack of diffusion time makes the carbide particles appear as extremely fine spheroids in a matrix of α-iron. As expected, such a structure possesses a desirable combination of strength and ductility. Isothermal heat treatment leading

FIG. 2.24 Time-temperature-transformation diagram for a steel of 0.8% C content.

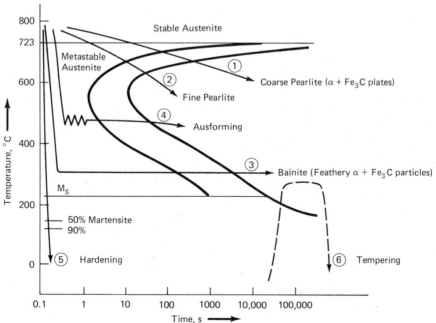

to bainite or very fine lamellar pearlite (*patenting*) has long been used to make the highest-strength wires, such as spring and piano wires. Full strength is attained by heavy cold reduction, such as wire drawing after heat treatment.

There is also the possibility of improving the properties of the final product by working the metastable austenite, thus introducing a high dislocation density which then causes a much finer precipitation of carbides in the transformation product. When such *thermomechanical processing* is carried out approximately 100–200°C below the transformation temperature of 723°C, it is described as *ausforming* (line 4 in Fig. 2.24). Even higher strengths, but usually reduced ductilities, are obtained if working is done at lower temperatures (*low-temperature thermomechanical treatment*). In either case, the need for rapid cooling shown in Fig. 2.24 would make the time available for such processing much too short; therefore these treatments are more practicable for alloy steels in which the alloying element moves the nose of the transformation curve to the right.

When cooling is fast enough to miss the nose in the time-temperature-transformation (TTT) diagram entirely (line 5 in Fig. 2.24), cooling (quenching) suppresses the separation of the carbide phase and the carbon atoms are retained in a supersaturated solid solution of the body-centered iron structure. The presence of carbon distorts the cubic into a tetragonal lattice and this highly stressed structure (*martensite*) is very hard and brittle. Ductility can be restored by reheating the martensite (line 6 in Fig. 2.24) so that carbide can precipitate in a very fine form. Such *tempered martensite* can be further strengthened by limited cold deformation, although ductility suffers.

The temperature at which martensite transformation starts (M_s) can be depressed by various alloying elements. Thus, a metastable austenitic structure can be retained at room temperature. When such material is subjected to tensile deformation, the greater mobility of atoms during deformation initiates the transformation to martensite. Therefore, an incipient neck will be stabilized by transforming the austenite to the much stronger martensite, and the onset of localized necking will be delayed until the entire volume of the tensile specimen is transformed. This provides, then, a third means (besides increasing n and m values) of increasing the ductility of a metal by delaying necking. Some stainless steels, as well as *transformation-induced plasticity* (TRIP) steels offer great strength with unusual ductility. The latter steels benefit from a high dislocation density induced in the austenite by warm-working.

It should be noted that the martensite need not be hard. If the carbon content is very low, as in *maraging steels,* the martensite is soft and readily worked, but subsequently can be greatly strengthened by precipitation of intermetallic compounds (such as Ni_3Ti or Ni_3Mo) at the numerous sites of high dislocation density induced by cold-working.

2.3 MULTIPHASE MATERIALS

We have assumed hitherto that the alloy had only two components. In a great many technically important materials more than two are frequently encountered.

2.3.1 Ternary and Polycomponent Alloys

In practice, very few truly pure metals or binary alloys are used. Even if only of the order of a few parts per million, contaminants are always present, while intentional additions may reach such high proportions that it becomes difficult to classify a material according to its base metal. Such ternary, quaternary, etc., alloys (*polycomponent systems*) still exhibit the same phases as binary alloys, thus one can find solid solutions, eutectics, peritectics, intermetallics, and their various combinations. Their behavior and properties can be derived from an analogy to two-phase materials, especially if the phase diagram is known or at least a section of the phase diagram (for a constant percentage of one of the alloying elements) has been established. The solidification behavior can be estimated approximately from the temperature range of solidification, and the deformation behavior from the principles outlined in Sec. 2.2.3.

2.3.2 Inclusions

The term *inclusion* is used to describe the unintentional presence of solid (often refractory) particles in a metallic structure. They find their way into the alloy usually during melting (for example, from the furnace lining, contaminations of the charge, or even as a result of reaction, usually oxidation, with the surrounding atmosphere), or during pouring (again through oxidation, or by washing of refractory particles into the melt from the ladle or mold). As with all second-phase particles, their effect depends greatly on whether they are wetted by the matrix or not.

When inclusions are wetted, their effects can be judged as those of any second-phase particle (Sec. 2.2.3): if they are strong and perhaps even ductile, and are dispersed inside the grains in an approximately globular or fibrous form, they are harmless and sometimes even useful. Arranged along grain boundaries they are likely to be harmful, unless they are extremely small and well distributed (which is the basis of *dispersion strengthening,* used, for example, in TD-nickel that has small quantities of thorium oxide interspersed). Brittle plates and, particularly, films (such as are formed by aluminum oxide) are detrimental in cast structures, even though they may be neutralized by breaking them up through hot-working.

Nonwetted inclusions are almost always harmful. The interface between inclusion and matrix acts as a premade crack which readily propagates, thus reducing the strength and ductility of the material. If gases are present, they tend to congregate on these interfaces and can build up such high pressures that a bubble (*blister*) is formed on inclusions close to the surface, particularly during heating for annealing or hot-working.

Like any other second-phase particles, inclusions orient themselves in the direction of major material flow and give rise to or accentuate directional properties (fibering, referred to in Sec. 2.2.4). In steels they contribute to outlining the flow lines discussed in Sec. 2.2.5.

2.3.3 Gases

The solubility of gases is much higher in the molten than in the solid state and, unless melting and pouring take place *in vacuum* or in a *protective atmosphere* of a gas that is not soluble in the particular metal, gases appear in the alloy as generally troublesome constituents.

If the melt was allowed to become saturated with gases, their excess is rejected during solidification (Fig. 2.25), frequently causing *gas porosity* to appear in the solidified structure. Such holes are usually round. If the holes contain a reducing gas, then they have a clean, bright surface and may weld during subsequent hot-working. They are, however, essentially like inclusions of zero strength and, if not eliminated, weaken the structure as discussed earlier. They can also cause blistering, as discussed in conjunction with nonwetted inclusions. The volume occupied by gas cavities

FIG. 2.25 Solubility of gases in metals as a function of temperature.

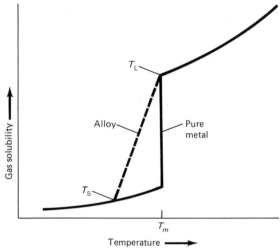

of various sizes may partially or fully counterbalance solidification shrinkage and may thus have useful side effects (e.g., in rimming steel).

Micropores in castings may or may not contain gases; in either case, the ragged edges of such pores cause stress intensification and are, therefore, harmful. One can say that, with very few exceptions, voids and gases are undesirable in any metallic structure, whether cast or worked. Indeed, one finds that the highest quality (combining greatest strength with greatest ductility and fracture toughness) in any composition is obtained only when it is free of gases, reaction products with gases, and of inclusions. There is, for example, dramatic improvement in the workability as well as the high-temperature creep performance of nickel-base superalloys when they are *melted in vacuum* rather than air, and further improvement is assured by remelting under a protective and refining slag (*electroslag melting*). Similar improvements have been achieved in many other critical materials such as die and tool materials and roller-bearing steels.

2.4 ADHESION

In Sec. 2.2.3 we discussed some properties of interfaces formed in two-phase materials. Interfaces, on a much larger scale, are also created by the manufacturing process itself.

In most instances, the part is shaped in contact with a mold, die, or tool. Even apparently clean surfaces are usually covered with invisible, adsorbed *contaminant films* that are surprisingly effective in separating contacting surfaces, at least at low pressures and temperatures. Manufacturing processes do, however, create high temperatures (as in casting) and pressures (as in cold-working) or a combination of both (as in hot-working). Displacement of the metal along the mold or die wall is very effective in removing adsorbed films, and some processes (notably machining) involve penetration of the tool under the protective surface.

The above effects, alone or in combination, may well bring atoms of the workpiece and of the die within typical interatomic distances. *Metallic bonds* are then established; in the terminology adopted by workers in tribology (the study of friction, wear, and lubrication), *adhesion* occurs. As might be expected, adhesion is greatest when the contacting metals possess similar crystal structures of similar interatomic spacings. No wonder then that a metal shows greatest adhesion against itself, and dissimilar metals adhere to each other if they also form a solid solution at the temperature of contact. Strong adhesion results in *pressure welding* and, as the workpiece moves against the die, parts of the softer material (usually the workpiece) are torn out and remain attached to the die. Apart from the ruined workpiece, such *die pickup* also damages subsequent workpieces and finally the tool itself will *wear* away.

Adhesion is sometimes purposely exploited for welding (Chap. 9), but, in the majority of instances, it is to be avoided. The most frequently used die (tool, mold) material is steel, and its major component, iron, shows a solubility for most metals. It must, therefore, be protected against adhesion and its consequences. Heat treating to a greater hardness helps, but additional protection is almost always necessary. This usually takes the form of specially developed surface films, such as refractory *mold washes* for the casting of metals, *parting agents* for the casting and molding of polymers, and *lubricants* (which act as parting agents while also reducing friction) for metalworking and machining processes.

Externally applied films are of little use once they fail. The ultimate in protection would be a tool material that shows no adhesion to the workpiece material. This usually means that the tool has to be made from a nonmetal, or the metal surface has to be transformed, to a depth of several tens of atoms, into a nonmetallic or intermetallic compound—one which does not adhere to the tool and also prevents the diffusion of metal atoms across the interface. In the following chapters many examples of applying these principles will be found.

2.5 SUMMARY

From the manufacturing point of view, the most important material properties are strength, ductility, and adhesion.

2.5.1 Strength

1 Strength increases with decreasing grain size in all materials, irrespective of whether they are in the cast or worked condition, and this offers a powerful control method both during processing and in the finished product.

2 Strength varies with crystallographic direction. Single crystals with the greatest strength in the loading direction find limited application (e.g., for turbine blades) but polycrystalline materials with a well-developed texture can offer advantages for processing (e.g., a high r value for deep drawing) or in use (e.g., for rocket-motor cases).

3 Strength increases with cold-working, as expressed by the strain-hardening exponent n of a power function.

4 Strength drops with temperature, and above $0.5T_m$ restoration processes (recovery and recrystallization) prevail over strain hardening. In such hot-working, strength is governed by strain-rate sensitivity, expressed by an exponent m.

2.5.2 Ductility

1   One facet of ductility is the ability to elongate without fracture. A high strain-hardening rate (high n) delays necking, while a high strain-rate sensitivity (high m) allows spreading of the neck. These will be important in stretching-type operations for both metals and polymers.

2   The other facet of ductility is resistance to triaxial tensile stresses, conveniently expressed by the reduction in area in the tensile test. It is a function of composition and also of cleanliness (freedom from inclusions, gases, and contaminants). By definition, elimination of triaxial tension (superposition of triaxial compression) delays fracture, and this is an important process-control method for bulk deformation.

2.5.3 Multiphase Materials

Almost all engineering materials consist of more than a single phase.

1   Solidification in solid solutions is dendritic and is accompanied by coring, segregation, and interdendritic porosity, choking off of liquid supply, and also by the appearance of nonequilibrium phases at the grain boundary.

2   Eutectic compositions solidify with a smooth front and little porosity and hence are the preferred casting alloys.

3   Mechanical properties depend on the properties, wetting, relative position, shape, size, and distribution of the various phases. A strong and brittle phase strengthens a solid solution when finely distributed, but reduces its ductility when in a coarse, platelike form. A weak phase on the grain boundaries destroys strength and creates the problems of hot shortness in casting, hot-working, and welding but it may impart desirable properties (internal lubrication in metal cutting, or for bearings) when dispersed within the grains.

2.5.4 Adhesion

Adhesion is a manifestation of interatomic bonds that develop when completely clean surfaces are brought within interatomic distances.

1   The strength of bond is greater when atomic registry is attained, typically between like materials or those having similar interatomic distances (and, therefore, forming solid solutions).

2 Adhesion can be undesirable, as when it leads to sticking in casting, tool pickup in plastic deformation processes and built-up edge in machining.

3 Adhesion can be beneficial, indeed essential, for ensuring strength in powder consolidation, solid-state welding and brazing, soldering, and adhesive bonding.

Further Reading

GENERAL, INTRODUCTORY TEXTS ON MATERIALS:

BARRETT, C. R., W. D. NIX, and A. S. TETELMAN: *The Principles of Engineering Materials,* Prentice-Hall, Englewood Cliffs, N.J., 1973.

CLAUSER, H. R.: *Industrial and Engineering Materials,* McGraw-Hill, New York, 1975.

HANKS, R. W.: *Materials Science Engineering: An Introduction,* Harcourt, Brace & World, New York, 1970.

HUTCHINSON, T. S., and D. C. BAIRD: *The Physics of Engineering Solids,* Wiley, New York, 1963.

JOHN, V. B.: *Introduction to Engineering Materials,* Macmillan, London, 1972.

KEYSER, C. A.: *Materials Science in Engineering,* Merrill, Columbus, Ohio, 1968.

ROSENTHAL, D.: *Introduction to Properties of Materials,* Van Nostrand, Princeton, N.J., 1964.

VAN VLACK, L. H.: *Elements of Materials Science,* 3d ed., Addison-Wesley, Reading, Mass., 1975.

VAN VLACK, L. H.: *A Textbook of Materials Technology,* Addison-Wesley, Reading, Mass., 1973.

WULFF, J., (ed.): *The Structure and Properties of Materials,* Wiley, New York, 1964 (in four volumes).

SELF-TEACHING TEXT ON METALS:

ROGERS, B. A.: *The Nature of Metals,* American Society for Metals, Metals Park, Ohio, 1964.

EMPHASIS ON MATERIALS SELECTION:

CLARK, D. S., and W. R. VARNEY: *Physical Metallurgy for Engineers,* 2d ed., Van Nostrand, Princeton, N.J., 1962.

MECHANICAL BEHAVIOR:

BIGGS, W. D.: *The Mechanical Behavior of Engineering Materials,* Pergamon, New York, 1965.

COTTRELL, A. H.: *The Mechanical Properties of Matter,* Wiley, New York, 1964.

DIETER, G. E., JR.: *Mechanical Metallurgy,* 2d ed., McGraw-Hill, New York, 1976.

MC CLINTOCK, F. A., and A. S. ARGON: *Mechanical Behavior of Materials,* Addison-Wesley, Reading, Mass., 1966.

POLAKOWSKI, N. H., and E. J. RIPLING: *Strength and Structure of Engineering Materials,* Prentice-Hall, Englewood Cliffs, N.J., 1966.

TEGGART, W. J. M.: *Elements of Mechanical Metallurgy,* Macmillan, New York, 1966.

WOOD, W. A.: *The Study of Metal Structures and Their Mechanical Properties,* Pergamon, Elmsford, N.Y., 1971.

ZACKAY, V. F. (ed.): *High Strength Materials,* Wiley, New York, 1965.

Ductility, American Society for Metals, Metals Park, Ohio, 1968.

Strengthening Mechanisms in Solids, American Society for Metals, Metals Park, Ohio, 1962.

PHASE DIAGRAMS AND MICROSTRUCTURES:

Metals Handbook, 8th ed., American Society for Metals, Metals Park, Ohio, vol. 8, *Metallography, Structures and Phase Diagrams,* 1973; vol. 7, *Atlas of Microstructures of Industrial Alloys,* 1972.

Problems

2.1 A tensile test is conducted on a sheet specimen (as in Fig. 2.5*a*) of $l_0 = 2$ in, $w_0 = 0.25$ in, and $h_0 = 0.040$ in. The test is interrupted before the onset of necking; at this time, $l_1 = 2.4$ in and $w_1 = 0.224$ in (the thickness h_1 is difficult to measure with sufficient accuracy). Calculate the r value.

2.2 From data given in Table 4.2 for cartridge brass (70/30 brass), construct (*a*) a true stress σ vs. true strain ϵ curve (on a linear scale); (*b*) a true stress vs. engineering tensile strain e_t curve. (*c*) Superimpose on (*b*) the σ_{eng} vs. e_t curve as far as defined by the data (note the relation between the two curves).

2.3 From data in Table 1.3, calculate and tabulate the temperature ranges for cold-, warm-, and hot-working of Sn, Pb, Al, Cu, and Ni.

2.4 Make a properly scaled sketch to show that the 12 slip systems of the FCC structure define an octahedron.

2.5 A tensile test (as in Fig. 2.7) is conducted at room temperature on annealed 2024 Al sheet specimens, with an extensometer clamped to the gage length. An *x-y* recorder is used to obtain a diagram of force P vs. gage length increase $l_1 - l_0$. A typical diagram is shown here in a solid line. (*a*) Calculate σ_{eng}, σ, e_t, and ϵ at five points selected to cover the range from initial yielding to necking (if you are not sure about the calculations, check App. A). (*b*) Plot, in separate diagrams, σ_{eng} vs. e_t, σ vs. e_t, and σ vs. ϵ. (*c*) Plot σ vs. ϵ on a logarithmic scale; if the plot or parts of it are reasonably linear, find n. (*d*) If (*c*) results in a definite n value, $n = \epsilon_u$. Convert this to e_t and check against the point of necking in the σ_{eng} vs. e_t diagram. (*e*) Calculate uniform elongation e_u. (*f*) In the same tensile test, the force vs. displacement curve was also recorded from the movement of the crosshead of the machine and the curve shown in a broken line was obtained. Explain why this should be different from the full line. (*g*) Compare the UTS and e_f values with those quoted in the literature (e.g., *Metals Handbook,* vol. 1).

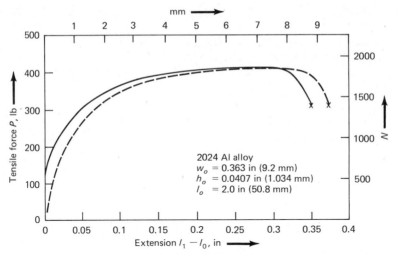

2.6 By taking σ_0 and k_y in Eq. (2.2) arbitrarily equal to unity, calculate and plot a generalized σ vs. d curve (with d varying from 10^{-2} to 10^2).

2.7 A pure metal is to be produced in strip form, in the annealed condition, but with the highest possible strength. What processing steps would assure this and why?

2.8 Tungsten carbide (WC) tools are made by bonding the very hard WC particles with cobalt. (*a*) Draw a single diagram to show, in principle, the trends in hardness and ductility with increasing Co content. (*b*) What could be done to increase both hardness and ductility?

2.9 A 10% Sn, 90% Cu binary alloy is made up in the laboratory under well-controlled melting conditions. When hot-rolling is attempted, the billet breaks up. (*a*) Review the equilibrium diagram of Cu-Sn alloys and (*b*) identify a possible cause of the problem. (*c*) Suggest a remedy.

2.10 The operator of a riveting machine reports that many rivets crack, and the riveting force also seems to be inadequate as judged by the many incompletely formed rivet heads. The rivets are of 2024 aluminum alloy and are riveted in the solution-treated condition, to attain their full strength by natural (room-temperature) aging after riveting. Determine what could have gone wrong, and suggest, step-by-step, what remedial action could be taken.

3

SOLIDIFICATION PROCESSES

As shown in Figs. 1.3 and 1.4, one of the shortest routes from raw material to finished part is that of melting the metal or alloy and pouring it into a shape that approaches that of the finished component. Essentially the same technique also produces the ingots that are subsequently subjected to plastic deformation processes. The latter application consumes approximately 85 percent of most metals produced, while the remainder is cast directly into shapes. The total quantities cast into shapes (Table 3.1) have held reasonably constant since 1970, at least in the industrially developed nations, but their value has gone up considerably because of the increasing complexity and highly improved quality of the product.

While there are a great many casting processes, some basic principles apply to all of them and these will be discussed first.

3.1 THE METAL

The general manufacturing properties of metals have been discussed in Chap. 2. It will now be necessary to expand our considerations with a specific view to the casting process.

3.1.1 Melting

The first step is obviously that of bringing the metal into a molten state. This requires heating it above its melting point (the liquidus temperature) in a *furnace*. While furnaces may vary greatly in construction, they all possess some common features (Fig. 3.1).

TABLE 3.1   SHIPMENTS OF CASTINGS* (U.S.A. 1972)

Type	Thousand metric tons
Gray iron	14,000
Malleable iron	860
Ductile iron	1,830
Steel	1,450
Aluminum alloys	850
Zinc alloys	460
Copper and brass	345
Magnesium alloys	21
Tin (+solder)	7 (+21)
Lead	7

* Compiled from *Metal Statistics 1974*, American Metal Market, Fairchild Publications, Inc., New York, 1974.

1   The *melt* is contained in a *crucible* or *furnace hearth* that is made of a material of substantially higher melting point than the metal, assuring minimum contamination of the melt by inclusions or dissolved elements. The furnace material may range from iron (for lead) to graphite or refractory crucibles, and to refractory-lined furnace structures.   Alternatively, the melt may be contained by maintaining a chilled outer zone that forms its own container.

FIG. 3.1   Elements of a melting system.

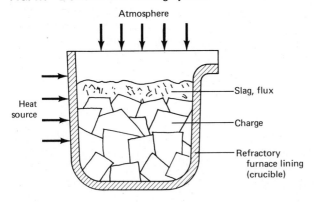

2 *Heat* is provided either externally (for example, by electric, gas, or oil heating) or internally (as by magnetic induction or by mixing the charge itself with the fuel—such as the coke in a blast furnace or cupola).

3 A *charge* is made up to yield the alloy of the desired composition upon complete melting. It is seldom practicable to have the charge made up entirely of *primary metals.* Alloying elements of much higher melting point than the base metal would be very slow in dissolving and would require excessive overheating; therefore they are added in the form of a *master metal (temper alloy, hardener)* which contains a higher concentration of the alloying element in the base metal. For economy of operation, it is most important that as much *scrap* as possible should be added, since its price may be substantially below that of primary metals. The aim is to produce a melt of the composition specified by relevant standards while holding contaminants below the allowed maximum levels, and to accomplish all this at the lowest possible cost. Because this necessitates an immediate check on the composition of each melt before *pouring,* extensive laboratory facilities are required. Smaller plants that do not possess these often find it more economical to use prealloyed material cast into ingots (*ingotted melting practice*), purchased from specialized companies.

4 An inevitable factor is the presence of an atmosphere. This may be air (with its humidity and various pollutants), a protective atmosphere (such as argon), or even vacuum, produced at some expense. Combustion products are also present in oil- and gas-fired furnaces. When the charge is mixed with the fuel (such as coke in the cupola), reactions of the fuel and its combustion products with the melt are inevitable. Thus, interactions with the atmosphere may range from simple dissolution of gases in the melt to reactions such as oxidation or, in the presence of reducing agents, reduction, and even carbon enrichment.

5 The charge is covered or mixed with elements that form *slags* (usually silicates of various metal oxides) and *fluxes* (various inorganic compounds that are capable of spreading on the surface and which either react with the melt to remove contaminants or mechanically entrap inclusions). These may also isolate the melt from the atmosphere and reduce vapor losses of metals of high vapor pressure.

3.1.2 Pouring and Solidification

When the melt reaches the desired temperature and composition, it is *tapped* by breaking through a refractory plug placed in a hole close to the bottom of a stationary furnace, by tilting the furnace or, with the lower-melting-point metals, even by siphoning or pumping out the melt. It may be transferred directly to the mold, or it may be tapped into a *ladle* (a refractory-lined vessel), which is then taken to the mold.

Solidification is achieved (Sec. 2.1.1) by cooling the melt (or part of it) below the liquidus temperature. In a very clean melt the first nuclei to form would consist only of the atoms of the metal (homogeneous nucleation), but in technical melts it is much more likely that tiny solid particles remain floating in the melt even after it is superheated to the pouring temperature. Such *nucleating agents* promote the rearrangement of atoms around themselves, leading to heterogeneous nucleation, especially if the melt wets the nucleating agent. In this instance much less undercooling is sufficient to initiate solidification. Sometimes nucleating agents are added on purpose to provide many sites on which solid grains can form, thus refining the grain size with attendant benefits in greater strength (Sec. 2.1.4). Finer grain may also be achieved by faster cooling rates (but this may lead to undesirable side effects such as the development of more shrinkage porosity and entrapment of gases) or by presolidification and forging of the melt (Sec. 2.1.1).

Ingot casting

Alloys destined for further plastic deformation are cast into *ingots,* that is, simple bodies of circular, square, or rectangular cross section of a size suitable for subsequent working (Fig. 3.2). The mold may be a metal (usually iron or steel) *permanent mold* (for all alloys, Fig. 3.2a) or a *water-cooled mold* (mostly for copper-base alloys, Fig. 3.2b). In all cases, solidification begins from the mold walls and proceeds toward the center, giving

FIG. 3.2 Ingot solidification: (*a*) permanent mold, (*b*) water-cooled mold, and (*c*) semicontinuous casting.

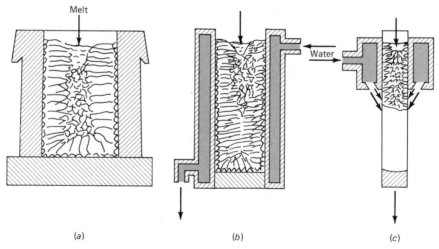

rise to the typical solidification patterns shown in Figs. 2.2 and 2.15. Piping is avoided by feeding molten metal either from the ladle or from a hot metal reservoir contained within a refractory-lined extension of the mold (*cap*). The metal may be kept hot even further by placing some exothermic compound on top of the melt (*hot top*). The smallest ingots cast by these techniques may be 1–2 in (25–50 mm) thick in nonferrous metals and 6–8 in (150–200 mm) thick in steels; they may range up to 20 tons in weight in nonferrous metals and 300 tons in steels.

In continuous casting processes (Fig. 3.2c) the solidification zone is localized in the water-cooled die, and the ingot is withdrawn gradually as solidification progresses while the ingot is further cooled with water sprays. The process may be periodically interrupted to allow removal of an ingot (*semicontinuous casting*). If it goes on almost indefinitely (*continuous casting*), the ingot is either cut up during its movement or is directly fed into a rolling mill (thus producing a completely continuous process). The latter approach is eminently practicable for aluminum and copper alloys, particularly in the wire and strip form.

A ceramic mold wash or a lubricant-type parting compound, often with graphite, prevents adhesion and welding of the melt to the mold. Further protection is obtained by oscillating the mold in the continuous casting of steel.

Casting of shapes

When the casting process is designed to produce a component of complex shape, the *mold* is prepared to provide that shape (with due allowance for dimensional changes after solidification). The melt has to be brought to this cavity in an orderly fashion and, while details may vary, some features (Fig. 3.3) are common to the various processes.

The *pouring basin* is a receptacle large enough to accommodate the stream of metal, and is often shaped to assure smooth flow of the melt. The *sprue,* usually of conical shape to minimize turbulence and aspiration (intake of air from the sand mold), transports the fluid downward, toward the distributing branches (*runners*), which are of larger cross section and often streamlined to slow down and smooth out the flow, and are designed to provide approximately uniform flow rates to various parts of the cavity. The runners are connected by *gates* to the mold cavity; at the junction to the cavity these gates are much reduced in thickness (*in-gates*) to allow easy separation from the solidified casting. At this point the metal stream is chocked so that it flows quietly into the cavity. The runner is often extended beyond the last gate and a *well* may also be provided at the base of the sprue to retain inclusions, scum, and various refractory materials that may have been washed along with the fluid stream. Positioning and dimensioning of the runner system vitally affects the soundness of the

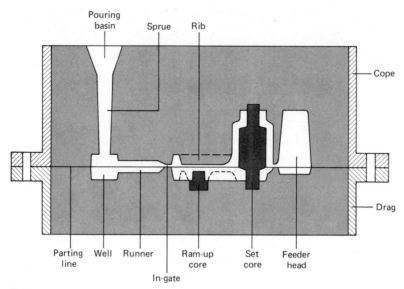

FIG. 3.3 Characteristics of a cope-and-drag sand mold.

casting, because it determines at what rates the melt is supplied to various parts of the cavity. Therefore, this subject has been greatly developed in recent years, both experimentally and theoretically, drawing on many of the principles of fluid flow.

Heat is extracted so as to move the solidification front from the remotest sections toward points of feeding. In expendable molds this can be aided by placing metal inserts (*chills*) into the refractory at points where maximum cooling is desired. In permanent molds localized cooling is achieved by placement of *cooling fins* or *pins* on external surfaces, or even by *air* or *water cooling* passages in the mold. In addition, it is necessary that liquid should be supplied to compensate for solidification shrinkage. Solidification time t_s is, as might be expected, directly proportional to volume over surface area (through which heat extraction occurs). Chvorinov[1] has shown for a large variety of shapes and sizes that the relationship is quadratic:

$$t_s \propto (V/A)^2 \tag{3.1}$$

Chunky portions of the casting must be fed with *risers* (*feeder heads*) that themselves are chunky, of a high *V/A ratio* (Fig. 3.4). An example is shown in Fig. 3.5.

---

[1] N. Chvorinov, *Proc. Inst. Br. Foundrymen, 32*:229, 1938–39.

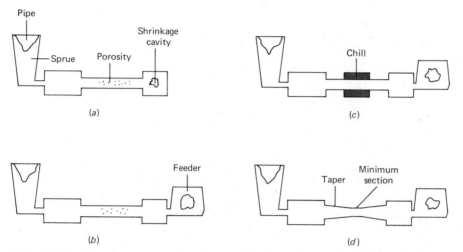

FIG. 3.4 Feeding of a casting: (a) no feeder, (b) cavity eliminated with feeder head, (c) porosity eliminated with chills, and (d) porosity eliminated by tapering the thinnest section.

Gases and solidification

As discussed in Sec. 2.3.3, melts are capable of accommodating large quantities of gas, but the solubility suddenly drops on solidification (Fig. 2.25). The excess could be retained in the melt in the atomic form or may appear as gas bubbles entrapped in the solidified metal (e.g., nitrogen in steel, hydrogen in aluminum). By and large, gases are undesirable in the solidified metal.

The excess gas may be removed before casting by various techniques:

1 Since the solubility S of a gas in the melt increases (or decreases) with the square root of the partial pressure p_g of the gas over the melt according to Sievert's law,

$$S \propto p_g^{1/2} \tag{3.2}$$

excess gas can be readily removed (although at some expense) by melting, holding, or even pouring in vacuum, or by *purging* with a gas that is insoluble in the melt (such as argon) or with a gas that also fulfills a fluxing function (chlorine for aluminum).

2 The gas may be removed by combining it with another element to form a compound (typically an oxide) by adding aluminum, silicon, calcium, or other metals to iron, or phosphorus to copper. This, of course, leads to inclusions formed by the *deoxidation* product in the metal which may or may not become trapped in the casting.

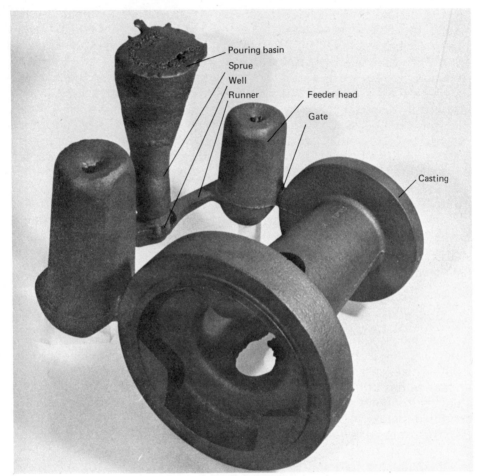

FIG. 3.5 A cored gray-iron casting showing sprue, runners, gates, and risers. (Massey-Ferguson Brantford Foundry, Brantford, Ontario.)

3.1.3 Casting Properties

The properties of metals that make them suitable for casting are partly fundamental and partly technological.

Fundamental properties

Among the fundamental properties the most obvious is *viscosity.* Since the pouring process is essentially a problem of fluid flow, lower viscosity is beneficial. The viscosity of a melt is a function of *superheat,* that is, the degree to which the metal was heated above its melting point. While information is somewhat sketchy, one might generalize by saying that viscosity is a function of composition and of superheat as expressed on the homologous temperature scale. For example, in eutectic systems one may find (Fig. 3.6) that the viscosity changes linearly with alloy composition, but it could also show marked variations with phase boundaries (maximum viscosity at the limit of solid solubility, minimum at the eutectic composition). Another fundamental property, surface tension, affects the wetting of inclusions (Sec. 2.3.2) and also limits the minimum radius that can be filled in the corner of a mold without applying external pressure (typically to 0.1 mm in gravity casting).

While one thinks of the melt as a fluid that is subject to the laws of fluid flow, it must be recognized that the surface usually gets very rapidly covered with an *oxide film,* and the nature of this film greatly influences casting behavior. Thus, the extremely dense and tenacious oxide of aluminum makes it flow as if it were inside a rather tough bag, and alloying elements that modify the oxide greatly affect the casting behavior of aluminum alloys.

FIG. 3.6   Probable variation of viscosity in a eutectic system.

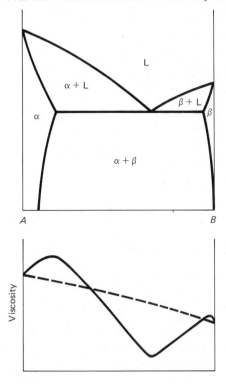

Technological properties

From the point of view of the foundryman, complex technological properties—those that may be more difficult to define but bear a closer relationship to the observed casting behavior—are much more important than the fundamental properties.

Foremost among the technological properties is that described as *fluidity.* This is a semiquantitative index showing the ability of the melt to fill out a long spiral-shaped or a thin platelike cavity (Fig. 3.7). It is obviously a property that is a function not only of the metal but also of the mold. For a given mold material and temperature, the fluidity index (length of the spiral) increases with increasing superheat because this lowers viscosity and delays solidification. There is, of course, a limit set by the furnace refractories. A solidification mechanism that allows orderly freezing of the melt, such as is found in pure metals and eutectic compositions, also helps. Dendrite arms growing into the path of the liquid supply cut off the flow gradually (Fig. 3.8); therefore the fluidity of alloys with a wide solidification range is generally lower (except in pressure die casting, Sec. 3.2.2). For a given alloy and pouring temperature, fluidity increases with increasing mold temperature, because solidification is slowed down; this benefit is, however, gained at the expense of slower solidification which may limit productivity (Fig. 3.9) and leads to coarser grain. The advantage of the fluidity index is that it reflects the effect of those variables that can also be employed in controlling solidification in actual castings.

FIG. 3.7   Determination of fluidity in (*a*) spiral and (*b*) plate molds.

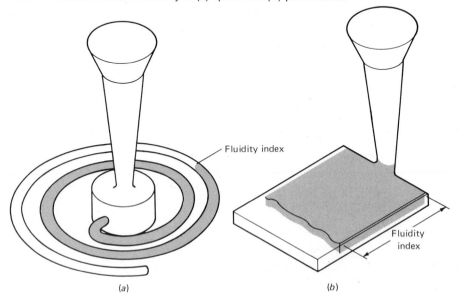

Fluidity index

Fluidity index

(*a*)   (*b*)

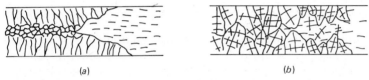

FIG. 3.8 Solidification in a channel of (a) pure metals and eutectics, and (b) solid-solution alloys.

An even broader technological index is the *castability* of a metal. This term incorporates, in addition to the technological concept of fluidity, the aspects that define the ease of producing a casting under average foundry conditions. Thus, an alloy classified as highly castable has not only a high fluidity but is also relatively insensitive to accidental changes in process conditions, is more tolerant of gating and risering, is less sensitive to wall-thickness variations, and, in general, will produce acceptable quality castings with less skill.

3.1.4 Casting Alloys

With the exception of metals and alloys that are produced directly by powder metallurgy or electrolytic techniques, all metals and alloys must first go through the melting stage (Fig. 1.3). It is, however, usual to distinguish between *wrought alloys* that possess sufficient ductility to permit hot or cold plastic deformation (these will be discussed in Chap. 4) and *casting alloys,* which excel in castability (such as the eutectics) or have a structure that cannot tolerate any deformation (for example, alloys with very high proportions of intermetallic compounds and other hard constituents). There are, of course, overlaps between the two groups, and the

FIG. 3.9 The effect of mold temperature on casting and freezing behavior.

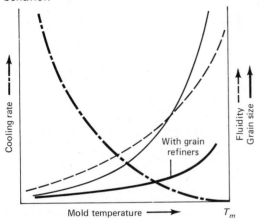

same material, because of its attractive service properties, may be produced both in the wrought and cast form.

The casting alloys are most conveniently classified according to the base metal, in the order of increasing melting points (Table 1.3), which also indicates increasing cost and difficulty in melting and superheating them to the appropriate pouring temperature. Some properties of the most popular alloys are given in Table 3.2; general properties are discussed in the following sections.

Tin-base alloys

Of the widely used metals, tin has the lowest melting point (232°C). It is highly corrosion-resistant and nontoxic; therefore a large proportion of the total consumption is for coating mild steel sheet (*tin plate*) for container purposes. Its low strength precludes its use as a construction material, except for bearing surfaces where the low shear strength and low adhesion to other metals assures low friction coefficients even when lubrication fails. Because of its low strength, it must be backed by a stronger material or, if the bearing layer is to be thicker, it must be strengthened by a duplex structure in which a hard compound is dispersed in the soft tin matrix. This is achieved by adding Sb to form the hard intermetallic compound SbSn, in the shape of small, hard cubic crystals (*cuboids*). These tend to rise to the surface of the melt; the addition of some copper improves the situation by forming a copper-tin intermetallic which solidifies as a spatial network of needles, thus trapping the cuboids.

Lead-base alloys

Lead too has a low melting point (327°C) and good corrosion resistance, but it is toxic and its use as a coating on mild steel (*terne plate*) is limited to applications where human contact is avoided. Its low strength qualifies it for a bearing material, although of somewhat lower quality than tin. Strengthening is again obtained by alloying, usually with tin and antimony, so that the SbSn cuboids are dispersed in a matrix of ternary Sn-Pb-Sb eutectic. These ternary alloys are not only hard but also possess a high fluidity imparted by the presence of tin; therefore they are also used as type metal with a higher antimony content that reduces solidification shrinkage to very small values, thus giving a clear, clean type face.

Zinc-base alloys

Zinc is the only low-melting (419°C) metal extensively used as a structural casting material. Its high fluidity and low melting point make it eminently suitable for casting into steel dies. Strengthening is obtained by solid-solution alloying with approximately 4% Al and 1–2% Cu. Its major weakness is low creep strength. Also, its corrosion resistance is low in the presence of contaminants such as Cd, Sn, and Pb. This characteristic is, however, useful in protecting mild steel by coating it with a sacrificial zinc coating (*galvanized steel*).

Magnesium-base alloys

The melting point of magnesium is substantially higher (649°C) but still low enough to allow casting into permanent molds made of steel. Therefore, structural parts may be cast of it by a variety of techniques. Its low density and reasonable strength, coupled with corrosion resistance, make it very attractive for structural applications. The major barrier is its cost. Casting alloys are solid-solution strengthened with up to 10% Al (the eutectic composition with 32% Al is too brittle to be practical), and some precipitation hardening may be obtained by adding Mn, Zr, or Zn. The fluidity is quite adequate because the oxide is not dense and does not hinder flow.

Aluminum-base alloys

Melting at an only slightly higher temperature than magnesium (660°C), and almost as light but considerably cheaper, aluminum and its alloys represent (beside nodular iron) the fastest-growing segment of the casting industry. Its corrosion resistance is excellent (except to alkali) and its strength is readily improved through solid-solution and precipitation-hardening mechanisms.

The oxide film formed on the melt surface is dense and tough and, as already mentioned, reduces fluidity. The ease of casting is greatly affected by the influence of alloying elements on this oxide film. Silicon is most beneficial, making silicon alloys the most castable aluminum alloys. The eutectic composition (around 12% Si) is, of course, the most favorable. Its properties are greatly improved by *refining* the eutectic structure through the addition of a small quantity of sodium to the melt just prior to pouring. Hypereutectic alloys contain the very hard and brittle silicon in a pro-eutectic form and are thus extremely wear-resistant (for making engine blocks without cast-iron cylinder liners). The hard and brittle silicon of the eutectic does, however, limit the ductility of the alloy; therefore lower Si contents are frequently used, sacrificing some of the castability in exchange for improved ductility.

Magnesium is a useful solid-solution strengthening element but creates the problems typical of a wide solidification range. The same applies to copper; however, smaller quantities suffice because the interme-

TABLE 3.2 PROPERTIES OF SELECTED CASTING ALLOYS*

Alloy		Typical composition, weight %	Preferred casting method	Liquidus (solidus), °C	Shrinkage allowance,† %	Mechanical Properties‡			
Name	ASTM No.					UTS, MPa	YS, MPa	Elongation (2 in), %	Hardness,§ BHN
Ferrous:									
Cast steel	60-30	≤0.25C	Expendable mold		1.5–2	420	210	24	
	175-145		Expendable mold		1.5–2	1200	1000	6	360
Gray iron	20	3.5C, 2.4Si, 0.4P, 0.1S	Expendable mold	1180	1	125		(<1)	170
	60	2.7C, 2.0Si, 0.1P, 0.1S, 0.8Mn	Expendable mold	1290	1	420		(<1)	270
Malleable iron	A47	2.5C, 1.4Si, 0.1P, 0.1S, 0.4Mn	Expendable mold		1	350	220	10	
Nodular iron	60-40-15	3.5C, 2.4Si, 0.1P, 0.03S. <0.8Mn	Expendable mold		0.8–1	420	320	15	160
Stainless steel	CF8	0.08C, 19Cr, 9Ni	Expendable mold		2.5	500	240	45	
Cu-base:									
Leaded red brass	4A	5Sn, 5Pb, 5Zn	All	1010(854)	0.8–1.8	250	105	32	62
High-lead tin brass	3A	10Sn, 10Pb	All	926(760)	1–2	280	125	30	67
Leaded yellow brass	6C	1Sn, 1Pb, 37Zn	All	940(912)	0.8–1.5	280	100	15	50

Al-base:

Alloy	Nominal composition, %	Casting process	Melting range	†				
108	4Cu, 3Si	Sand	627(521)	1.5	150	100	2.5	55
D132	3.5Cu, 9Si, 0.8Mg, 0.8Ni	Permanent mold	582(520)	1	250	195	(1)	105
380	3.5Cu, 8Si	Die	593(538)	0.6	300	170	(2)	

Mg-base:

AZ 91	9Al, 0.7Zn, 0.2Mn	All	596(468)	1.5 (die 0.6)	280	135	5	70
EZ 33A	2.7Zn, 0.5Zr, 3 rare earths	Sand and permanent mold	643(543)	1.2	160	110	3	50

Zn-base:

AG40A	4Al, 0.04Mg	Die	387(381)	0.3–0.6	285		10	82

Pb-base:

Type metal	3Sn, 11Sb	Die(type)	248(240)					

Sn-base:

Babbitt Alloy 1	4.5Sb, 4.5Cu		371(223)		65		2	17

* Data compiled from *Metals Handbook*, 8th ed., vol. 1, American Society for Metals, Metals Park, Ohio, 1961.

† Patternmakers' allowance.

‡ Minimum properties in the as-cast condition, except malleable and nodular cast iron (annealed) and D132 and EZ 33A (precipitation hardened). To convert MPa into 1000 psi, divide by 7.

§ Load: 3000 kg for ferrous, 500 kg for nonferrous materials.

tallic compound is very hard and is effective as a precipitation-hardening phase. This is true also of zinc, but the fluidity is reduced.

Copper-base alloys

The melting point of copper (1083°C) is too high for steel dies (unless protected by heavy coatings), but other casting methods are practiced because copper has attractive corrosion resistance and high electrical conductivity. The majority of castings are, however, made of alloys that combine good fluidity with reasonably high strength. Strengthening is usually achieved by solid-solution alloying, as in tin bronzes, aluminum bronzes, and the brasses. Because copper alloys have been around for such a long time, many of them were given proprietary names and sometimes misleading designations (some brasses are commonly called bronzes). The technologically important features, however, can always be deduced from their composition. Many casting alloys are hardened by adding solid-solution elements up to the limit of solubility (e.g., 10% Sn bronze) for maximum strength without undue embrittlement by excessive intermetallic particle content.

There are few copper alloy systems with useful eutectics; therefore most alloys tend to be solid solutions and, whenever the solidification range is wide, fluidity suffers. This is particularly true of tin bronzes, in which fluidity is then increased by adding phosphorus which forms a low-melting ternary eutectic (*phosphor bronzes*). The addition of zinc with its low vapor pressure also increases fluidity. Lead is often added, primarily to improve machinability, but it also benefits fluidity.

Nickel- and cobalt-base alloys

The high melting point of nickel (1435°C) and of cobalt (1495°C) and their good corrosion resistance make them eminently suitable for many critical applications. Their strength and, particularly, hot strength can be greatly increased with solid-solution and precipitation-hardening alloying elements. Some of these *superalloys* have such high second-phase content that they are not deformable, and such cast superalloys can outperform other materials in high temperature applications, particularly as jet-engine parts.

Iron-base alloys

In the familiar iron-carbon diagram, carbon appears in the form of the intermetallic compound Fe_3C. Up to a carbon content of 1.7% (the maximum solubility in the γ-iron) one speaks of *steels.* Because of their high melting point and fairly wide solidification range, steels are not favored as casting alloys, and their use is limited to cases where the high strength and ductility of steels is essential.

The majority of ferrous castings is in the class of *cast irons,* in which the carbon content is high enough to make the eutectic appear. Because the intermetallic compound Fe_3C is very hard and brittle, the so-called *white cast irons* are also hard and brittle and have only limited use.

If sufficient time is available and other conditions are favorable, the carbon separates in the more stable form of graphite. The Fe-C equilibrium diagram, representing the more stable form, shows the eutectic temperature somewhat higher and the eutectic composition somewhat lower than in the $Fe-Fe_3C$ system. To obtain the stable Fe-C form during solidification, it is necessary to alloy the system with graphite-stabilizing elements such as silicon or nickel; such *gray iron* has excellent casting properties. Within the maximum temperatures (typically 1450°C) allowed by the usual refractory furnace linings, increasing the carbon content toward the eutectic composition allows greater superheat (160°C at 2.5% C to 280°C at 3.6% C) and therefore higher fluidity. Solidification in the graphitic form counterbalances much of the solidification shrinkage of the iron thus assuring soundness and relative absence of porosity. A disadvantage is, however, that the graphite separates in large petallike plates that reduce the strength of the structure and practically eliminate ductility.

Much of the ductility can be regained by bringing the graphite into a less detrimental, globular form. This is achieved by inoculating the melt just prior to casting with a small amount of magnesium which, through a mechanism that is only partially understood, causes the graphite to separate in well-defined, roughly spherical particles, distributed in the iron or pearlite matrix. Such *nodular (spheroidal) cast iron* combines the advantages of gray iron with some of the ductility of steel.

An alternative and older method of arriving at roughly the same end result uses *malleable iron,* which is cast as a white iron but is then subjected to prolonged heating which causes the iron carbide to dissociate into carbon (which congregates into somewhat diffused globules) and a ductile ferrite matrix.

Stainless steels are used because of desirable service properties but their high melting point and solid-solution structure present great challenges to the foundryman.

High-temperature materials

Some of the higher-melting alloys are used in only very specific circumstances to produce castings. Thus, titanium alloys combine high elevated-temperature strength with low weight and high corrosion resistance, but the great affinity of titanium for oxygen, the high melting point, and the low fluidity place great demands on the skills of the melter and caster.

The very rapid oxidation of the refractory metals (molybdenum, tungsten, and niobium) necessitates special melting and casting techniques.

3.2   CASTING PROCESSES

While the number of casting processes is very large, all can be grouped into two broad categories (Fig. 3.10): casting into expendable and into permanent molds. The *expendable mold* is used only once and is broken up to free the solidified casting, while the *permanent mold* is expected to last for several hundreds or thousands of castings and must be of such a construction as to release the solidified casting. The competitive position of casting processes is greatly influenced by dimensional tolerances, surface finishes, machining allowances, and so forth. These aspects are dealt with in Chap. 11.

3.2.1   Expendable-Mold Casting

The large variety of techniques coming under this heading can be further subdivided by considering the pattern that creates the shape of the mold cavity.

FIG. 3.10   Classification of casting processes.

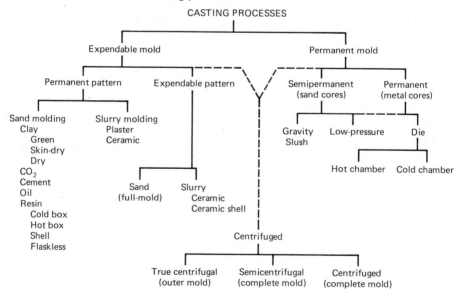

Patterns

The expendable mold is prepared by consolidating a refractory material (sand or other refractory powder or slurry) around a pattern that defines the shape of the cavity and also incorporates, in most instances, the gates, runners, sprue, and risers required to fill the mold (Fig. 3.3).

The *pattern* can be *permanent* so that it may be reused repeatedly; therefore it must have a shape that allows its withdrawal from the consolidated refractory mold. Cavities, undercuts, and recesses must be shaped by the insertion of stronger *cores* which allow greater complexity of shape but at a higher cost.

Alternatively, the pattern may be *expendable,* made up of a material that is melted out before or burnt up during casting, and need thus not be removed from the mold. Shape limitations are then much fewer and the only criterion is that the refractory can be shaken out from all cavities and intricate details of the finished casting.

Permanent patterns are made of wood or, for greater durability and dimensional stability, from a metal or, sometimes, a strong plastic. Their shape and dimensions differ in several respects from those of the finished casting (Fig. 3.11). To allow withdrawal of the pattern from the mold, a *parting plane* is selected that conveniently divides the shape into two or more parts. Vertical surfaces are sloped (have *draft*) to allow easy pattern removal, and the dimensions are increased to allow for contraction of the solidified casting from the solidus to room temperature (*shrinkage allowance*). The pattern must also provide nesting holes (*core prints*) for the accurate location of cores.

The simplest pattern for producing the shape shown in Fig. 3.11 would be in *one piece,* and the gates, runners, etc., would be added during molding. For higher productivity, these elements are usually incorporated into the pattern, which is then parted (split) along the parting line and the two

FIG. 3.11 Pattern for a cast connecting rod; finished part shown in broken lines.

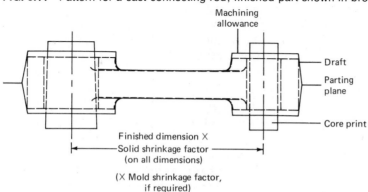

halves are either formed on two surfaces of the same plate (*match plate*) or are formed on separate upper and lower plates (*cope and drag patterns*). The production rates are further increased if several pieces are molded and cast simultaneously in the same mold; for this, *multipiece pattern plates* are prepared. If a very large part of fairly simple configuration is needed in small numbers, *skeleton patterns* suitable for hand molding may be economical. Simple shapes, especially of rotational symmetry, may also be formed by rotating a cross-section board (a *sweep pattern*) in the sand.

While patterns are required to make cavities in the mold, cores themselves are formed in the cavities of *core boxes.* Cores are placed in the mold after the pattern has been withdrawn from it. Their correct location is assured by core prints. If their weight cannot be supported, the cores are placed on *chaplets* (small, often perforated metal supports that will melt into the casting alloy).

The refractory material will have to be contained around the pattern and this container is traditionally called a *flask.* When it is in two pieces to accommodate the upper and lower halves of the pattern, they are called the *cope* and *drag halves,* respectively. Very large molds may be formed in a pit in the ground.

Because the mold and cores are collapsed after casting, all expendable-mold processes offer a freedom in part geometry that often can only be matched by other processes at great extra expense. This feature, coupled with the ability to choose the heat resistance (refractoriness) of the mold material to suit the pouring temperatures, makes for the widest applicability of the processes.

Sand casting

Of all refractory materials, *sand* (SiO_2) has the lowest cost and is satisfactory for rather high casting temperatures, including that of steel (although it is now supplanted by zircon for the latter purpose). Sand in itself flows freely and must be bonded temporarily to form a mold of sufficient strength to withstand the pressure of and erosion by the melt. The bond, however, must be sufficiently destroyed by the heat of the solidifying metal so that it allows shrinkage of the metal and, finally, removal of the sand without damage to the solid casting. The bonded sand must have adequate permeability to allow escape of gases from the melt and also of any gases that may be produced in the binder itself by the heat of the melt. Processes are often described according to the *bonding agent* used with the sand.

Green sand molds are the cheapest because they are bonded with *clay.* Clay is an aluminosilicate with a layered structure which is fairly strong in the dry state, but becomes readily deformable when water is added that adsorbs on the platelets. Some sands already contain the required few percent clay, but superior qualities are usually obtained when a quality clay (e.g., 6–8 percent bentonite) is added to pure quartz sand. With 2–3 percent water and thorough mixing (by *mulling*) a readily transportable and moldable *sand mix* is obtained. The mold may be left in the damp condition (*green sand*); it may be partially dried around the cavity to improve the surface quality of the casting and reduce pinhole defects in it (*skin-dried sand*), or the entire mold may be dried out (*dry sand*). Dry sand gives better surface finish, but may cause tearing in hot-short materials and it places greater demands on floor space and equipment. Facing materials (coal or graphite) added to the sand or applied to the mold surface generate gases on contact with the hot metal, thus reducing metal penetration and reaction with the sand.

The CO_2 *process* uses a silica gel as the bonding agent, which is formed by mixing the sand with 3–5 percent water glass ($Na_2O \cdot xSiO_2 + nH_2O$), a liquid. On completion of molding, CO_2 is bubbled through, forming the reaction products Na_2CO_3 and a gel of $xSiO_2 \cdot nH_2O$ composition. This gives a firmer sand mold with less wall movement and, therefore, more accurate castings.

Cement binder (which also forms a gel of great strength) is used occasionally, mostly for steel.

Oil sands are mixed with a drying-type vegetable oil and some cereal flour. These oils are unsaturated, and their double and triple bonds allow cross-linking and polymerization on heating to temperatures around 230°C. Thus, the sand is bonded with what could be regarded a flour-filled polymer and acquires high strength, suitable for core applications.

Resin-bonded sands are a more recent development with thermosetting resins as the bonding agent. The resin may be set by heating the core box to 200–250°C (*hot-box method*) or by curing with an airborne catalyst (*cold-box method*).

Shell molding is a variant of the resin-bonded sand technique. The pattern must be made of metal and heated to 200–260°C and coated with a parting agent. It is then placed in a box that contains sand, coated with a heat-curing resin; when the box is inverted, the sand settles on the pattern and a thin shell cures *in situ,* faithfully reproducing details of the pattern. After the excess sand is removed, the shell is stripped, combined with the other half, placed in a flask, and backed with some inert material such as steel shot to provide support. The greater strength of the mold often allows forming of cores integral with the mold. Floor space and sand quantity are reduced but recycling of the sand is more expensive.

An interesting variant of sand casting combines an expendable mold with an expendable pattern made of polystyrene foam or similar material. This *full-mold* process allows great freedom in shapes, because the pattern is left in the mold to burn up during casting.

The production rates of sand-casting processes can be adjusted to fit the needs. For only a few parts, the sand may be shoveled into the flask around a one-piece pattern and *rammed* by hand. For mass production, the sand is conveyed to the mold and dropped, blown, or slung into it (by having it flow over a fast-rotating wheel). Compaction in the flask is obtained by mechanical means, such as *jolting* or *squeezing* over match or cope- and drag-pattern plates. When pressures are high enough (around 1000 psi, 7 MPa), a properly bonded sand acquires enough strength to maintain its integrity without a supporting flask (*flaskless molding*); production rates can reach 250–1000 molds per hour, stacked end to end, in a single production unit.

Slurry-molding processes

Instead of the mold being compacted of a granular refractory material with an appropriate binder, a finer-grained refractory may be made into a *slurry* with water and poured around the pattern. A smoother surface finish is assured and, if so chosen, the refractory may be more heat-resistant than the bonded-sand variety. Since the shrinkage of mold and casting can be closely controlled, these processes are often called *precision-casting* processes.

Plaster molding relies on the well-known ability of a plaster of paris (gypsum) slurry to flow around all details of a pattern. Various inorganic fillers may be added to improve strength and permeability. After a rather complex baking sequence, the mold is assembled and the metal poured. Since gypsum is destroyed at 1200°C, the process is not suitable for ferrous castings. This limitation also applies to a patented variant in which sand is bonded with gypsum.

Ceramic-mold casting is suitable for all materials, because the slurry is made up of selected refractory powders, such as zircon ($ZrSiO_4$), alumina (Al_2O_3), or fused silica (SiO_2), with various patented bonding agents. The fine-grained ceramic slurry is applied as a thin facing to the pattern and is backed up with lower-cost fire clay. The higher cost of these processes is well justified by their success in producing quality castings of highly alloyed, high-melting-point metals.

Expendable-pattern molding

The full-mold process has already been mentioned. Much more widespread is investment casting.

Investment casting (the *lost-wax process*) is capable of producing the most complex shapes, because the pattern—complete with the feeding system—is made of wax (but sometimes of a plastic, or even frozen mercury), and the refractory slurry is poured around this. Because there is no need to remove the pattern, cores are not needed either, and the only limitation on shape is that the ceramic shall be removable after the casting has solidified. The disposable wax patterns are readily produced in large quantities by injection molding into metal dies.

Individual patterns are assembled with sprues, runners, and gates into a so-called *tree.* This is then precoated by dipping it into a refractory slurry, dusting it with refractory sand, and placing it in a flask where a thick refractory slurry is poured around it. When the slurry has jelled by drawing off excess water, the mold is dried in an oven in an upside-down position to allow the wax to run out. Before casting, the mold is fired at 700–1000°C to increase the fluidity of the melt that will be poured. The cost of the mold may be reduced and the rate of production increased by dispensing with the solid mold. In the *ceramic shell-casting process* the tree is prepared as before, but is then covered with refractory in a fluidized bed. Several layers of gradually coarsening coats are applied to reach sufficient thickness. The shell mold is dried and fired, if necessary supported by a granular material, and the metal is cast.

3.2.2 Permanent-Mold Casting

While in the processes described above the mold was destroyed after the solidification of the casting, the mold is reused repeatedly in the *permanent-mold casting* processes. This requires a mold material that has a sufficiently high melting point to withstand erosion by the liquid metal at pouring temperatures, a high enough strength not to deform in repeated use, and a high thermal-fatigue resistance to resist premature *crazing* (the formation of thermal fatigue cracks) that would leave objectionable marks on the finished casting. Finally, and ideally, it should also have low adhesion (Sec. 2.4) to the melt.

The *mold material* may be cast iron, although alloy steels are the most widely used. For higher-melting alloys (brasses and ferrous materials) the mold steel must contain large proportions of stable carbides; more recently refractory metal alloys (particularly molybdenum alloys) have found increasing application. *Graphite molds* can also be used for steel although only for relatively simple shapes. The resistance of the mold to the melt can be increased with refractory coatings (mold washes) and adhesion can be reduced by graphitic, silicone, or other films (parting compounds).

A prime requirement of permanent-mold casting is that the solidified casting be readily removed from the die cavity. Therefore, shapes that can be produced are more limited than those for casting in an expendable mold, although complexity can be attained at the expense of making more complicated dies with *fixed* or *movable metal cores.* Or, when casting is accomplished under gravity or low-pressure feed, sand cores may be inserted into the permanent mold (*semipermanent molds*). *Ejector pins* are necessary to remove the solidified casting, particularly if the process is mechanized. The permanent mold works as a heat exchanger and, to ensure the heat balance for optimum casting conditions, provision is made for radiating pins or fins or water cooling.

Since the mold can be preheated to a desirable temperature (typically 150–200°C for Zn, 250–275°C for Mg, 225–300°C for Al, and 300–700°C for Cu), the wall thickness of castings can be reduced, even though at the expense of slower solidification and a prolonged casting cycle (Fig. 3.9). Nevertheless, solidification rates are much higher than in refractory molds; therefore the grain size tends to be finer and the supply of liquid is more rapidly cut off. It is imperative, therefore, that proper feeding be provided, and even so, porosity can be and usually is higher than in similar castings made in expendable molds. The permanent molds have no permeability and *venting* is critical; otherwise, gas porosity compounds the problems. The permanent mold is always stronger than the solidifying casting; therefore hot shortness is more of a danger and casting alloys prone to this defect (long freezing range, low-melting matrix) are avoided.

While the term permanent-mold casting should really be used as a generic one, everyday usage has made the terminology rather confusing.

Gravity-fed permanent-mold casting (usually referred to simply as *permanent-mold casting*) builds on the same principles as expendable-mold casting, except that the mold is made of an appropriate permanent material. The casting machine is simply a bed that takes the mold halves and is equipped with a long handle or a hydraulic actuator to move one of the mold halves (Fig. 3.12*a*). In conjunction with sand cores, it is a fairly versatile process. The mold is protected from the melt by a thick (up to 0.040 in or 1 mm) *ceramic coating,* thus producing a fairly rough surface finish which is nevertheless comparable to the better expendable-mold methods. In return, the process allows casting of even iron and steel, at least in smaller weights, but its main application is to aluminum and copper alloys.

A variant of permanent-mold casting, the *slush-casting* process, is used mostly for nonstructural, decorative purposes; the mold is filled with the melt and, after a short time, it is inverted to drain off much of the melt, leaving behind a hollow casting.

Low-pressure permanent-mold casting is applied primarily to aluminum alloys. The mold is situated right above the melting or holding furnace and metal is fed by air pressure into the mold cavity (Fig. 3.12*b*). Solidification is directed from the top downward and the air pressure is released as soon as the cavity is filled with solid metal, thus material losses are minimized. Thinner coatings (mold dressings) assure an acceptable surface quality. The die halves must be held together under sufficient force to resist the pressure in the cavity.

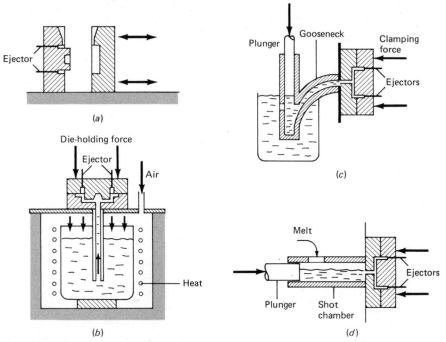

FIG. 3.12 Permanent-mold casting processes: (*a*) gravity, (*b*) low-pressure, (*c*) hot-chamber, and (*d*) cold-chamber die casting.

A variant of the process is used for casting steel into graphite molds.

Die-casting methods rely on filling the mold cavity under moderate to high pressures, thus forcing the metal into intricate details of the cavity. Solidifying dendrites of solid-solution alloys are torn off the cavity walls and the resulting solid-liquid mixture flows more readily than a pure metal or eutectic would; therefore pure metals are seldom cast by this process. The die halves are held together by a correspondingly high force; therefore the *die-casting machines* resemble a press, in which the die-holding force is exerted by the press movement while the liquid is forced into the die by a separate plunger. The machines are rated by the die-holding force.

In the *hot-chamber process* the metal is transferred to the mold directly from the *holding pot* by a submerged *pump* (cylinder and plunger, Fig. 3.12*c*) made of refractory materials, at pressures up to 6000 psi (40 MPa). The range of alloys that can be handled is limited by the pump material. At present, temperature and erosion problems limit the process to zinc (and lower-melting) alloys.

The *cold-chamber process* uses separate melting facilities, and quantities of melt sufficient for one shot are individually transferred to the cylinder (*shot chamber*) from which the plunger squirts it into the cavity (Fig.

3.12*d*) at pressures often exceeding 6000 psi (40 MPa). The alloys that can be handled are limited by the mold material; the process has long been established for Mg and Al alloys and is finding increasing use for copper-base alloys.

3.2.3   Centrifugal Casting

When a mold is set in rotation during pouring, the melt is thrown out by the centrifugal force under sufficient pressure to assure better die filling. Solidification progresses from the outer surface inwards; thus porosity is greatly reduced and any contaminants present are also pushed toward the center (which is often machined out anyway). The grain is refined by the forced movement of the melt. Centrifuging can be applied to all processes with a mold strong enough to stand rotation; nevertheless, it is customary to distinguish between various processes according to the shape of the mold.

True centrifugal casting employs molds of rotational symmetry made of steel (with a refractory mold wash or even a green- or dry-sand lining) or of graphite. The melt is poured while the mold rotates, giving a hollow product such as a tube or ring. By proper control of flow rates and movement of the pouring orifice, long and large tubes of very uniform quality and wall thickness can be cast. If desired, the outer contour of the casting can be varied, while the inside remains cylindrical.

Semicentrifugal and centrifuged casting are terms applied to the production of castings in molds prepared by any of the techniques discussed under expendable and permanent molds. When only one piece of approximately rotational symmetry (e.g., a wheel with spokes and central hub) is cast, the term semicentrifugal is applied. When odd-shaped parts are placed around a central sprue in a balanced manner (e.g., by investment molding), the term centrifuged casting is usual.

3.3   FINISHING PROCESSES

The solidification process seldom leads to a component that may be immediately used.

When casting is performed in expendable molds, first of all the casting must be freed from the mold. For green and dry sand, *shaking* is a most effective and quite adequate procedure; the clay-bonded sand is then recycled and, with suitable additions, reused. With other molding materials *reclamation* is a matter of economy because it often requires special equipment or processes. This is one of the main reasons for the economy of sand casting and for its survival as the dominant process for making larger parts even in large quantities.

Whatever the process by which the casting was made, the gates, runners, risers, and sprue must be removed. In brittle materials this is accomplished simply by breaking off, but in more ductile materials sawing or grinding becomes necessary. The remnants of the gate as well as the fin (that may have flowed between the two mold halves at the parting line or at cores) are removed through various trimming operations. The overall surface is cleaned by processes such as blasting with glass beads or steel shot, tumbling (either in a dry or wet medium of some refractory material or steel shot), or by chemical pickling. Any defects detected may often be repaired by welding without jeopardizing the function of the finished part.

Quality control and inspection at all stages of production are vital to the success of the modern casting process.

Many of the *casting defects* can be detected by visual inspection. Improperly designed melt flow or molding practices may cause sand erosion and embedment (*sand skin, scab*) in expendable mold casting. In the extreme case, the melt penetrates into the sand and causes sand and metal to fuse into one mass. Shifting of mold halves and particularly of cores is a common cause of exceeding tolerance specifications. Poorly controlled sand compaction may cause the dimensional tolerance limits to be exceeded by allowing too much movement of the mold walls (*swell*) in expendable-mold casting. The melt may break out from the flask if the mold is inadequately weighted or the mold wall is too thin. An incipient *breakout* shows up as a thin fin at the parting plane.

Insufficient cavity filling may occur with any process and can be caused by inadequate metal supply, improperly designed gating, or too low mold or melt temperatures. The latter can also lead to *cold shuts,* that is, surface depressions showing that two streams of metal met without complete fusion. This can be particularly troublesome when the metals flow inside an oxide cover, especially with alloys containing aluminum.

The greatest advances have recently been made by the extensive use of nondestructive testing techniques, including x-ray, ultrasonic, and eddy-current inspection for internal soundness. While these measures can be costly, they make the casting process competitive in applications hitherto reserved for forgings or other plastically deformed components, some of the prime examples being the crankshaft and connecting rod of the internal combustion engine. Nondestructive techniques are particularly important in detecting porosity, whether it be caused by solidification shrinkage, internal hot tearing, or gas porosity.

The finished casting is sometimes heat-treated to impart desirable properties (provided, of course, that it is made of a heat-treatable material). Even if the part is going to be used in the as-cast condition, *distortion* from residual stresses may occur in service or during machining. If this is undesirable, a *stress-relief anneal* is given, although the residual stress level cannot be reduced below the yield stress of the material at the annealing temperature.

3.4 DESIGN FOR CASTING

While the skill of the foundryman may make it possible to cast a very large variety of shapes with relatively thin walls and sudden changes in direction and cross section, design practices that violate the rules of sound casting principles can make production impossible or possible only at the expense of accepting the high cost of large rejection rates. There are several books available that detail the usual design practices for castings; however, it is possible to indicate the major principles in a general manner:

   1   As with all design, it is essential that preconceived ideas about the final shape of the part should not intrude at early stages of design. Instead, the basic function of the part must be considered, and alternative shapes that satisfy these functions must be kept under consideration as long as possible.

   2   Stress levels should be kept consistent with the casting-alloy strength readily achieved. If the material is brittle (e.g., gray and, particularly, white cast iron), one must remember that the tensile strength is much less than the compressive strength. In many other materials, the as-cast structure is likely to contain porosity, inclusions, and unfavorably located second phases resulting from nonequilibrium solidification. Therefore, the tensile and, particularly, fatigue strengths are lower than in a thoroughly worked material.

   3   The shape of the casting must ensure that the pattern can be removed from the mold and the casting from the permanent mold (except, of course, with the lost-wax technique). The extra material required by the draft should preferably contribute to load bearing, and it should not interfere with positive clamping and location in subsequent machining. Locations of drilled holes should be strengthened by bosses, shaped preferably so as to make the drill enter perpendicular to the cast surface.

   4   The shape of the casting should allow orderly solidification by moving the solidification front from the remotest parts toward the feeding end. When wall-thickness variations are essential, transition must be made by generous radii (Fig. 3.13a). Small radii or sharp corners act as stress raisers in the finished casting and prevent proper feeding during solidification. Localized heavy cross sections, such as result when the appropriate radius is applied to only the inside surface of a corner or when two ribs cross each other (Fig. 3.13b), create *hot spots* where the melt so-

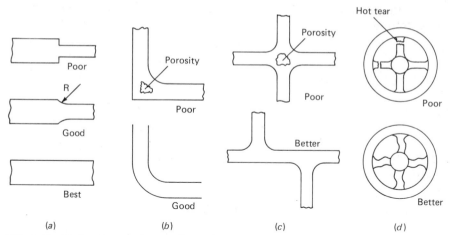

FIG. 3.13 Design features to alleviate casting problems due to (a) wall thickness variation, (b) hot spot in corner, (c) hot spot at cross-ribs, and (d) hot tearing.

lidifies only after adjacent zones have frozen; therefore shrinkage cavities form. Applying a radius to the outer surface or offsetting the ribs alleviates the problem. Otherwise, it would be necessary to reduce the cross section by placing extra cores in the thickest section.

5 A material that has a high solidification shrinkage and contains an equilibrium or nonequilibrium low-melting phase is susceptible to hot shortness. To avoid *hot tears,* such materials must be cast in molds that are either of simple enough shape not to develop tensile stresses during solidification, or in mold materials that collapse or give sufficiently to accommodate shrinkage. Whenever possible, the casting shape should allow deformation without moving large mold masses; for example, in a spoked wheel straight spokes would tear even in a sand mold, but S-shaped spokes can straighten somewhat by displacing relatively little sand and thus accommodate the required shortening on and after solidification (Fig. 3.13c). Even if fracture is avoided, distortions and internal (residual) stresses may result from restrained shrinkage.

3.5 CRYSTAL GROWING

A solidification process of greatest importance is the growing of single crystals directly from the melt, without any mold. The resulting crystal is, typically, of cylindrical shape. If of very small diameter, the crystal can be free of dislocations, and such *whiskers* possess enormous strength. Hopes for their economic, industrial-scale production and use have not materialized yet.

Much larger diameter [2 in (50 mm)] single crystals of very high purity are the foundation of the *semiconductor* industry. Germanium crystals are usually grown from the melt, by gradual withdrawal of a single-crystal seed onto which atoms grow in the same crystallographic orientation. Silicon single crystals are formed from a polycrystalline bar, by resting it on a single-crystal seed and slowly passing a molten zone (created by high-frequency heating) from the bottom up. Again, solidification proceeds by the arrangement of atoms in the lattice orientation determined by the seed crystal.

The tetravalent single crystal may be doped with small amounts of tri-valent or pentavalent elements to impart an excess hole or electron popu-lation (*p*- and *n*-type semiconductor), respectively, and is then cut with diamond-tipped saws into 0.01 in (0.25 mm) thick wafers. After chemical etching to less than half this thickness, further localized doping of a small (say 2 mm square) chip creates a *transistor* which, when wired into an elec-tronic circuit, replaces the bulky electron tube of yesterday.

By masking off selected areas on the chip surface, the location of *p* and *n* zones on an undoped silicon substrate can be accurately controlled. Resistors, capacitors, insulating oxide films, and conducting paths may be deposited in a succession of steps and *integrated circuits* created that may contain up to 2000 transistors on a chip only 3 by 4 mm.

3.6 SUMMARY

Solidification processes are involved in the production of the vast majority of all metallic materials. Almost all wrought materials are cast before de-formation. A significant portion of metals is cast into a shape directly, for applications as diverse as machine-tool bases, automotive engine blocks and crankshafts, turbine blades, plumbing fittings, and decora-tive hardware. With the appropriate molding process, parts of a com-plexity unmatched by any other process may be cast.

A sound casting results only if the limitations imposed by the solidifi-cation process are recognized:

1 For solidification to occur, the mold must be colder than T_m of the metal. Problems of fluid flow and of heat transfer limit the minimum attainable wall thickness, especially if the alloy solidifies with the growth of dendrites which choke off fluid flow.

2 Gates, runners, and risers must assure smooth, complete filling of the die cavity followed by orderly solidification, with a liquid metal supply sufficient to feed the pipes that would otherwise form.

3 Heat transfer must be locally controlled to prevent starvation of late-solidifying portions of the casting and resultant porosity.

4 Grain-size control is a most powerful means of improving mechanical properties, usually by aiming for a fine grain, but exceptionally (for high-temperature creep resistance) by directional solidification of a single grain.

5 Moldless solidification of semiconductor materials yields a very simple, round bar which is, however, the essential starting material of solid-state technology.

Further Reading

DETAILED PROCESS DESCRIPTIONS:
    BRIGGS, C. W. (ed.): *Steel Castings Handbook,* 3d ed., Steel Founders' Society of America, Cleveland, Ohio, 1960.
    Metals Handbook, 8th ed., American Society for Metals, Metals Park, Ohio, vol. 5, *Forging and Casting,* 1970.

PROCESS DETAILS, MOSTLY FOR THE LABORATORY SCALE:
    BUNSHAH, R. F. (ed.): *Techniques of Metals Research,* Interscience, New York, 1968, vol. 1, part 2, pp. 627–1188.

TEXTS:
    CHALMERS, B.: *Principles of Solidification,* Wiley, 1964.
    COOK, G. J.: *Engineering Castings,* McGraw-Hill, New York, 1961.
    FLEMINGS, M. C.: *Solidification Processing,* McGraw-Hill, New York, 1974.
    FLINN, R. A.: *Fundamentals of Metal Casting,* Addison-Wesley, Reading, Mass., 1963.
    HEINE, R. W., C. R. LOPER, JR., and C. ROSENTHAL: *Principles of Metal Casting,* McGraw-Hill, New York, 1967.
    KONDIC, V.: *Metallurgical Principles of Founding,* Arnold, London, 1968.
    WINEGARD, W. C.: *An Introduction to the Solidification of Metals,* Institute of Metals, London, 1964.

DESIGN OF CASTINGS:
    CAINE, J. B.: *Design of Ferrous Castings,* American Foundrymen's Society, Des Plaines, Ill., 1963.
    Casting Design Handbook, American Society for Metals, Metals Park, Ohio, 1962.

Problems

3.1 Look up the Al-Si equilibrium diagram and select compositions with 2% and 12% Si. Which alloy is (*a*) more ductile in the equilibrium condition, (*b*) more prone to coring, (*c*) more susceptible to hot shortness, (*d*) of higher fluidity, (*e*) easier to feed, and (*f*) more prone to microporosity?

Support your considerations with sketches, and then (g) state which alloy is more favorable for casting.

3.2   The part shown in the figure for Example 11.1 will be sandcast of steel. (a) Using data on machining allowances, draft angles, and radii (e.g., from *Metals Handbook,* vol. 1), design and draw the cast shape. (b) Taking shrinkage into consideration, make a drawing of the pattern. (c) For greater productivity, eight rings will be cast in each mold. Sketch a possible layout complete with sprue, runners, and gates; provide for streamlined flow; assure equal feeding rates to each ring by taking the runner cross section proportional to the flow rates at various points.

3.3   The part shown is to be cast of 10% Sn bronze, at the rate of 100 parts per month. Review Fig. 11.7 for surface finish. To find an appropriate casting process, consider all, then reject those that are (a) technically inadmissible or (b) technically feasible but too expensive for the purpose, and (c) identify the most economical one (make common-sense assumptions about costs, reinforced by data from Table 11.2).

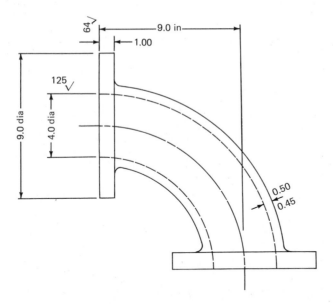

3.4   The part of Prob. 3.3 is machined at the flange face. Utilizing data on machining allowances, drafts, shrinkage, etc. (from the literature, e.g., *Metals Handbook,* vol. 1), choose an appropriate parting line and design a properly dimensioned pattern and a corebox.

3.5   A wheel is cast of low-carbon steel with straight spokes (as in Fig. 3.13, "poor"). The length of each spoke is 100 mm. The mold is refractory material, unyielding, and changes its dimensions insignificantly during heating or cooling.   The spoke cools from 1100 to 900°C in 10 min. Calculate (a) the strain, if the thermal expansion coefficient is 23×10^{-6} per °C; (b) the average tensile strain rate; (c) the flow stress at 1000°C (from Table 4.2).   (d) Assuming that the material at this temperature behaves like an ideal elastic-plastic body, and Young's modulus is 60 percent of the room-temperature value, determine whether shrinkage will be accommo- dated by the development of elastic (residual) stresses or by plastic defor- mation.   (e) Subject the above problem statement to a detailed critique regarding the validity of the simplifying assumptions.

3.6   With the aid of Chvorinov's rule, calculate the relative solidification times for castings of identical volumes and of the following shapes: (a) sphere of diameter d_s; (b) cylinder with $h/d = 1$; (c) cylinder with $h/d = 10$; (d) cube; (e) right rectangular prism with $h/a = 10$; (f) flat plate of the same length as (e) but of ⅓ the thickness.   (g) Plot the results to illustrate the ef- fect of shape changes.

3.7   If the part discussed in Prob. 3.2 were to be die-cast of an Al alloy, what size machine would be required?

4

BULK DEFORMATION PROCESSES

Most metallic materials are subjected, at one or another stage of processing, to plastic deformation. This means that the shape of the workpiece is changed without its volume being changed and without it being brought into the liquid state. The properties defined in Sec. 2.1.2 through 2.1.9 become important; in other words, it is essential that the material should be able to undergo plastic deformation without fracture.

4.1 CLASSIFICATION OF PROCESSES

A great many systems for classifying deformation processes can be and are often applied but, fundamentally, the soundest distinction is made according to the extent of the plastic deformation zone. In *bulk deformation processes* the thickness of the workpiece is substantially changed; in *sheet-metalworking processes* any change in sheet thickness is fairly limited. In bulk deformation it is useful to distinguish between *steady-state* processes in which all parts of the workpiece are subjected to the same mode of deformation (Fig. 4.1) and *non-steady-state* processes in which the geometry changes continually. Some processes have a transitionary character; for example, deformation is non-steady-state at the beginning and end of extrusion, but acquires steady-state characteristics while the greater part of a billet is extruded. The same principles apply whether deformation is conducted hot or cold, on the large or small scale.

A further useful distinction may be made according to the purpose of deformation. If the process aims at destroying the cast structure by

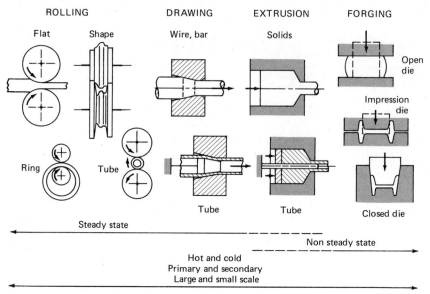

FIG. 4.1   Classification of bulk deformation processes.

successive deformation steps and the resulting semifabricated product is destined for further shaping or forming, it is customary to speak of *primary processes* (as shown in Fig. 1.3). *Secondary processes* take the products of some primary process and further transform them into a finished part, as indicated in Fig. 1.4. In the present context, secondary processes are of first importance. These include specific variants of the bulk deformation processes (Fig. 4.1) and all the sheet-metalworking processes to be discussed in Chap. 5.

Apart from the individual characteristics of processes, there are some features common to many of them:

Complex geometries can be obtained equally well by hot- or coldworking, but hot-working often allows more extensive deformation and complexity. However, small dimensions, tighter tolerances, and better surface finish can be attained in cold-working.

Cold-working is required when the finished part is used in the strainhardened condition (Sec. 2.1.7). A soft state is obtained most often by sequences of cold-working and annealing (Sec. 2.1.8) but can also be obtained by controlled hot-working (Sec. 2.1.9). Intermediate or unusual properties are imparted by controlled hot-working (Sec. 2.1.9) or by thermomechanical working (Sec. 2.2.8).

Output rates and costs are highly variable and can be adjusted according to need, as discussed in Chap. 11.

4.2 THE METAL

In all plastic metalworking processes, the workpiece shape is formed from the solid metal or alloy by the displacement of material from unwanted locations into the positions required by the part shape. This demands that the material should possess a property rather vaguely described as ductility—that is, the ability to sustain substantial plastic deformation without fracture. A material of a given ductility may fare very differently in various processes, depending on the conditions imposed on it. Therefore, our main concern is a more complex property, usually called *workability* in bulk metalworking operations and *formability* in sheet-metalworking operations (to be discussed in Chap. 5). In both groups of metalworking processes, failure occurs by the mechanism of *ductile fracture* and is induced by tensile stresses. The difference is that in sheet metalworking the tensile stresses are intentionally imposed in order to achieve deformation. In bulk deformation processes the imposed stresses are usually compressive; tensile stresses appear only as a result of inhomogeneous deformation and, for this reason, they are usually called *secondary tensile stresses.*

4.2.1 Bulk Workability

The technological concept of *bulk workability* has two components:

1 The basic *ductility* of the material (Sec. 2.1.6) allows it to deform without fracture even in the presence of tensile stresses. The reduction in area, Eq. (2.9), measured in the tensile test is therefore a useful (but not universally applicable) measure of basic ductility. Other possible measures are the number of turns to fracture in a torsion test, or the reduction in height in axial upsetting with sticking friction at the end face, which causes severe barreling and thus surface cracking in a material of low ductility (Sec. 4.4.1).

2 The *stress state* induced by the process modifies ductility. If the process maintains compressive stresses in all parts of the deforming workpiece, cavity formation cannot begin and ductile fracture does not occur. If, however, the process allows secondary tensile stresses to develop, cavity formation can begin and may lead to fracture. At what point this fracture should occur is predicted by *workability criteria,* the most useful of which has been formulated by Cockroft and Latham[1] who state that, for a given metal, the work done by the highest local tensile stress must reach a critical value characteristic of the basic workability of that material. It follows that when the development of secondary tensile stresses can be

[1] M. G. Cockroft and D. J. Latham, *J. Inst. Metals, 96:*33–39, 1968.

suppressed, deformation can be taken much further, just as fracture in the tensile test is delayed by hydrostatic pressure (Fig. 2.9a). The practical exploitation of this principle can be found in many processes that are designed to minimize secondary tensile stresses.

4.2.2 Wrought Alloys

Because bulk workability combines elements both of basic ductility and of effects imposed by the deforming process itself, it is difficult to set a lower limit of basic ductility that is essential for the success of plastic deformation. In a favorable, fully compressive system, a material of virtually zero ductility (zero reduction in area) can be successfully deformed, but the same material will fail very quickly in a process that allows secondary tensile stresses to develop. Therefore, all one can say is that wrought alloys

TABLE 4.1 SHIPMENTS OF WROUGHT PRODUCTS* (U.S.A. 1972)

Alloy group	Thousand metric tons
Steel	
Sections, rails	5,900
Plate	7,300
Hot-rolled sheet and strip	14,200
Cold-rolled sheet and strip	17,800
Galvanized sheet	4,900
Tinplate	5,000
Hot-rolled bar	11,800
Cold-finished bar	1,600
Wire	2,700
Tube, pipe	6,900
Forgings	1,200
Copper and brass	5,100
Aluminum	4,100
Lead (incl. battery)	480
Zinc	45
Magnesium	16

* Compiled from *Metal Statistics 1974*, American Metals Market, Fairchild Publications Inc., New York, 1974.

must possess a minimum ductility commensurate with the contemplated process.

This requirement is amply satisfied by all pure metals with a sufficient number of slip systems (Sec. 2.1.2) and also by most solid solution alloys of the same metals (Sec. 2.2.1). Two-phase (Sec. 2.2.3) and multiphase (Sec. 2.3) materials are deformable if they meet certain minimum requirements. There must be no liquid or brittle phase on the grain boundaries or across several grains (thus, gray iron, white cast iron, or a hypereutectic Al-Si alloy is not deformable). Excessive amounts of brittle constituent are not permissible even in a ductile matrix, especially if the brittle constituent is also coarse or lamellar. The greater the quantity of brittle constituents and the lower the ductility of the matrix, the more important it is that the material should be free of other weakening features such as inclusions, voids, or grain-boundary contaminants (Sec. 2.3.2 and 2.3.3).

Steels represent the largest portion of wrought products (Table 4.1) but, in line with the system adopted in Chap. 3 for casting, wrought materials too will be discussed in order of increasing melting point of the base metal. Apart from ductility, an immediate concern is the flow stress of the material, since this determines the magnitude of deforming stresses and forces and thus can often set a limit to attainable deformation. Therefore, relevant data are given in Tables 4.2 and 4.3 for some of the most frequently encountered materials. Every effort was made to use reliable data, and the two C and m values entered for Cu show the worst of the extreme variations occasionally found in published data. As noted in Sec. 2.1.9, C and m values are often a function of strain. The data given here for $\epsilon = 0.5$ should be used only for approximate calculations.

Tin alloys

The low strength of tin makes it unsuitable as a structural material, except for collapsible tubes and foil when its corrosion-resistant properties are desired.

Of the tin alloys, modern *pewter* contains no lead but 2–3% Cu and 1–3% Sb. It is equally well suited to casting and plastic deformation and is enjoying a revival because of its aesthetic appeal.

Lead alloys

Even though lead has low strength, its corrosion resistance amply justifies its use in the forms of sheet, tube, and cable sheathing. It can be strengthened by a number of elements (As, Sn, Bi, Te, and Cu). Antimonial lead with 6–7% Sb is widely used as flashing and roofing in the building industry. It also serves as an excellent sound, vibration, and radiation

TABLE 4.2 MANUFACTURING PROPERTIES OF STEELS AND COPPER-BASE ALLOYS[a]
(Annealed condition)

Designation and composition, %	Hot-working						Cold-working						Annealing temp.,[e] °C
	Liquidus/solidus, °C	Usual temp., °C	Flow stress,[b] MPa			Workability[c]	Flow stress,[d] MPa		$\sigma_{0.2}$, MPa	UTS, MPa	Elongation, %	q R.A. %	
			at °C	C	m		K	n					
Steels:													
1008 (0.08C), sheet		<1250	1000	100	0.1	A	600	0.25	180	320	40	70	850–900(F)
1015 (0.15C), bar		<1250	800 1000 1200	150 120 50	0.1 0.1 0.17	A	620	0.18	300	450	35	70	850–900(F)
1045 (0.45C)		<1150	800 1000	180 120	0.07 0.13	A	950	0.12	410	700	22	45	790–870(F)
~ 8620 (0.2C,1 Mn, 0.4Ni; 0.5Cr 0.4Mo)			1000	120	0.1	A			350	620	30	60	
D2 tool steel (1.5C, 12Cr, 1Mo)		900–1080	1000	190	0.13	B	1300	0.3					880(F)
H13 tool steel (0.4C, 5Cr 1.5Mo,1V)			1000	80	0.26	B							
302 SS (18 Cr, 9 Ni)	1420/1400	930–1200	1000	170	0.1	B	1300	0.3	250	600	55	65	1010–1120(Q)
410 SS (13 Cr)	1530/1480	870–1150	1000	140	0.08	C	960	0.1	280	520	30	65	650–800

Copper-base alloys:

Alloy	Melting (°C)	Hot-working temp (°C)	Temp (°C)	Hot flow stress	m	Rating	Cold flow stress	n					Annealing range (°C)
Cu (99.94%)	1083/ 1065	750– 950	600 900	130 (48) 41	0.06 (0.17) 0.2	A	450	0.33	70	220	50	78	375– 650
Cartridge brass (30 Zn)	955/ 915	725– 850	600 800	100 48	0.24 0.15	A	500	0.41	100	310	65	75	425– 750
Muntz metal (40 Zn)	905/ 900	625– 800	600 800	38 20	0.3 0.24	A	800	0.5	120	380	45	70	425– 600
Leaded brass (1 Pb, 39 Zn)	900/ 855	625– 800	600 800	58 14	0.14 0.20	A	800	0.33	130	340	50	55	425– 600
Phosphor bronze (5Sn)	1050/ 950		700	160	0.35	C	720	0.46	150	340	57		480– 675
Aluminum bronze (5 Al)	1060/ 1050	815– 870				A			170	400	65		425– 750

[a] Compiled from various sources; most flow stress data from T. Altan and F. W. Boulger, *Trans. ASME, ser. B, J. Eng. Ind.*, 95:1009, 1973.

[b] Hot-working flow stress is for a strain of $\epsilon = 0.5$. To convert to 1000 psi, divide calculated stresses by 7.

[c] Relative ratings, with A the best, corresponding to absence of cracking in hot rolling and forging.

[d] Cold-working flow stress is for moderate strain rates, around $\epsilon = 1/s$. To convert to 1000 psi, divide stresses by 7.

[e] Furnace cooling is indicated by F, quenching by Q.

TABLE 4.3 MANUFACTURING PROPERTIES OF VARIOUS NONFERROUS ALLOYS[a]
(Annealed condition)

Designation and composition, %	Liquidus/ solidus, °C	Hot-Working				Cold-Working						An-nealing temp.,[f] °C
		Usual temp., °C	Flow Stress,[b] MPa		Work-ability[c]	Flow stress,[d] MPa		$\sigma_{0.2}$, MPa	UTS[e], MPa	Elonga-tion[e], %	q R.A. %	
			at °C	C	m		K	n					
Light metals:												
1100 Al (99%)	657/ 643	250– 550	300 500	60 14	0.08 0.22	A	140	0.25	35	90	35		340
~2017 Al(3.5Cu,0.5Mg, 0.5Mn)	635/ 510	260– 480	400 500	90 36	0.12 0.12	B	380	0.15	100	180	20		415(F)
5052 Al(2.5Mg)	650/ 590	260– 510	480	35	0.13	A	210	0.13	90	190	25		340
~7075 Al(6Zn,2Mg, 1Cu)	640/ 475	260– 455	450	40	0.13	B	400	0.17	100	230	16		415
Mg alloy (1 Mn)	649/ 648	290– 540	400	14	0.3	A			130	230	10		370
Low-melting metals:												
Sn (99.8%)	232	100– 200				A				15	45	100	150
Pb (99.7%)	327	20– 200	100	10	0.1	A				12	35	100	20– 200
Zn (0.08% Pb)	417	120– 275	75 225	260 40	0.1 0.1	A				130/ 170	65/ 50		100

High-temperature alloys:

Material											
Ni (99.4 Ni + Co)	1446/1435	650–1250				A		140	440	45	65	650–760
Hastelloy X (47Ni, 9Mo, 22Cr,18Fe,1.5Co,0.6W)	1290	980–1200	1150	~140	0.2	C		360	770	42		1175
Ti (99%)	1660	750–1000	600 / 900	200 / 38	0.11 / 0.25	C / A		480	620	20		590–730
Ti-6Al-4V	1660/1600	790–1000	600 / 900	550 / 140	0.08 / 0.16	C / A		900	950	12		700–825
Zirconium	1852	600–1000	900	50	0.25	A		210	340	35		500–800
Uranium (99.8%)	1132	~700	700	110	0.1			190	380	4		10

[a] Empty spaces indicate unavailability of data. Compiled from various sources; most flow stress data from T. Altan and F. W. Boulger, *Trans. ASME, ser. B, J. Eng. Ind.*, 95:1009, 1973.

[b] Hot-working flow stress is for a strain of ϵ = 0.05. To convert to 1000 psi, divide calculated stresses by 7.

[c] Relative ratings, with A the best, corresponding to absence of cracking in hot rolling and forging.

[d] Cold-working flow stress is for moderate strain rates, around $\dot{\epsilon}$ = 1/s. To convert to 1000 psi, divide stresses by 7.

[e] Where two values are given, the first is longitudinal, the second transverse.

[f] Furnace cooling is indicated by F.

absorber. Electrical storage batteries still rely, almost exclusively, on lead plates.

Zinc alloys

Pure zinc has wide use as the material for drawn battery cans, corrugated roofing, and weather stripping (usually with 1% Cu for the last two applications). Because of its hexagonal structure it is cold-worked above 20°C.

A eutectoid transformation in the zinc-aluminum system allows commercial production of extremely fine-grained material that exhibits superplasticity (Sec. 2.1.9). Binary alloys with 22% Al and further-alloyed variants are gaining acceptance because they can be deformed almost like plastics at elevated temperatures and attain a substantial strength at room temperature.

Magnesium alloys

The hexagonal structure of magnesium (Sec. 2.1.2) makes it rather brittle at room temperature, but it is worked very readily at only slightly elevated temperatures, typically above 220°C. Such low temperatures create no tool or lubrication problems and yield the benefit of great ease of forming. Both solid-solution alloying and precipitation hardening are exploited to obtain material of greater strength.

Aluminum alloys

The fastest-growing segment of the metalworking industry has been the working of aluminum alloys. An FCC material, aluminum is readily deformable at all temperatures. With the aid of solid-solution and precipitation-hardening mechanisms, materials of great strength can be produced with an often unsurpassed strength-to-weight ratio. Aluminum alloys have been the main constructional material for aircraft and are beginning to make larger inroads into the construction of land vehicles. Corrosion resistance and light weight make them attractive for a great many household, food industry, container, marine, and chemical plant applications. Equivalent electrical conductivity may be obtained at a cost often below that of copper and, especially in large cross sections, there are no installation problems.

Most alloys are formed in the annealed (O) condition. Non-heat-treatable alloys acquire useful strength through cold-working (H condition) although at the expense of ductility. Heat-treatable alloys may be worked in the annealed condition, then subjected to solution treatment and natural aging (T4 condition) or artificial aging (T6 condition). Even greater strength is obtained by cold-working a solution heat-treated material, since on subsequent natural aging (T3 condition) or artificial aging (T8 condition), the precipitates become extremely fine and well distributed (Sec. 2.2.7).

Copper-base alloys

Copper is one of the most ductile materials, and its solid-solution alloys with zinc (brass), tin (tin bronze), aluminum (aluminum bronze), beryllium (beryllium bronze), and nickel (cupronickel) and ternary alloys (such as *nickel silver,* a Cu-Ni-Zn alloy) preserve most of these desirable qualities, often with much-enhanced strength, fatigue strength, corrosion resistance, and high-temperature properties. Pure copper has the highest electrical conductivity after silver, and its high thermal conductivity and easy joining by soldering and brazing methods make it the main constructional material for radiators and for electrical wiring. Further alloying allows strengthening by precipitation-hardening mechanisms.

Most copper-base alloys are readily hot- and cold-worked, although some require considerable skill. Thus, the wide solidification range of tin bronzes and the phosphorus eutectic formed as a result of their deoxidation make them hot-short unless homogenization is assured prior to hot-working. We have already mentioned that minute amounts of lead destroy the hot-workability of alpha brasses, while substantial quantities of lead are added to α-brass and β-brass to improve their machinability (Sec. 2.2.6). The presence of lead makes these brasses less workable at room temperature.

The warm glow of copper and of its alloys has appealed to humans over millenia and their aesthetic appeal is often enhanced by corrosion products (patina).

Steels

Steels are produced in sheet, plate, wire, tube, and other wrought forms in vast quantities (Table 4.1). Low- and medium-carbon steels represent the greatest tonnages and are utilized either in the as-rolled condition or after annealing. The latter procedure, annealing, is essential for severe cold deformation, and a *spheroidizing anneal* is usual for the most demanding applications such as cold extrusion. As indicated in Sec. 2.2.3, spheroidal distribution of the carbide increases ductility while lowering strength.

Heat-treatable alloy steels are most readily worked in the annealed condition, even though increasing carbide contents increase the forming forces and die wear and reduce ductility. These materials are usually hot-worked, since in the austenitic temperature range their flow strength is not much higher than that of the carbon steels.

Most stainless steels can be hot-worked with proper precautions. Those containing both nickel and chromium are among the most cold-formable materials because of their high strain-hardening rate. Maraging steels contain nickel as the main alloying element and are readily deformable; in the finished shape, they are heat-treated to strengthen the soft martensite by precipitation of intermetallic compounds (Sec. 2.2.8).

Titanium alloys

Hexagonal titanium, stable at room temperature, is not particularly workable, but the BCC form (over 880°C) is most ductile. Titanium alloys have become indispensable for critical aircraft components and the compressor stages of jet engines. Titanium and its alloys have also made considerable gain in chemical applications because of their superior corrosion resistance. For control of finished properties, they are often worked just below the transformation temperature, and then they rank among the more difficult-to-work materials.

High-temperature alloys

Nickel in its pure form is readily deformable, just like a low-carbon steel. Some of its alloys, particularly those with copper, present no problem, while others, particularly the heavily alloyed nickel-base superalloys, have very high hot strength and a ductility limited to a narrow temperature range. Sophisticated melting and pouring techniques must be coupled with a thorough knowledge of the metallurgy of the alloys and with processing technologies that prevent the development of secondary tensile stresses or localized cooling.

The refractory metal alloys (molybdenum, tungsten, and niobium) are somewhat problematic because they readily oxidize at high temperatures. Tungsten, which is used extensively in a wire form in incandescent lamps, is processed from bars compacted by powder-metallurgy, first by hot-working and then by working at gradually lower temperatures as the ductile-to-brittle transformation temperature drops with increasing deformation. Most development in refractory metal alloys was spurred by space-age technology, which necessitated materials that would function at very high temperatures.

4.3 FORCES IN BULK DEFORMATION PROCESSES

Once it is ascertained that a wrought alloy satisfies service requirements and that plastic deformation is feasible for obtaining the required shape, it will inevitably be asked whether tooling can be made to withstand the pressures and a machine found that can exert sufficient force and power. Indeed, the calculation of *interface pressures* and *deformation forces* is the primary preoccupation of books dealing with plastic deformation processes. Very often, the emphasis is on the relative accuracy of various competing theories. For our purpose, it is much more important that any estimate of the pressures and forces should be truly relevant to the process conditions. The simple approach presented here will be accurate to plus or minus 20 or 30 percent; more sophisticated theories could improve the accuracy by a few percent, but in choosing tooling and equipment an appropriate safety factor must be applied anyway. In order to get a meaningful estimate of pressures and forces, three points must be observed: a relevant flow stress must be found, the effect of friction must be judged, and inhomogeneous deformation must be allowed for.

4.3.1 The Relevant Flow Stress

The stress at which plastic deformation can be maintained, called in the following the *flow stress* σ_f, must be taken for the temperatures, strains, and strain rates prevailing in the process. The yield stress $\sigma_{0.2}$ found in many handbooks has seldom anything to do with this and must never be used, because errors of 200–300 percent magnitude could be committed.

It cannot be sufficiently emphasized that our interest is not just in initiating but also in maintaining plastic flow. Thus the flow stress traverses the true stress-strain curve (Figs. 2.8a and 2.14a), within the strain limits given by the condition of the starting material and the end strain. In cold-working it can be assumed that the power law, Eq. (2.7), holds for most materials and the K and n values can be taken whenever available, as from Tables 4.2 and 4.3. For a non-steady-state process, such as forging, the *instantaneous flow stress* σ_f at the end of deformation is taken (Fig. 4.2a); for a steady-state process such as extrusion or wire drawing, the work-

FIG. 4.2 Determination of relevant flow stress for (a) non-steady-state and (b) steady-state processes, and (c) its relation to the UTS. (A strain of 0.5 is used for illustrative purposes.)

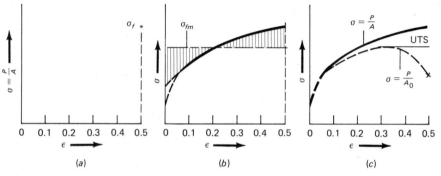

piece gradually strain-hardens within the deformation zone, therefore a *mean flow stress* σ_{fm} is needed. This is found by integration, or by calculating the value of the flow stress at selected strains (e.g., $\epsilon = 0.1, 0.2, 0.3$, etc.) up to the deformation to be obtained; the curve thus defined is then extrapolated to zero strain (Fig. 4.2*b*) and a mean is found by visual averaging.

A problem arises when the *K* and *n* values are not known. If equipment is available, they can be determined (App. A) fairly rapidly. Otherwise, the only guide could be the UTS, from Tables 4.2 and 4.3 or other source. Paradoxically, the basically nonsensical method of calculating the UTS (Sec. 2.1.5) happens to give a reasonable approximation of the mean true flow stress σ_{fm} (Fig. 4.2*c*). Since the UTS is measured at the point of necking, where $n = \epsilon_u$, some reasonable correction can be made for smaller or larger strains (see Example 4.4 at the end of this chapter).

For hot-working, the flow stress must be calculated from the power function, Eq. (2.11), with the appropriate *C* and *m* values (Tables 4.2 and 4.3). It is quite inadmissible to use hot-strength values determined in conventional slow tensile tests, because they are often only a fraction of the true flow stress prevailing at the much higher (typically, 1/s to 100/s) strain rates attained in deformation processes. If the *C* and *m* values are not available for various strains, one has to assume that the flow stress remains constant throughout deformation (as in the curve for a strain rate of 1/s in Fig. 2.14). If no *C* and *m* data are available, one is obliged to make a test (App. A); extrapolation from low-strain-rate tensile tests is much too hazardous.

4.3.2   The Effects of Friction

Deformation is most frequently accomplished by bringing the workpiece in contact with a tool or die. Consequently, *friction* between the two contacting bodies is unavoidable.

In most engineering and physical situations, friction effects are described by the *coefficient of friction* μ,

$$\mu = \frac{F}{P} \tag{4.1}$$

where P is the normal force and F the frictional force (Fig. 4.3*a*). It is well established that contact between two bodies is limited to only a few microscopic high points (*asperities*); nevertheless, it is customary to calculate stresses by assuming that the forces are distributed over the total (*apparent*) *area A*. Thus, the interface pressure is $p = P/A$, the frictional stress (the *interface shear stress*) is $\tau_i = F/A$, and

$$\mu = \frac{\tau_i}{p} \qquad\qquad (4.2)$$

In many engineering situations (e.g., in bearings) the interface pressure p is low relative to the flow stress σ_f of the contacting materials and Eqs. (4.1) and (4.2) hold. With increasing pressure p the interface shear stress τ_i increases linearly (Fig. 4.3b), and μ could assume any constant value.

In plastic deformation processes one of the contacting materials (the workpiece) deforms and in doing so also slides against the harder surface (the tool or die). A frictional stress τ_i is again generated, but this time there is a limit to the coefficient of friction. The interface shear stress τ_i cannot rise beyond a maximum given by the *shear flow stress* τ_f of the workpiece material; at this point the workpiece refuses to slide on the tool; instead, it deforms by shearing inside the body. Since the flow stress in shear τ_f is approximately one-half of the flow stress σ_f in tension or compression, it is often said that the maximum value of $\mu = \tau_f/\sigma_f = 0.5$ (Fig. 4.3b). This statement is true only when $p = \sigma_f$ but, in general, it is much more accurate to say that the coefficient of friction becomes meaningless when $\tau_i \geq \tau_f$, since there is no relative sliding at the interface. This is often described as *sticking friction* (even though the workpiece does not actually have to stick to the die surface).

Because of the conceptual difficulties introduced by the coefficient of friction, it is often preferable to use the actual value of τ_i, especially when interface pressures are very high, and we will follow this practice for extrusion calculations (Sec. 4.5.5).

We will soon see that friction increases pressures and forces and could easily limit the attainable reduction. With a few exceptions noted later, every effort is made to reduce friction by applying a suitable lubricant (Table 4.4). A good lubricant accomplishes much more: it separates die

FIG. 4.3 Interface friction: (*a*) definitions, and (*b*) possible variations when workpiece deforms plastically.

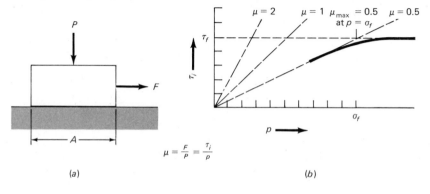

TABLE 4.4 TYPICAL LUBRICANTS* AND FRICTION COEFFICIENTS IN PLASTIC DEFORMATION

Workpiece material	Working	Forging Lubricant	μ	Extrusion† lubricant	Wire drawing Lubricant	μ	Rolling Lubricant	μ	Sheet metalworking Lubricant	μ
Sn, Pb, Zn alloys		FO-MO	0.05	FO or soap	FO	0.05	FA-MO or MO-EM	0.05 0.1	FO-MO	0.05
Mg alloys	Hot or warm	GR and/or MoS₂	0.1– 0.2	None			MO-FA-EM	0.2	GR in MO or dry soap	0.1– 0.2
Al alloys	Hot	GR or MoS₂	0.1– 0.2	None			MO-FA-EM	0.2		
	Cold	FA-MO or dry soap	0.1 0.1	Lanolin or soap on PH	FA-MO-EM, FA-MO	0.1 0.03	1–5% FA in MO (1–3)	0.03	FO, lanolin, or FA-MO-EM	0.05– 0.1
Cu alloys	Hot	GR	0.1– 0.2	None (or GR)			MO-EM	0.2		
	Cold	Dry soap, wax, or tallow	0.1	Dry soap or wax or tallow	FO-soap-EM, MO	0.1 0.03	MO-EM	0.1	FO-soap-EM or FO-soap	0.05– 0.1
Steels	Hot	GR	0.1– 0.2	GL (100– 300), GR			None or GR-EM	ST‡ 0.2	GR	0.2
	Cold	EP-MO or soap on PH	0.1 0.05	Soap on PH	Dry soap or soap on PH	0.05 0.03	10% FO-EM	0.05	EP-MO, EM, or soap on polymer	0.05– 0.1

Material							ST‡	GR	
Stainless steel, Ni and alloys	Hot	GR	0.1–0.2	GL (100–300)		None		GR	0.2
	Cold	CL-MO or soap on PH	0.1 / 0.05	CL-MO or soap on PH	Soap on PH 0.03 / or CL-MO 0.05	FO-CL-EM or CL-MO	0.1 / 0.05	CL-MO, soap, or polymer	0.1
Ti alloys	Hot	GL or GR	0.2	GL (100–300)				GR, GL,	0.2
	Cold	Soap or MO	0.1	Soap on PH	Polymer 0.1	MO	0.1	Soap, or polymer	0.1

* Some more frequently used lubricants (hyphenation indicates that several components are used in the lubricant):

CL = chlorinated paraffin.
EM = emulsion; the listed lubricating ingredients are finely distributed in water.
EP = "extreme-pressure" compounds (containing S, Cl, and P).
FA = fatty acids and alcohols, e.g., oleic acid, stearic acid, stearyl alcohol.
FO = fatty oils, e.g. palm oil and synthetic palm oil.
GL = glass (viscosity at working temperature in units of poise).
GR = graphite; usually in a water-base carrier fluid.
MO = mineral oil (viscosity in parentheses, in units of centipoise at 40°C).
PH = phosphate (or similar) surface conversion, providing keying of lubricant.

† Friction coefficients are misleading for extrusion and are therefore not quoted here.

‡ The symbol ST indicates sticking friction.

Source: Data extracted from J. A. Schey (ed.): *Metal Deformation Processes: Friction and Lubrication,* Dekker, New York, 1970.

and workpiece and thus prevents adhesion (Sec. 2.4) with its undesirable side effects of pickup, workpiece damage, and die wear; it controls the surface finish of the part produced; and it cools the system in cold-working and helps to prevent heat loss in hot-working. The lubricant must not be toxic or allergenic, it must be easy to apply and remove, and residues must not interfere with subsequent operations or cause corrosion.

4.3.3 Inhomogeneous Deformation

There is also a second important source of high interface pressures and forces which has nothing to do with interface friction and can be best understood from the example of indenting a semi-infinite body with a narrow anvil. Inspection of Fig. 4.4a will show that a small tool cannot possibly deform the entire bulk of a large (semi-inifinite) workpiece. If the tool is to penetrate, we must assume that localized *indentation* with highly *inhomogeneous material flow* takes place.

This could occur by the mechanism shown: a part of the workpiece immediately under the indenter (1) remains immobile relative to the indenter and moves with it as though it would be an extension of the indenter itself. This rigid wedge then pushes aside two triangular wedges (2), which in turn push up two outer wedges (3), thereby pushing up the volume displaced by the indenter alongside it. The rest of the workpiece (4) is only elastically loaded, and it is the difficulty of moving the material purely locally, against the *restraint* given by the surrounding elastic material, that raises the required interface pressure. More detailed theory shows that the *average interface pressure* will be the uniaxial flow stress σ_f times a *pressure-multiplying factor* Q_i (where the subscript i refers to inhomogeneity), which in the limit reaches

$$p_{i(max)} = Q_{i(max)}\sigma_f = 3\,\sigma_f \tag{4.3}$$

It will be recognized that while physically the situation shown in Fig. 4.4a is very different from a *hardness test,* the *strain state* is actually very similar. In the hardness test an indenter of spherical or pyramidal configuration penetrates the material that is, for all intents and purposes, infinite in the width, length, and thickness directions; thus the indenter has to push the material out. Therefore, the *indentation hardness* of a material is approximately 3 times its uniaxial (compressive) flow strength. As the highly localized deformation causes rather severe strain hardening, the indentation hardness is 3 times the mean flow stress σ_{fm} prevailing in the *shear zones* and, for the reasons mentioned in Sec. 4.3.1, the UTS is a good approximation of this mean value. It is for this reason that the indentation hardness is often taken as $3 \times$ UTS (in consistent units).

In many bulk deformation processes a workpiece of finite thickness is

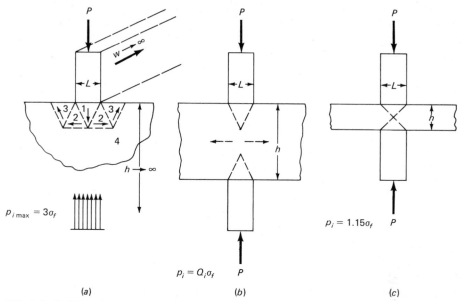

FIG. 4.4 Deformation modes and interface pressures in indenting (a) a semi-infinite body, (b) a thick workpiece ($h/L > 1$), and (c) a workpiece with $h/L = 1$.

deformed simultaneously from two sides (Fig. 4.4b). The effects of inhomogeneous deformation then depend on how far the two *deformation zones* are separated, and this is most conveniently expressed by the h/L ratio, that is, the ratio of height to *contact length.* It is found from both theory and experiment that when $h/L \geqslant 8.7$, the two deformation zones are entirely separated and the material between the zones, being only elastically deformed, exerts the same restraining effect as though it were of infinite thickness. Therefore, the maximum value of the pressure-multiplying factor $Q_{i(max)} = 3$ is again obtained. As the thickness h becomes smaller, the two deformation zones gradually interact, requiring less and less force to maintain plastic deformation. Therefore, the pressure-multiplying factor also diminishes (as given in Fig. 4.5). As might be expected, at a ratio of $h/L = 1$ the two deformation zones fully cooperate (Fig. 4.4c) and the material flows at a minimum pressure. If the ratio were to diminish further, deformation would be *homogeneous,* but the effects of friction would increase the pressures.

Inspection of Fig. 4.4b indicates that the two wedges penetrating the workpiece from top and bottom tend to pry the workpiece apart; in other words, inhomogeneous deformation generates secondary tensile stresses. These may not be high enough to fracture the workpiece during deformation. They will, nevertheless, result in a *residual stress* pattern (*internal stresses*) that may cause subsequent elastic deformation (warping) of the

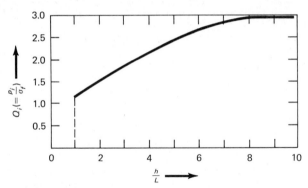

FIG. 4.5 Pressure-multiplying factor for average interface pressure in indentation (for both low and high friction, inclusive of the effects of plane strain deformation). (R. Hill, *The Mathematical Theory of Plasticity,* Clarendon Press, Oxford, 1950.)

workpiece, particularly on heating, or can combine with other effects to cause *delayed failures* (stress-corrosion cracking in the presence of a corrosive medium). In general, therefore, the aim of process development is to make deformation as homogeneous as possible. If harmful residual stresses remain, a stress-relief heat treatment is given (as for castings, Sec. 3.3). *Compressive residual stresses* concentrated in a thin surface layer greatly improve the fatigue-resistance of the workpiece in service. Highly inhomogeneous compressive deformation is then purposely applied, e.g., by roller burnishing or shotblasting the surface. The surface compressive stresses are balanced by internal tensile stresses but the latter are spread over such a large cross-sectional area that their level is harmless.

4.3.4 Calculation of Forces

Each deforming process has its own peculiarities, but there are some basic steps to be followed for all of them:

1 Calculate the strain imparted to the material. In everyday processing situations, this is expressed as engineering strain. If, for example, the height of a workpiece is reduced from h_0 to h_1 the *engineering (compressive)* strain is

$$e_c = \frac{h_0 - h_1}{h_0} \qquad (4.4a)$$

or $\quad e_c = 100\dfrac{h_0 - h_1}{h_0} \quad$ (percent) $\qquad (4.4b)$

For calculation purposes, the logarithmic or natural strain will be needed:

$$\epsilon = \ln \frac{h_1}{h_0} = -\ln \frac{h_0}{h_1}$$

The negative sign is usually ignored, since only the absolute value of the strain is of importance for force calculations:

$$\epsilon = \ln \frac{h_1}{h_0} = \left| \ln \frac{h_0}{h_1} \right| \tag{4.5}$$

2 Calculate the strain rate prevailing during hot deformation (as mentioned in Sec. 2.1.9, strain rate effects are minor in cold-working). For simple compression or tension, Eq. (2.10) can be used; for other processes, appropriate equations will be given.

3 Find the relevant flow stress (Sec. 4.3.1)—σ_f for non-steady-state processes (Fig. 4.2a), and σ_{fm} for steady-state processes (Fig. 4.2b).

4 Determine if the process could give rise to inhomogeneous deformation. Find the appropriate L and h; if $h/L > 1$, use Fig. 4.5 to determine Q_i. If deformation is reasonably homogeneous ($h/L < 1$), friction effects will predominate and, after the magnitude of friction is estimated (Table 4.4), appropriate multiplying factors Q can be found for various processes.

5 The average interface pressure p is simply

$$p = \sigma_f Q \tag{4.6}$$

or $\qquad p = \sigma_{fm} Q \tag{4.7}$

The deforming force P is the pressure multiplied by the contact area A. The work and power required for deformation can be easily calculated.

Most of the work expended in the course of deformation is *transformed into heat* and if this heat cannot be conducted away quickly, the temperature of the workpiece may rise to undesirably high levels.

4.4 FORGING

Forging processes are among the most important manufacturing techniques. As shown in Fig. 4.1, three broad groups can be distinguished: open-die forging allows free deformation of at least some workpiece surfaces, while deformation is much more constrained in impression-die forging and is fully constrained in closed-die forging.

4.4.1 Open-Die Forging

Because at least one of the workpiece surfaces deforms freely, *open-die forging* processes produce workpieces of lesser accuracy than impression- or closed-die forging; however, tooling is usually simple, relatively inexpensive, and allows the production of a large variety of shapes.

Upsetting of a cylinder

In the *axial upsetting* of a cylinder, a workpiece of cylindrical shape is placed between two flat parallel *platens* (*anvils*) and is reduced in height by force applied to the platens. In the absence of friction, deformation is homogeneous; i.e., the cylinder becomes shorter and assumes a greater diameter (to preserve the *constancy of volume*) but it remains a true cylinder (Fig. 4.6a). Since the volume V remains constant while the cross-sectional area changes,

$$V = A_0 h_0 = A_1 h_1 \tag{4.8}$$

The compressive strain (engineering strain) is usually calculated from the height change, Eq. (4.4), but because of the constancy of volume, cross-sectional areas can be used equally well:

$$e_c = \frac{h_0 - h_1}{h_0} = \frac{A_1 - A_0}{A_1} \tag{4.9}$$

The natural or logarithmic strain is obtained from Eq. (4.5). Again, because of constancy of volume, the areas can be substituted

$$\epsilon = \left| \ln \frac{h_0}{h_1} \right| = \ln \frac{A_1}{A_0} \tag{4.10}$$

Since we assumed that friction is absent (a somewhat overoptimistic assumption even with a very good lubricant), the interface pressure at any point of the compression process is quite simply the uniaxial flow stress σ_f for the given material, at the strain given by Eq. (4.10) and strain rate

$$\dot{\epsilon}_1 = \frac{v}{h_1} \tag{2.10}$$

The strain rate is that attained at the particular height h_1 (note that strain rate must always be expressed in units of reciprocal seconds, s^{-1}).

Since the platen overlaps the workpiece, no indentation effect is possible and one need not worry about any h/L ratio. Friction is, however, almost certain to develop on the platen face. In the presence of friction, the interface pressure rises because, in the course of upsetting, the diameter of the cylinder has to increase by sliding over the platen surface. This additional frictional stress retards displacement of the cylinder end face over the platen, thus creating an increased pressure that rises symmetrically from the edges of the cylinder towards the center (Fig. 4.6b). The greater the friction (expressed as, say, a coefficient of friction), the steeper this *friction hill* and the higher the average pressure, our main concern. Also, from the example shown in Fig. 4.6c, it is obvious that, for the same coefficient of friction, a cylinder of the same height but of larger diameter gives rise to a taller friction hill and, therefore, higher average interface pressure.

The average interface pressure is conveniently expressed as a multiple of the uniaxial flow stress σ_f that now prevails only at the edges of the workpiece. The *multiplying factor* Q_a (the subscript a signifying axial symmetry) must take into account both the effects of friction (the coefficient of friction) and workpiece geometry (the *d/h ratio,* which characterizes the

FIG. 4.6 Interface pressures in upsetting a cylinder with (a) no friction, (b) high friction, (c) high friction and larger *d/h* ratio, and (d) folding over of the sides in compressing with sticking friction.

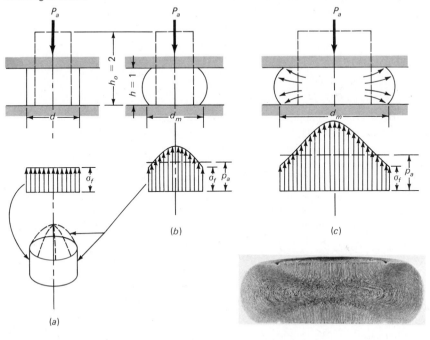

squatness of the cylinder). Without the appropriate formulas being derived, the relevant multiplying factor can be taken from Fig. 4.7.

Because friction retards movement of the cylinder end faces, the deformed cylinder assumes a *barrel shape,* and deformation within the cylinder is not homogeneous any more. The error committed through the use of Fig. 4.7 (which was derived on the assumption of homogeneous deformation) will, however, be rather small if an *average diameter* is calculated from the constant volume of the cylinder

$$V = A_0 h_0 = A_m h_1 = \frac{d_m{}^2 \pi h_1}{4} \tag{4.11}$$

In the extreme case, when the platen surface is rough and no lubricant is used, the interface shear stress τ_i may reach or exceed the shear flow stress τ_f of the workpiece material (Sec. 4.3.2) and movement of the end face is totally arrested. All deformation now takes place by internal shear in the cylinder (the sides of which *fold over,* Fig. 4.6d), and the friction hill rises in a cone shape, resulting in a linear increase of Q_a (Fig. 4.7) for sticking friction.

In evaluating the feasibility of a proposed upsetting operation, the first concern will be whether the *tooling* can stand the interface pressure and the total force. The *average pressure* is

$$p_a = \sigma_f Q_a \tag{4.12}$$

where both σ_f and Q_a must be calculated for the geometry existing at the

FIG. 4.7 Interface pressure-multiplying factor for the axial upsetting of cylinders. (After W. Schroeder and D. A. Webster, *Trans. ASME, 71*:289–294, 1949.)

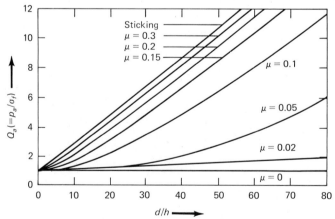

end of the stroke, since the pressure rises continuously (because of strain hardening and an increasing d/h ratio). The *force* at this point is

$$P_a = p_a A_m \qquad\qquad\qquad (4.13)$$

where A_m is the average cross-sectional area of the cylinder at the end of the stroke, calculated from Eq. (4.11).

Whether the tooling will stand this pressure and force depends not only on the hardness of the tooling but also on its shape. If the tool is a slender, long *punch,* it could buckle elastically and the maximum safe load must be calculated from buckling theory. When a stubby punch is loaded over most of its end face, it is subjected essentially to uniaxial compression; therefore the forging force should not exceed a safe fraction of the strength, calculated from the yield stress $\sigma_{0.2}$ of the die material. In many operations, the punch can be made into a platen that is substantially larger than the contact zone. Then the locally loaded contact area is backed up by the nondeforming bulk of the platen and, for the reasons elaborated in Sec. 4.3.3, deformation of the platen will be prevented unless the pressures exceed 3 times the yield strength $\sigma_{0.2}$ of the die material. If even higher pressures must be sustained, the working part of the platen may be set into a larger *shrink ring* which precompresses the insert and thus delays its yielding.

For production purposes it is important not only that the material can be deformed with feasible pressures and forces, but also that deformation should be uniform and free of defects. A very slender cylinder may *buckle* instead of upsetting uniformly; therefore it is advisable to limit the h_0/d_0 ratio to 2 (Fig. 4.8a). When friction on the anvil is very low, h_0/d_0 should be

FIG. 4.8 Limits set by buckling of long workpiece (*a*) between flat platens, (*b*) in heading, (*c*) with conical preform, and (*d*) in cold header.

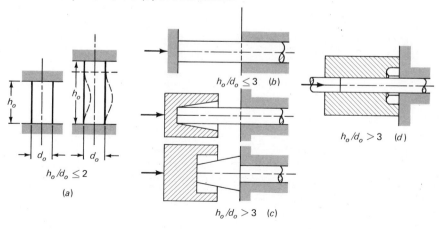

less than 1.5 to prevent skewing of the billet.   Very often, upsetting is conducted in a *heading* operation; i.e., only the end of a cylindrical workpiece is upset.   The longer part of the workpiece, firmly clamped in die halves, becomes fixed and the increased resistance to buckling allows somewhat greater free lengths (Fig. 4.8*b*).   An even longer length can be upset, either by *progressive upsetting* in stages to contain deflections in a *die cavity* (Fig. 4.8*c*) or an arrangement typical of cold-headers can be used: the long overhanging part of the wire or rod is now supported in the bore of a die (Fig. 4.8*d*) and upsetting is accomplished by a punch that moves the material into the space made available in this die.   Because the workpiece is guided on both of its ends, buckling is prevented and larger heads can be accumulated.

If deformation is truly homogeneous (Fig. 4.6*a*), most ductile materials can take relatively large strains in upsetting before their ductility is exhausted and fracture occurs by shearing at 45° to the application of the

FIG. 4.9 Cracking of deformed workpieces caused by (*a*) exhaustion of ductility in cold-working and (*b*) secondary tensile stresses generated by barreling in hot-working.

(*a*)

(*b*)

compressive stress (Fig. 4.9a). In practice, however, the presence of friction leads to barreling (Figs. 4.6b to 4.6d). It is readily seen that the material in the bulge is not directly compressed; instead, it is deformed indirectly, by the radial pushing action of the centrally located material. This expanding action creates circumferential as well as axial secondary tensile stresses on the free (barreled) surface, which may cause *cracking* of the barreled surface, either at a 45° angle or in the axial direction (Fig. 4.9b), depending on the relative magnitude of the circumferential and axial secondary tensile stresses. Since barreling is the primary culprit, improved lubrication (which reduces friction and thus barreling) may alleviate the problem. It is, nevertheless, quite common that one has to accept a limited deformation in a single stroke. Process anneals are then used to restore ductility and deformation can be taken further.

Upsetting may be practiced hot (and often on a very large scale in primary processing), cold, or perhaps even in between (warm-working), to take advantage of a somewhat lower flow stress and increased ductility. The process is readily automated, and vast numbers of nails, screws, bolts, and similar components are made in special-purpose machines (Sec. 4.6.3).

Forging of rectangular workpieces

A *rectangular workpiece* upset between two overhanging platens (Fig. 4.10) creates conditions different from those found in upsetting a cylinder.

FIG. 4.10 Interface pressure in upsetting a flat, rectangular workpiece with friction.

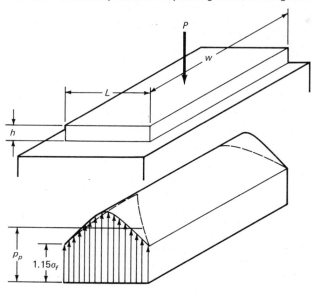

With a very wide ($w > 10L$) part, frictional constraints prevent movement in the width direction; therefore it is customary to regard this as deformation in *plane strain*. The stress that prevents movement in the width direction raises the interface pressure by some 15 percent over the uniaxial flow stress even at the edges of the workpiece (Fig. 4.10). The cross section of the friction hill is the same as would be found for a similar coefficient of friction in axial upsetting, but the average pressure is now higher, as will be obvious from comparing the shape of the friction hill in Fig. 4.10 with that of Fig. 4.6a. Therefore, the average interface pressure is now taken as the uniaxial flow stress σ_f multiplied by the plane-strain pressure-intensifying factor Q_p (Fig. 4.11, which includes also the effect of the intermediate stress).

$$p_p = \sigma_f Q_p \tag{4.14}$$

The surface area of a rectangular workpiece can be very large, resulting in an impracticably high total force; therefore it is customary to deform only one part of a large workpiece at a time. Properly sequenced individual *bites* gradually reduce the height of the entire length of the workpiece by the process of *cogging* or *drawing out* (Fig. 4.12). Successive bites must be spaced close enough to produce an even surface, but too short a bite will just fold the material down instead of deforming the entire cross section. Just as with a cylinder, a very tall and narrow workpiece would buckle rather than compress; therefore the ratio h/w is usually kept below 2.5.

Upsetting between overhanging platens is practiced in manufacturing when, for example the end of a pin is flattened. Drawing out is more a pri-

FIG. 4.11   Interface pressure-multiplying factor for the plane-strain compression of a rectangular slab. (After J. F. W. Bishop, *Quart. J. Mech. Appl. Math.*, 9:236–246, 1956.)

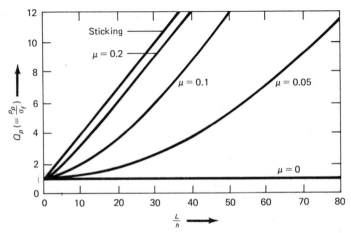

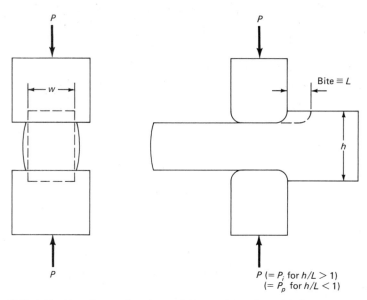

FIG. 4.12   Cogging or drawing out of a rectangular workpiece.

mary or preparatory process, sometimes used as a substitute for rolling when the quantities to be deformed are small.

The calculation of stresses and forces follows the principles described in Sec. 4.3.4.   The mean height is now readily seen; the active contact length L is the shorter dimension of a slab (Fig. 4.10) or the bite in cogging (Fig. 4.12); in other words, it is measured in the direction of major material flow.   When $h/L > 1$, inhomogeneous deformation prevails and the interface pressure is found from Q_i (Fig. 4.5); when $h/L < 1$, friction predominates and Q_p (Fig. 4.11) should be used (provided that $w/L > 10$; for narrower pieces the multiplying factor is smaller and equals Q_a when $w/L = 1$).

Piercing

Impressions or holes are made in a workpiece by *piercing*.   The simplest case to consider is that of piercing in a *container,* so that the workpiece is supported at its base and around its sides (Fig. 4.13).   As far as pressures are concerned, this is really indentation, equivalent to hardness testing, and the average pressure under the punch p_i is 3 times the uniaxial flow stress, Eq (4.3).   When the punch penetrates to any depth into a strain-hardening material, the pressure is 4 to 5 times the mean flow stress σ_{fm}.   The material displaced by the punch flows in a direction opposite to that of the punch movement, and friction on the container surface should be minimized, otherwise the piercing pressure will further rise.

When the workpiece is unconstrained, the deformation pattern depends on the ratio of workpiece diameter d_0 to punch diameter d_p.   When

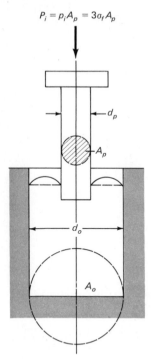

$$P_i = p_i A_p = 3\sigma_f A_p$$

FIG. 4.13 Punch pressure for piercing in a container.

$d_0/d_p > 3$, the situation is similar to that of piercing in a container. Piercing with smaller d_0/d_p ratios (Fig. 4.14) results in a complex deformation pattern; pressures drop roughly linearly to reach the value of the uniaxial flow stress at $d_0/d_p = 1$. Piercing with two punches from opposing ends of a cylindrical workpiece can be used for the preparation of a through-hole (the remaining *web* has to be removed in a separate operation).

The most frequent application of piercing for secondary processes is indenting the heads of screws and bolts. Since this is done mostly cold and in a container, pressures on the indenting tool can become excessive. Another limitation is indicated by the appearance of cracks, a result of either secondary tensile stresses set up by the expansion of an unrestrained head, or exhaustion of the ductility of the material in the prior heading operation.

4.4.2 Impression-Die Forging

More complex shapes of greater accuracy cannot be formed by open-die forging techniques. Specially prepared dies are required that contain the negative of the forging to be produced.

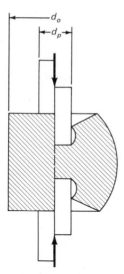

FIG. 4.14 Deformation of cylindrical workpiece pierced by two opposing punches.

In one variant of the process (Fig. 4.15) the shape is obtained by filling
out the die cavity defined by the upper and lower dies. Excess material is
allowed to escape into the *flash;* since the die is not fully closed, it should
properly be called an *impression die.* The term closed-die forging is, nevertheless, often applied, while the term drop forging is sometimes used to

FIG. 4.15 Sequence of impression-die forging: (*a*) blocker and (*b*) finishing-die cross section for an I-beam-like part.

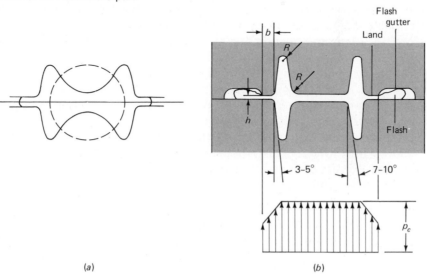

(*a*)   (*b*)

denote forging conducted upon a hammer; however, this distinction has no particular technical merit.

The die cavity must be filled without defects of *material flow,* such as could occur when parts of the workpiece material are pinched, folded down, or sheared through. A complex shape cannot be filled simply by forging a round or rectangular bar into the die cavity, and some *preforging* steps are necessary. These can be free-forging (open-die forging) operations, sometimes performed with the aid of specially shaped surfaces in the impression-die *forging blocks* themselves (Fig. 4.16) or on separate forging equipment, or even by other preparation methods such as rolling.

The preform may be further shaped to bring it closer to the final configuration in a so-called *blocker die,* which assures proper distribution of material but not the final shape (Fig. 4.15a). Excess material is allowed to run out between the flat die surfaces and this flash is sometimes removed (*trimmed*) prior to forging in the *finishing die.* The excess material is again allowed to escape into a flash, which must now be thin to assure die filling and close tolerances. A thin flash means a large L/h ratio and thus high pressures (Fig. 4.11); therefore the flash is reduced to its minimum thickness over only a small width (*flash land*), and the rest is allowed to flow freely into the *flash gutter* (Fig. 4.15b). The flash is removed either hot or cold, in a separate die resembling a blanking die (Sec. 5.2).

The shape of the forging must promote smooth material flow. Therefore, a *parting line* is chosen with proper consideration of the fiber structure of the finished forging. Fibers (caused by alignment of inclusions, second-phase particles, and microsegregation, Secs. 2.2.4, 2.2.5, and 2.3.2) should follow the contour of the forging as far as possible, because this ensures greatest toughness or ductility (Fig. 4.17). At the parting line the fibers are unavoidably cut through when the flash is trimmed; therefore the parting line is best placed where minimum stresses arise in the service of the forging. After the parting line is located, the cavity walls are given sufficient draft to allow removal of the forging from the die cavity. The *internal draft* is greater than the *external draft* (Fig. 4.15) because the forging tends to shrink onto bosses of dies prior to its removal from the die. Fillets and corners must be given appropriate radii to assure both smooth material flow and reasonable die life.

There is no simple yet satisfactory method of calculating impression-die pressures and forces, partly because the strain rate varies tremendously in various parts of the workpiece. To obtain a very approximate estimate, an *average strain rate* is calculated first:

$$\dot{\epsilon}_m = \frac{v}{h_m} = \frac{vA_t}{V} \tag{4.15}$$

where V is the volume, A_t is the total *projected area* of the workpiece

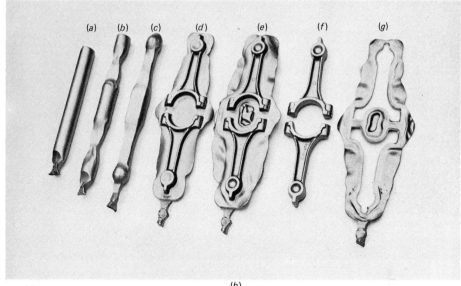

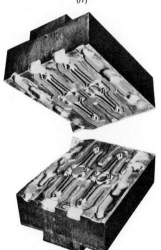

FIG. 4.16 Hammer forging two connecting rods: (a) bar stock; after (b) fullering, (c) "rolling," (d) blocking, (e) finishing, (f) trimming; (g) the flash; and (h) the forging dies. (Forging Industry Association, Cleveland, Ohio.)

(including the flash), h_m is the average height, and v is the deformation velocity.

Similarly, an *average strain* is found (the negative sign is ignored):

$$\epsilon_m = \ln \frac{h_0}{h_m} = \ln \frac{h_0 A_t}{V} \qquad (4.16)$$

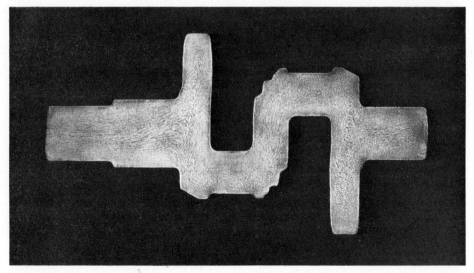

FIG. 4.17 Grain flow (flow lines) in a forged steel workpiece revealed by macroetching. (Forging Industry Association, Cleveland, Ohio.)

From these, the relevant flow stress σ_f is determined, which is then multiplied by a factor Q_c to allow for shape complexity. Its value is taken (with some experience and faith) according to Table 4.5.

Impression-die forging is practiced both in the hot and cold state, on hammers, presses and special purpose equipment. Presses have a rated maximum force, and this should be well over the estimated value of *forging force*

$$P = \sigma_f Q_c A_t \tag{4.17}$$

Hammers deliver a certain amount of energy (Sec. 4.6) and this should be greater than the *energy requirement of forging,* estimated with the aid of the multiplying factor Q_{fe} from Table 4.5 by

$$E = \sigma_f Q_{fe} V \epsilon_m \tag{4.18}$$

TABLE 4.5 MULTIPLYING FACTORS FOR ESTIMATING FORCES (Q_c) AND ENERGY REQUIREMENTS (Q_{fe}) IN IMPRESSION-DIE FORGING

Forging shape	Q_c	Q_{fe}
Simple, no flash	3–5	2.0–2.5
With flash	5–8	3
Complex (tall ribs, thin webs), with flash	8–12	4

Production rates vary greatly depending on the size of the part, the number of cavities that are needed to produce it, and the number of parts that can be forged simultaneously. Cooling—and thus forces—in hot-forging are much reduced with heated dies. With the dies at the work-piece temperature (*isothermal forging*), very slow forging speeds are permissible, and complex, thin-walled parts can be forged at low pressures.

4.4.3 Closed-Die Forging

In true *closed-die forging* the workpiece is completely *trapped* in the die and no flash is generated. Economy of forging is thus increased, but die design and process variables must be very carefully controlled. At the end of the stroke the cavity is completely filled with an incompressible solid and die pressures rise very steeply; this becomes a critical factor in setting up the equipment (Sec. 4.6). Forces are as in an impression-die forging.

A special case of closed-die forging is *coining,* in which a three-dimensional surface detail is imparted to a preform. The largest application is, of course, that of minting coins, but it is useful for improving the dimensional accuracy, surface finish, or detail of other parts too. The forging pressure is at least $p_i = 3\sigma_f$ but filling of fine details calls for pressures up to $5\sigma_f$ or even $6\sigma_f$.

4.4.4 Forge Rolling and Rotary Swaging

These are two of the more specialized forging processes.

Forge rolling performs an impression-die forging operation, but this time the die-half contours are machined into the surfaces of two rolls. Reciprocating roll motion is suitable for the rolling of short pieces while unidirectional rotation is used in high-production lines. Forge rolling often replaces open-die forging for preforming but is suited also for finishing more or less flat forgings such as cutlery and scissors.

A very special form of a hammer is the *rotary swager.* The workpiece is usually stationary, while the hammer itself rotates. The construction resembles a roller bearing (Fig. 4.18): the anvils are free to move rapidly in a slot of the rotating shaft and are thus hurled against the rollers, which in turn knock them back. A rapid sequence of blows is assured and the workpiece, fed axially, is reduced in diameter by a drawing-out process. While, strictly speaking, swaging should be regarded (and sometimes is used) as an open-die forging process, it is capable of producing exceptionally smooth surfaces to close tolerances. The process can be employed for pointing, assembling a bar and collar, or shaping the internal contour of a tube on a mandrel (Fig. 4.18*b*).

Manufactured components of complex shape are often produced by a combination of forging and extrusion; therefore extrusion processes will be discussed next.

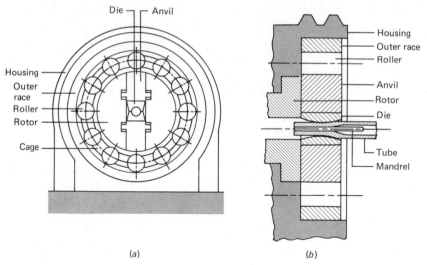

FIG. 4.18   The principal components of a rotary swager (*a*) and the swaging of a tube on a mandrel (*b*).

4.5 EXTRUSION

In *extrusion* the workpiece is pushed against the deforming die while it is being supported in a container against uncontrolled deformation.   The extrusion process therefore offers the possibility of heavy deformations coupled with a wide choice of extruded cross sections.

4.5.1   The Extrusion Process

To initiate extrusion, a *cylindrical billet* is loaded inside a container and is pushed against a die held in place by a firm support.   The press force is applied to the punch and, after the billet has upset to fill out the container, the product emerges through the die (Fig. 4.19).   Initially, deformation is non-steady-state but once the product has emerged, steady-state conditions prevail until close to the end of extrusion when continuous material flow is again disturbed.   In *direct* or *forward extrusion,* the product emerges in the same direction as the movement of the punch (Fig. 4.19*a*) while in *indirect* (*reverse* or *back*) *extrusion* the product travels against the movement of the punch (Fig. 4.19*c*).   By definition, piercing in a container (Fig. 4.13) may be regarded as a case of back extrusion.

Extrusion may be carried out without a lubricant, and then the material seeks a *flow pattern* that results in minimum energy expenditure.   With a die of *flat face* the material cannot follow the very sharp directional changes that would be imposed on it; instead, the corner between the die

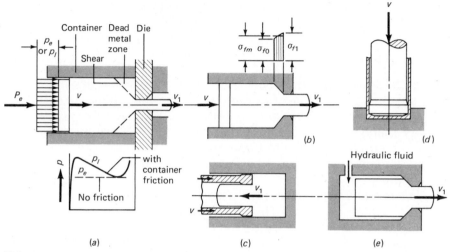

FIG. 4.19 Extrusion processes: (a) forward or direct with no lubrication, (b) forward with full lubrication, (c) reverse, (d) reverse can (impact), and (e) hydrostatic extrusion.

face and container is filled out by a stationary, *dead-metal zone* and material flow takes place by shearing along the surface of this zone (*unlubricated extrusion,* Fig. 4.19a). Thus, the extruded product acquires a completely freshly formed surface which is highly desirable in materials such as aluminum extruded at hot-working temperatures. In cold-working it is much better to apply a very effective lubricant to assure complete sliding on the die face (*lubricated extrusion*); accordingly, the die is now provided with a conical entrance zone that, ideally, corresponds to the angle of minimum-energy flow (Fig. 4.19b). A similar situation prevails in the hot extrusion of steel with *glass* as a *lubricant.* Even though the die has a flat face, the glass pad applied to it melts off to give a die shape of optimum profile.

The movement of the punch must be stopped before the conical die entry is touched or, in unlubricated extrusion, before material from the dead-metal zone is moved, since this would create internal defects. When the purpose of extrusion is to produce a bar of uniform cross section, the remaining material (*butt*) is scrap which is removed by taking it out with the die. After the butt is cut off, the extrusion can be extricated from the die, and the die is returned for inspection, conditioning, and reuse. When, however, the purpose of extrusion is to produce a finished component, with the butt forming the head of the component, it is customary to *eject* the extrusion by pushing it back through the extrusion die and lifting it out from the container. Since ejector actuation can be mechanically synchronized with the punch movement, high production rates are achieved, provided, of course, that the extruded product is strong enough to take the ejection force.

4.5.2 Hot Extrusion

While hot deformation is often typical of primary processes, the *hot extrusion* of shapes, particularly of nonferrous materials, offers such a wide scope for custom design that this process can justifiably be regarded as a secondary manufacturing technique. *Shapes* are usually classified according to their complexity into three groups (Fig. 4.20):

Solid shapes are produced by extruding through a suitably shaped stationary die.

Hollow products necessitate the use of a die insert that forms the cavity in the extruded product. This insert may be a *mandrel* fixed to the punch (Fig. 4.21a) or moving inside the punch (Fig. 4.21b), or it may be a *bridge section* attached to the die (Fig. 4.21c). The last method is permissible only if the material flow can be divided and then reunited prior to leaving the die, with complete pressure-welding of the separated streams. This is practicable only for the unlubricated hot extrusion of aluminum and lead; in lubricated extrusion the lubricant would prevent rewelding.

Semihollow products appear to be solid sections, except that their shape makes the use of a single-piece die impracticable. The die tongue forming the internal shape is connected to the external contour by such a small cross section that it would break off; therefore techniques similar to the extrusion of hollow sections must be used.

In designing complex sections it should be kept in mind that the *circumscribed circle* must be less than the container diameter, and that wall

FIG. 4.20 Extruded sections of (a) solid, (b) semihollow, and (c) hollow configurations.

Increasing
circumference/area
ratio

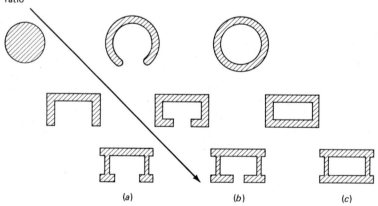

(a)   (b)   (c)

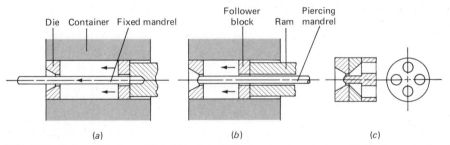

FIG. 4.21 Tube extrusion with (*a*) fixed and (*b*) piercing mandrel, and (*c*) schematic drawing of a bridge die. (After J. A. Schey, in *Techniques of Metals Research,* R. F. Bunshah (ed.), vol. 1, Pt. 3, Interscience, 1968, p. 1494.)

thicknesses should be kept as uniform as possible in order to equalize material flow. Increasing complexity is often expressed as a perimeter/weight or perimeter/cross-sectional area *shape factor;* the higher its value, the more skill is required to produce the part. The thinnest possible wall is 1 mm for aluminum, and the value increases with increasing section complexity.

Copper and brass are usually extruded unlubricated, but cooling on the colder container and die limits the complexity and thinness of shapes. This is true also of the hot extrusion of steel, conducted mostly with a glass lubricant, although shorter lengths and thinner sections are sometimes produced with graphitic lubricants.

In the hot extrusion of as-cast material, the extrusion ratio should be at least 4 to 1, to assure adequate working.

4.5.3 Cold Extrusion

The purpose of *cold extrusion* is mostly that of producing a finished part. In most instances, the residue (butt) in the container becomes an integral part of the finished product (e.g., in the forward extrusion of a screw shank or the back extrusion of a toothpaste tube).

The low flow strengths of tin and lead facilitated their early cold extrusion for collapsible tubes (often called *impact extrusion,* Fig. 4.19*d*). With sufficient lubrication aluminum can be similarly treated. Only smaller extrusion ratios are permissible with copper and brass, and the cold extrusion of steel would be quite impossible without a lubricant capable of withstanding very high pressures while following the extension of the surface. The most successful commercial approach converts the steel surface into a zinc-iron phosphate (*phosphate coating*); this porous surface, integrally joined to the metal surface, is then impregnated with a suitable lubricant (usually a soap, Table 4.4).

4.5.4 Hydrostatic Extrusion

In an extensively investigated variant of the process, the billet is extruded by pressurizing a *liquid medium* inside a closed container (Fig. 4.19e). This helps to reduce friction on the container wall, but does not fundamentally change the stress state inside the deforming workpiece (reduced die friction can even increase the tendency to internal crack formation). The absence of *container friction* permits extrusion of very long billets or even wires, and large reductions can be taken. Nevertheless, the process has not yet attained broad practical application mostly because of the long cycle time imposed by the need for pressurizing the container.

4.5.5 Extrusion Forces

Let us first assume that there is no friction between billet and container. The *extrusion pressure* that must be developed is a function of die geometry and of the *extrusion ratio*

$$R_e = \frac{A_0}{A_1} \tag{4.19}$$

where A_0 is the cross-sectional area of the container and A_1 is that of the extruded product.

Deformation tends to be inhomogeneous and for approximate calculations the following formula may be adopted

$$p_e = \sigma_{fm}Q_e = \sigma_{fm}(0.8 + 1.2 \ln R_e) \tag{4.20}$$

A mean flow stress σ_{fm} must be used because the workpiece material strain hardens during its passage through the die. In hot-working the effect of strain rate becomes important and the average strain rate may be calculated from

$$\dot{\epsilon}_m = \frac{6vd_0^2 \tan \alpha \ln R_e}{d_0^3 - d_1^3} \tag{4.21}$$

where v is the ram velocity, d_0 is the diameter of the billet, and d_1 is the diameter of the extruded product. If the extruded section is other than a solid round, its area is calculated and an *equivalent diameter* is substituted in the above equation. The *half angle* α is given by the cone angle of the die or, in unlubricated extrusion with a dead-metal zone, it may be taken as 45°.

The total *extrusion force* is simply

$$P_e = p_e A_0 \tag{4.22}$$

A word of warning is in order here. We have already observed (Sec. 4.5.1) that back extrusion of a can is similar to piercing in a container (Fig. 4.13). The extrusion formula, Eq. (4.20), gives the pressure over the base area A_0 but the process limit may really be given by the punch (indenter) pressure p_i which, as discussed in Sec. 4.4.1 under piercing, can never be less than $3\sigma_f$ [Eq. (4.3)]. It is advisable, therefore, to calculate the extrusion force from both Eq. (4.22) and from the *punch force P_i*

$$P_i = p_i A_p \tag{4.23}$$

and take the larger of the two values. It does not matter whether the indenting punch is solid as in Fig. 4.13 or hollow as in Fig. 4.19c. The uniaxial yield strength $\sigma_{0.2}$ of the punch material must not be exceeded; more likely, the permissible stress is even lower because of the danger of buckling.

In direct extrusion the billet is pushed forward against the frictional resistance developed on the *container wall*. Correspondingly, the extrusion pressure is higher at the beginning of the stroke when a long length rubs against the container wall (Fig. 4.19a). Interface pressures can be very high and the use of a coefficient of friction value could be misleading (Sec. 4.3.2). Therefore, it is better to estimate the shear strength of the interface τ_i and add the corresponding pressure to the calculated extrusion pressure

$$p_1 = p_e + \frac{4\,\tau_i l}{d_0} \tag{4.24}$$

where l is the length of the billet at the point considered. Data for τ_i are scarce but an upper limit is given by sticking when $\tau_i = \tau_f$ or $0.5\,\sigma_f$. With a truly effective lubricant the pressure will drop toward the basic pressure, Eq. (4.20).

4.5.6 Process Limitations

Extrusion tends to create inhomogeneous deformation conditions, especially at extrusion ratios below 4. Inhomogeneity (Sec. 4.3.3) is, in general, a function of the ratio h/L or mean height h over compressed length L. Even though the h/L ratio is measured on a rectangular slab and extrusion is usually conducted with axial symmetry, the same principles apply (Fig. 4.22), except that a *mean diameter* $(d_0 + d_1)/2$ is now substituted for h.

As before, deformation is inhomogeneous when the h/L ratio is large (Fig. 4.22a), in other words, when the extrusion ratio is small and the die half angle α is large. Deformation is now concentrated in the outer zones which are, therefore, directly elongated. The center of the extrusion is not

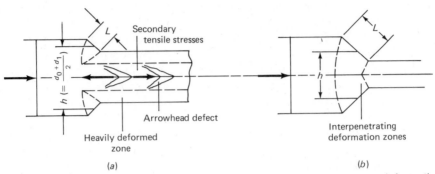

FIG. 4.22 Homogeneity of deformation in extrusion: (*a*) nonhomogeneous deformation and center-burst (arrowhead) defect with large *h/L* ratio, and (*b*) homogeneous deformation with *h/L* $\simeq$ 1.

directly deformed; instead, it is dragged along by the surface material. This generates secondary tensile stresses in the core which may ultimately suffer a characteristic *arrowhead fracture* (also described as *center-burst defect*). The danger is greatest at an *h/L* ratio of 2 and over. The situation can be remedied by lowering the *h/L* ratio, which implies either a smaller die half angle α or a heavier reduction and thus smaller *h* (Fig. 4.22*b*). With a component of fixed geometry, neither of these remedies may be allowable and the only hope is then the use of a more ductile material.

In very special instances the workpiece material is kept in the compressive stress state throughout the extrusion process, even at critically low extrusion ratios, by extruding the material into a pressurized space, a process usually described as *extrusion against back pressure,* not to be confused with hydrostatic extrusion.

In hot extrusion, the heat generated during extrusion may cause the workpiece temperature to rise above the solidus temperature of the material. This causes circumferential surface cracks to appear (*speed cracking*) which can be eliminated by slowing down the press, thus reducing the strain rate and the rate of heat generation.

4.6 FORGING AND EXTRUSION EQUIPMENT

Forging and extrusion are closely related processes. Sometimes they are difficult to distinguish (e.g., piercing in a container vs. back extrusion); at other times a distinctly forging-type process is combined with extrusion (e.g., in making a bolt by upsetting the head and extruding the shank). They also share many types of equipment (Table 4.6).

TABLE 4.6 CHARACTERISTICS OF HAMMERS AND PRESSES*

Equipment type	Energy,† kN·m	Ram mass, kg	Force,‡ kN	Speed m/s	Strokes/ min	Stroke, m	Bed Area m × m	Mechanical efficiency
Hammers:								
Mechanical	0.5–40	30–5,000		4–5	350–35	0.1–1.6	0.1 × 0.1 to 0.4 × 0.6	0.2–0.5
Steam and air	20–600	75–17,000 (25,000)		3–8	300–20	0.5–1.2	0.3 × 0.4 to (1.2 × 1.8)	0.05–0.3
Counterblow	5–200 (1250)			3–5	60–7		0.3 × 0.4 to (1.8 × 5)	0.2–0.7
Herf	15–750			8–20	<2			0.2–0.6
Presses:								
Hydraulic, forging			100–80,000 (800,000)	<0.5	30–5	0.3–1 (3)	0.5 × 0.5 to (3.5 × 8)	0.1–0.6
Hydraulic, sheet m.w.			10–40,000	<0.5	130–20	0.1–1	0.2 × 0.2 to 2 × 6	0.5–0.7
Hydraulic, extrusion			1000–50,000 (200,000)	<0.5	<2	0.8–5	0.06 to 0.6 dia. container	0.5–0.7
Mechanical, forging			10–80,000	<0.5	130–10	0.1–1	0.2 × 0.2 to 2 × 3	0.2–0.7
Horizontal upsetter			500–30,000 (1–9 in dia.)	<1	90–15	0.05–0.4	0.2 × 0.2 to 0.8 × 1	0.2–0.7
Mechanical, sheet m.w.			10–20,000	<1	180–10	0.1–0.8	0.2 × 0.2 to 2 × 6	0.3–0.7
Screw			100–80,000	<1	35–6	0.2–0.8	0.2 × 0.3 to 0.8 × 1	0.2–0.7

* From a number of sources, chiefly A. Geleji: *Forge Equipment, Rolling Mills and Accessories,* Akademiai Kiado, Budapest, 1967.

† Multiply number in column by 100 to get m·kg, by 0.73 to get 10^3 lbf·ft.

‡ Divide number by ~ 10 to get tons. Numbers in parentheses indicate the largest sizes, available in only a few places.

4.6.1 Hammers

Hammers are impact devices, in which a mass (the *ram*) is accelerated by gravity and/or compressed air, gas, or steam.   For a ram mass M and impact velocity v the hammer energy E_h is

$$E_h = \frac{Mv^2}{2} = \frac{Wv^2}{2g} \tag{4.25}$$

where W is the weight of the ram and g is the gravitational acceleration. The *striking velocity v* increases with the *stroke (drop height)* H_d and acceleration ξ

$$v^2 = 2\xi H_d = 2H_d \left(g + \frac{Ap_m}{M} \right) \tag{4.26}$$

where A is the cross-sectional area of the driving piston, p_m is the mean indicated pressure of the pressurized medium (air, gas, or steam), and M is the accelerated mass.

The energy of impact is absorbed mostly by the deformation of the workpiece, Eq. (4.18).   The excess energy is transmitted to the die, the foundation, the ground, and also the hammer components.   These latter undesirable effects are avoided in *counterblow hammers.*   High impact velocities and short contact times minimize cooling, therefore hammers are used mostly for open-die forging and for impression-die forging of intricate shapes.   Except for counterblow and high-energy-rate forging (HERF) hammers, the forging is produced by several blows in any one die cavity; therefore, the total energy requirement [Eq. (4.18)] can be delivered by a relatively small hammer.   Hammer forging does require, however, considerable operator skill.

4.6.2 Presses

Presses are powered mechanically or hydraulically.   *Hydraulic presses* stall out when their load limit is reached and can be used with dies that make contact (*kiss*) at the end of the stroke.   *Mechanical presses* (with the exception of *screw presses*) have a preset stroke and develop an infinite force at the end of the stroke; therefore, if the die is completely closed, it must allow escape of excess material or must be set with extreme care. Because of the lower speeds and longer contact times, workpieces must be preformed carefully if complex parts are made by hot-forging.

4.6.3 Upsetters and Cold-Headers

A special class of presses comprises *horizontal upsetters* for hot- and cold-working and *cold-headers* for cold-working.   Both start with coiled or straight lengths of bar or wire. The material is fed with indexing *pinch rollers* (sometimes through a multiroll straightener and even a draw-die that assures tight tolerance). The end of the rod or wire is deformed in successive steps ranging from simple upsetting to the most complex combined forging-extrusion operation. Auxiliary movements are synchronized with the main ram movement and are used to open and close *clamping dies,* to actuate auxiliary punches and shearing dies, and to transfer the workpiece from one die cavity to another. The workpiece material is cut off the bar or wire either at the beginning or end of the sequence, and either one workpiece may go through the die sequence at a time or a workpiece may reside in each die during each stroke.

The construction of these machines is often very ingenious and their production rates are difficult to match with other techniques.

An example of a hot-upsetting sequence is given in Fig. 4.23 and one of cold extrusion in Fig. 4.24.

4.6.4 Tools and Dies

Bulk deformation processes are characterized by high interface pressures coupled with high temperatures in hot-working. *Tool* and *die materials*

FIG. 4.23 A typical hot-upsetting sequence showing the development of the workpiece from the bar, and the associated tooling. (National Machinery Co., Tiffin, Ohio.)

Gripper dies   Heading tools (punches)

FIG. 4.24   A typical cold-upsetting sequence in a seven-station cold former, producing hose connectors at a rate of 60 per minute. (National Machinery Co., Tiffin, Ohio.)

are selected and manufactured with the greatest care.   In general, ductility is sacrificed for greater hardness in cold-working dies but a compromise between hardness and ductility must be struck for hot-working dies that are exposed also to thermal shock (Table 4.7).

The yield stress $\sigma_{0.2}$ of a fully heat-treated die steel is in excess of

TABLE 4.7 TYPICAL DIE MATERIALS FOR DEFORMATION PROCESSES*

Process	Die material and hardness (R_C) for working	
	Al, Mg, and Cu alloys	**Steels and Ni alloys**
Hot forging	6G 30–40 H12 48–50	6G 35–45 H12 40–56
Hot extrusion	H12 46–50	H12 43–47
Cold extrusion: Die	W1, A2 56–58 D2 58–60	A2, D2 58–60 WC
Punch	A2, D2 58–60	A2, M2 64–65
Shape drawing	O1 60–62 WC	M2 62–65 WC
Cold rolling	O1 55–65	O1, M2 55–65
Blanking	Zn alloy W1 62–66 O1 57–62 A2 57–62 D2 58–64	As for Al, and M2 60–66 WC
Deep drawing	W1 60–62 O1 57–62 A2 57–62 D2 58–64	As for Al, and M2 60–65 WC
Press forming	Epoxy/metal powder Zn alloy Mild steel Cast iron O1, A2, D2	As for Al

* Compiled from *Metals Handbook,* 8th ed., vol. 1, American Society for Metals, Metals Park, Ohio, 1961.

Die materials mentioned first are for lighter duties, shorter runs.

Tool steel compositions, percent (representative members of classes):
6G (prehardened die steel): 0.5C, 0.8Mn, 0.25Si, 1Cr, 0.45Mo, 0.1V
H12 (hot-working die steel): 0.35C, 5Cr, 1.5Mo, 1.5W, 0.4V
W1 (water-hardening steel): 0.6–1.4C
O1 (oil-hardening steel): 0.9C, 1Mn, 0.5Cr
A2 (air-hardening steel): 1C, 5Cr, 1Mo
D2 (cold-working die steel): 1.5C, 12Cr, 1Mo
M2 (Mo high-speed steel): 0.85C, 4Cr, 5Mo, 6.25W, 2V
WC (tungsten carbide)

280,000 psi (1900 MPa) for cold-working and between 170,000 and 250,000 psi (1100 and 1700 MPa) for hot-working. In simple compression they are loaded to 60–70 percent of these values, while up to 3 times this pressure is allowable before they suffer indentation.

Extrusion tooling presents a special case since the container is subjected to internal pressure and would burst on overloading.

A single-piece container made of high-strength die steel can take up to 150,000 psi (1000 MPa) pressure, while pressures up to 250,000 psi (1700 MPa) can be accommodated when the internal part of the container (*liner*) is shrunk into a larger outer shrink ring (container). Special container constructions permit pressures of up to 400,000 psi (2700 MPa). Steel punches are limited to approximately 180,000 psi (1200 MPa) in simple compression, and cobalt-bonded WC punches operate up to 500,000 psi (3300 MPa). In all instances, the tooling is surrounded by heavy *shielding* because a fractured die part becomes a potentially deadly projectile.

4.7 DRAWING

Long components of uniform cross section can be produced not only by extrusion but also by *drawing*. Instead of being pushed, the material is pulled through a stationary die of gradually decreasing cross section.

4.7.1 The Drawing Process

The material is again deformed in compression (Fig. 4.25), but the deformation force is now supplied by pulling the deformed end of the wire; therefore it is often said that the deformation mode is that of *indirect compression*.

The stationary *draw die* may be replaced with two, three or four idling

FIG. 4.25 Drawing of a round wire (a) with reasonably homogeneous deformation and (b) with the same reduction but large die half angle α and high h/L ratio leading to center-burst defect.

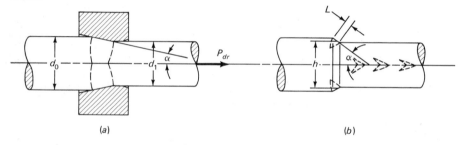

(a)   (b)

rollers, all with their axes in a common plane. A *Turk's head* is a tool containing four rollers with adjustable positions.

Tubes are sometimes drawn simply through draw dies, either to reduce their diameter (*sinking*) or to change their shape (say, from round to square). If their wall thickness is to be reduced, an internal die is also needed. This could be a short, conical *plug* held by a long bar from the far end, or a plug shaped so as to stay in the deformation zone (*floating plug*), or it could be a full-length *bar* of tool steel.

The productivity of the drawing process is high, because speeds of up to 10,000 fpm (50 m/s) are possible on thin wire. Much slower speeds, of the order of a few hundred fpm (around 1 m/s) are common in the drawing of heavier bar. Sections that cannot be bent around a *draw drum* (*bull block*) must be drawn in straight lengths, on *draw benches* at low speeds and, because of the batch-type operation, reduced production rates.

4.7.2 Forces

The *maximum reduction* taken is limited by the strength of the deformed product, calculated from the yield stress $\sigma_{0.2}$ after deformation and the issuing cross section A_1. This should be greater than the *drawing force,* which can be approximately calculated as

$$P_{dr} = A_1\sigma_{exit} = A_1 Q_{dr}\sigma_{fm} \tag{4.27}$$

where Q_{dr} is a *stress multiplying factor* attributable to drawing and can be taken from

$$Q_{dr} = (1 + \mu \cot \alpha)\,\Phi\,\ln\frac{A_0}{A_1} \tag{4.28}$$

where μ is the coefficient of friction between workpiece and die, α is the half angle of the draw die, and Φ is a factor that takes the inhomogeneity of deformation into account. For reasons discussed in Sec. 4.3.3 and 4.5.6, this factor is a function of the h/L ratio. For drawing wire of circular cross section, h is taken as the mean diameter and L is the length of the contact zone, and the factor is

$$\Phi = 0.88 + 0.12\frac{h}{L} \tag{4.29}$$

For deformation in plane strain, e.g., in shaping a rectangular cross section, the factor is

$$\Phi = 0.8 + 0.2\frac{h}{L} \tag{4.30}$$

Just as in extrusion, a mean flow stress σ_{fm} must be used because the material strain hardens during its passage through the draw die.

4.7.3 Process Limitations

The requirement of keeping the draw force below the strength of the issuing wire puts a limit of approximately 50 percent area reduction on most drawing operations. Breaking of the wire severely limits productivity since the end of the wire must then be reduced (*pointed*) again so that it can be rethreaded. This is obviously time-consuming, and it is usually more profitable to limit reductions to approximately 30 percent per die. As seen from an inspection of the formula for the draw force, Eq. (4.28), friction increases the draw stress and limits reduction; therefore good lubricating practices are essential (Table 4.4).

A second limitation arises from possible nonuniformity of deformation. Just as in extrusion (Sec. 4.5.6), the depth of the compression zone may not be sufficient to assure homogeneous deformation. This is again governed by the h/L ratio: when $h/L>2$, secondary tensile stress can lead to the typical arrowhead (center-burst) defect in less ductile materials (Fig. 4.25*b*).

A further possibility of secondary tensile stresses arises when deformation is limited to one part of a section. This will be discussed in more detail for the rolling of shapes (Sec. 4.8.2). Suffice it to say here that cracking of drawn sections occurs when some part of the cross section is not directly subjected to deformation.

4.8 ROLLING

Of all bulk deformation processes, *rolling* occupies the most important position. Over 95 percent of all material that is ever deformed is subjected to rolling (see Table 4.1).

4.8.1 Flat Rolling

The process of reducing the thickness of a *slab* to yield a thinner and longer but only slightly wider product is commonly referred to as *flat rolling*. It is the most important primary deformation process, because it allows a high degree of automation and very high speeds, and thus provides starting material for various secondary sheet-metalworking processes at a low cost.

The cast structure is first destroyed and defects healed as far as possible by *hot-rolling*. The hot-rolled product has a relatively rough surface finish and dimensional tolerances are not very tight; nevertheless, hot-rolled *plate,* over 1/4 in (6 mm) thick and 72–200 in (1800–5000 mm) wide, in weights up to 150 tons, is an important starting material in shipbuilding, boilermaking, and the manufacture of pipes and miscellaneous welded machine structures. Since the smallest economical weight of a rolled slab is around 1 ton, the long lengths of thinner *sheet* issuing from the rolling mill are *coiled* up. Such hot-rolled sheet of typically 0.048 to 0.25 in (1.2 to 6 mm) thickness, in widths up to 90 in (2300 mm), is an important starting material for the cold pressing of structural parts of vehicles, heavy equipment, and machinery, and also for making welded tube.

Thinner gages, better surface finish, and tighter tolerances are obtained by *cold-rolling.* Again, the product is made in the form of coils which are then slit into narrower widths or cut into shorter lengths, or both, depending on the handling facilities of the manufacturing operations. Standard surface finishes and tolerances are obtained at no extra expense; however, exceptionally smooth finish or tight tolerances can also be produced, often at only a slight premium. The cost, of course, goes up as the gage decreases, especially if the thinner gage necessitates an extra pass through a single-stand or *multistand (tandem)* mill. It is possible to roll steel and other materials to a gage of 0.0001 in (3 μm), and large quantities of aluminum foil are produced at a 0.0003 in (8 μm) gage at relatively low cost.

The process of flat rolling looks deceptively simple (Fig. 4.26). Two

FIG. 4.26   The principle of flat rolling.

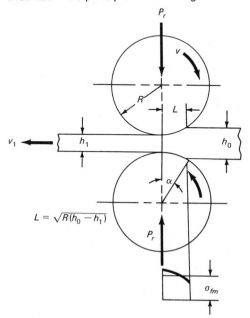

$$L = \sqrt{R(h_0 - h_1)}$$

driven rolls of cylindrical shape reduce the flat workpiece to a thinner gage. The finished product must have a uniform thickness in length and width, a flat shape, a controlled and uniform surface finish, and reproducible mechanical properties. Satisfying these requirements taxes the ingenuity of the production engineer, equipment designer, control specialist, and theoretician, and makes the process one of the most complex ones.

4.8.2 Shape Rolling

The rolling of shapes has a long history, beginning with the rolling of channels of lead for stained-glass windows (Table 1.1). The largest industrial application is now in the hot-rolling of *structural shapes,* which is a specialized primary deformation process practiced in special-purpose mills. The same techniques, however, can be applied to the cold-rolling of shapes to tight tolerances and excellent surface finish, and these specialized secondary manufacturing processes are gaining popularity as alternatives to drawing.

The starting material for cold shape rolling is a wire of square, rectangular, or circular cross section, and the finished shape is approached through a number of *passes* that gradually distribute the material in the desired fashion.

The crucial issue is always that of avoiding *nonuniform elongation.* As seen from the simple example of Fig. 4.27, those parts of the cross section that are *directly compressed* elongate as required to maintain constant volume, while parts not subject to direct compression elongate only because of their physical attachment to the deforming portion. Elonga-

FIG. 4.27 Secondary tensile stresses and edge cracking during nonhomogeneous rolling of a section.

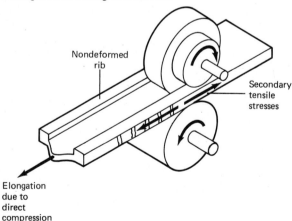

Nondeformed rib

Secondary tensile stresses

Elongation due to direct compression

tion in these noncompressed portions generates secondary tensile stresses which, as remarked before, easily lead to crack formation. Therefore, *roll pass design* aims at equalizing reductions in all portions of the cross section. This aim can be attained by moving the material sideways, especially in the early passes and, if necessary, by the use of *vertical rolls* that compress the section from its sides. Several rolling stands may be placed in tandem and it is then customary to alternate the axes of rolls from vertical to horizontal.

4.8.3 Ring Rolling

Seamless rings are important constructional elements, ranging from the steel tires of railway car wheels to rotating rings of jet engines and races of ball bearings.

After making a hole by any suitable technique, the thick-walled ring is rolled out by reducing its thickness and increasing its diameter as indicated in Fig. 4.1. Larger rings are rolled hot in specialized factories but smaller rings, especially those of small cross-sectional area, are increasingly rolled cold. In addition to simple rectangular profiles, rings of a fairly complex cross-sectional profile can be rolled.

4.8.4 Transverse Rolling

When a workpiece is placed between two counterrotating rolls with its axis parallel to the roll axes, it suffers plastic deformation (essentially, localized compression) during its rotation between the rolls. The consequences of this deformation depend on the shape and angular alignment of the rolls and, as in all compression (Sec. 4.3.3), on the h/L ratio. The height is now the workpiece diameter, and L is the length of contact with the roll (equivalent to L of an indenter in plane strain, Fig. 4.4b).

When $h/L > 1$, deformation is inhomogeneous and the plastic zones penetrating from the point of contact literally try to wedge the workpiece apart; in other words, high secondary tensile stresses are generated in the center of the workpiece. This is the principle of making thick-walled tubes by the *rotary tube-piercing* methods. As shown in Fig. 4.1, a *mandrel* or plug placed against the center of the billet helps in opening up and smoothing out the internal surface. Angular misalignment of the deforming rolls makes the tube progress in a helical path; thus its whole length is pierced through. Such tube-piercing methods are practiced in specialized plants equipped for hot-working.

The secondary deformation processes based on the same principle have the roll axes aligned and the workpiece rotates in the same plane (*transverse rolling,* Fig. 4.28). The rolls are shaped so as to avoid the generation of large tensile stresses while a sound workpiece of axial symmetry

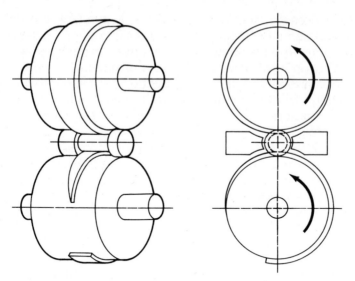

FIG. 4.28 Cross-rolling of a dumbbell-shaped workpiece. (J. Holub, *Machinery* (London), *102*:131, Jan. 16, 1963.)

is formed. For example, a dumbbell shape can serve either as a finished part or as a preform for further forging.

The rolls may be shaped to roll a thread on the workpiece. Large threads are rolled hot, but most *thread-rolling* operations are conducted cold. The same principle is utilized in thread-rolling machines equipped with so-called *flat dies* (Fig. 4.29). One of the dies is stationary, the other reciprocates; at an appropriate point of the stroke, a workpiece (typically, a cold-headed screw blank) dropped into the gap is grabbed by the moving die and is rotated against the stationary die; thus the screw thread profile is gradually developed. Rolled threads have a continuous grain flow and

FIG. 4.29 Thread rolling with reciprocating flat dies.

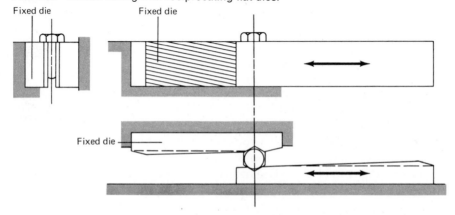

Fixed die   Fixed die

Fixed die

are, therefore, more fatigue-resistant than threads cut on a lathe. The productivity of the process is high. Even large, slow machines roll 60 screws per minute while smaller screws are produced at rates of 500 per minute. The "flat" dies may also be slightly curved and placed around the circumference of a rotating cylinder moving against stationary dies placed into an outer ring. For each revolution of the cylinder as many screws are made as there are tool pairs; thus production rates of 2000 per minute are achieved. The good quality and high productivity of thread rolling has eliminated thread cutting as a competitive process for all mass-production purposes.

Very large internal threads could be rolled but, apart from cutting, a more practical way is *cold form tapping*. The tool looks like a screw, except that its diameter changes periodically within the screw envelope, so that the protruding portions displace material from the roots into the threads.

4.8.5   Calculation of Forces and Power Requirements

A very rough estimate of the rolling forces can be obtained if rolling is regarded as a continuous forging (cogging) process. Thus, the *projected length of contact* between roll and workpiece (Fig. 4.26) is regarded as L of the forging tool, and the average height is taken as h. When $h/L > 1$, the inhomogeneity of deformation predominates and the pressure-multiplying factor Q_i is found from Fig. 4.5. When $h/L < 1$, friction effects are overriding and the pressure-intensification factor Q_p is found from Fig. 4.11.

The *roll force* is estimated from:

$$P_r = L \, w \, Q_p \, \sigma_{fm}$$
$$\text{or} \qquad P_r = L \, w \, Q_i \, \sigma_{fm} \qquad (4.31)$$

where w is the width of strip and σ_{fm} is the mean flow stress, used because the strip hardens while it is deformed in the roll gap (a steady-state process). In hot-working, the flow stress must be taken at the typical average strain rate

$$\dot{\epsilon}_m = \frac{v}{L} \ln \frac{h_0}{h_1} \qquad (4.32)$$

and the strain is simply calculated from its definition, Eq. (4.5).

The *torque* required to rotate the rolls can be obtained by assuming that the rolling force acts in the middle of the *arc of contact*. Since there are two rolls to be driven, the total torque will be

$$M_T = \frac{2 P_r L}{2} = P_r L \qquad (4.33)$$

and the *horsepower requirement* is readily calculated from

$$\text{hp} = \frac{2P_rL\pi N}{33,000} \tag{4.34a}$$

where N is rpm, L is in units of feet, and P is in units of pounds. In the SI system the *power requirements* are calculated in kilowatt units:

$$\text{kW} = \frac{2P_rL\pi N}{60,000} \tag{4.34b}$$

where P is the roll force in newtons, L is in meters, and N is rpm.

4.8.6   Process Limitations

In secondary manufacturing rolling operations, the major concern is soundness of the product, which requires that secondary tensile stresses be kept to a minimum. If this cannot be assured, cracking can be delayed or avoided if smaller deformation steps are taken, interspersed with frequent intermediate anneals. An undesirable side effect of annealing may be coarse grain in those parts of the rolled cross section that were exposed to only slight (critical) deformation (Sec. 2.1.8).

When thin sections are rolled in hard materials, elastic deformation of the roll may limit the rollable thickness. *Flattening* of the rolls can be minimized with a good lubricant, a small roll diameter, and a roll made of a material with a high elastic modulus, such as WC.

While good lubrication is essential to reduce roll forces, some minimum friction is still needed because it is the frictional component of the roll force that pulls the workpiece into the roll gap. The maximum *angle of acceptance* α_{max} (Fig. 4.26) is a function of the coefficient of friction

$$\tan \alpha \leqslant \mu \tag{4.35}$$

A heavier reduction can be taken by pushing the workpiece into the roll gap.

Inhomogeneous deformation, whether from a large h/L ratio or the absence of direct compression, is always harmful. There is one instance, however, when it is purposely induced. In *roller burnishing* the surface of a thick workpiece is superficially rolled; the deformation zone is very shallow and, in the absence of bulk plastic flow, the material of the surface is put in compression, making the part more resistant to fatigue.

4.9   SUMMARY

Bulk deformation processes have retained their importance over thousands of years of technological development. They not only provide the starting material for subsequent sheet metalworking, wire and tube bending, and most welding applications, but also assure the availability of finished components of great structural integrity. The products include hot-forged parts, from turbine blades and gear blanks to garden hoes; cold-forged parts, from nails, screws, and rivets to finished gears; cold-extruded parts, from automotive half axles and sparkplug bodies to toothpaste tubes; hot-extruded constructional sections and valve bodies; hot- and cold-rolled rings and sections for all purposes.

The plastic flow of material follows many of the rules of viscous flow, and sharp directional changes such as tight radii are avoided by the process-conscious designer. A great variety of shapes may nevertheless be produced economically if some basic process limitations are recognized:

1 Interface (die) pressures rise with increasing deformation (strain) and friction. Therefore, friction is, as a rule, minimized by lubrication, and deformation in cold-working and strain rates in hot-working are limited according to the n and m values of the material, respectively.

2 High pressures can be accommodated by suitable die design, usually based on the principle of prestressing.

3 Inhomogeneous deformation increases the deformation pressures and, more importantly, may lead to external or internal fracture in materials of limited ductility. In general, therefore, the process should ensure penetration of the deformation zones through the entire workpiece thickness. The principle is that of increasing the hydrostatic-pressure component of the induced stress state.

Further Reading

DETAILED PROCESS DESCRIPTIONS:
    Metals Handbook, 8th ed., American Society for Metals, Metals Park, Ohio, vol. 3, *Machining,* 1967, pp. 105–107 (Roller Burnishing); pp. 130–145 (Thread Rolling); pp. 145–146 (Spline Rolling).
Vol. 4, *Forming,* 1969, pp. 78–88 (Coining); pp. 322–333 (Straightening); pp. 333–346 (Rotary Swaging); pp. 465–496 (Cold Heading and Extrusion).
Vol. 5, *Forging and Casting,* 1970.

PROCESS DETAILS, MOSTLY FOR THE LABORATORY SCALE:
    HANKS, G. S., and D. J. SANDSTROM: High Energy Rate Forging and Impact Extrusion, in *Techniques of Metals Research,* R. F. Bunshah (ed.), Interscience, New York, 1968, vol. 1, pt. 3, pp. 1567–1587.
    SCHEY, J. A.: The More Common Fabrication Processes, *ibid.,* pp. 1409–1538.

GENERAL INTRODUCTORY TEXTS:
    COOK, N. H.: *Manufacturing Analysis,* Addison-Wesley, Reading, Mass., 1966.
    KALPAKJIAN, S.: *Mechanical Processing of Materials,* Van Nostrand, Princeton, N.J., 1967.

LISSAMAN, A. J., and S. J. MARTIN: *Principles of Engineering Production,* The English Universities Press, London, 1964.

PARKINS, R. N.: *Mechanical Treatment of Metals,* George Allen and Unwin, London, 1968.

SACHS, G.: *Fundamentals of the Working of Metals,* Pergamon, New York, 1954.

GENERAL, MORE ADVANCED TEXTS:

AVITZUR, B.: *Metal Forming: Processes and Analysis,* McGraw-Hill, New York, 1968.

BACKOFEN, W. A.: *Deformation Processing,* Addison-Wesley, Reading, Mass., 1972.

JOHNSON, W., and P. B. MELLOR: *Engineering Plasticity,* Van Nostrand, London, 1973.

ROWE, G. W.: *An Introduction to the Principles of Metalworking,* St. Martin's Press, New York, 1965.

SPECIALIZED BOOKS:

ALEXANDER, J. M., and B. LENGYEL: *Hydrostatic Extrusion,* Mills and Boon, London, 1971.

EVERHART, J. E.: *Impact and Cold Extrusion of Metals,* Chemical Publishing Co., New York, 1964.

FELDMAN, H. D.: *Cold Forging of Steel,* Chemical Publishing Co., New York, 1962.

GELEJI, A.: *Forge Equipment, Rolling Mills and Accessories,* Akademiai Kiado, Budapest, 1967 (in English).

GRAINGER, J. A.: *Flow Turning of Metals,* The Machinery Publishing Co., Brighton, England, 1969.

JENSON, J. E. (ed.): *Forging Industry Handbook,* Forging Industry Association, Cleveland, Ohio, 1970.

PEARSON, C. E., and R. N. PARKINS: *The Extrusion of Metals,* Wiley, London, 1961.

SABROFF, A, M., F. W. BOULGER, and H. J. HENNING: *Forging Materials and Practices,* Reinhold, New York, 1968.

SCHEY, J. A. (ed.): *Metal Deformation Processes: Friction and Lubrication,* Dekker, New York, 1970.

SHERIDAN, S. A.: *Forging Design Handbook,* American Society for Metals, Metals Park, Ohio, 1972.

WILSON, F. W. (ed.): *Die Design Handbook,* 2d ed., American Society of Tool and Manufacturing Engineers, McGraw-Hill, New York, 1965.

Forging: Equipment, Materials, and Practices, Battelle Memorial Institute, Metals and Ceramics Information Center, Columbus, Ohio, 1973.

Impact Machining, Verson Allsteel Press Co., Chicago, 1969.

Examples

4.1 A 302 stainless steel pin is to be produced from a square wire. One end is flattened, and the center is pinched (as shown in the illustration). Calculate die pressures and forces, assuming that no lubricant is used.

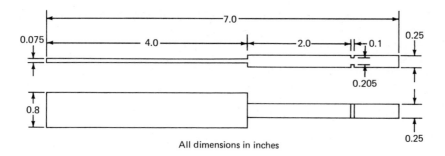

All dimensions in inches

Solution. For the *flattening* operation, consider upsetting a rectangular workpiece (Secs. 4.3.4 and 4.4.1):

(a) The engineering strain for $h_0 = 0.25$ in, $h_1 = 0.075$ in

$$e_c = (0.25 - 0.075)/0.25 = 0.7 \text{ (or 70\%)}$$

(b) The natural strain

$$\epsilon = \ln (0.25/0.075) = \ln 3.34 = 1.21$$

(c) For cold-working, strain-rate sensitivity can be ignored. At the end of the stroke (from Table 4.2 after conversion)

$$\sigma_f = K\epsilon^n = 190 \times 1.21^{0.3} \simeq 200{,}000 \text{ psi}$$

(d) The 4-in length of the pin increases very little during flattening, and this dimension must be regarded as the width w during plane-strain compression (Fig. 4.10). Most of the material gets displaced in the width of the pin; in terms of analysis, this becomes L. Check if friction or indentation predominates:

$$h/L = 0.075/0.8 = 0.094 \quad (<1)$$

Thus, friction is important. From Fig. 4.11, for

$$L/h = 10.6$$

and sticking friction (in the absence of a lubricant)

$$Q_p \simeq 4$$

Then $p_p = Q_p\sigma_f = 4 \times 200{,}000 = 800{,}000$ psi

This is high for any tool material, and a suitable lubricant (Table 4.4)

should be used. Assuming that the coefficient of friction is reduced to $\mu = 0.1$, from Fig. 4.11

$$Q_p \simeq 2$$

and $p_p = 2 \times 200{,}000 = 400{,}000$ psi

which is still high but feasible.

(e) The upsetting force is

$$P_p = Ap_p = wLp_p = 4 \times 0.8 \times 400{,}000 = 1{,}280{,}000 \text{ lb or 640 tonf.}$$

Note the large size of press needed for this seemingly minor operation even with the application of a lubricant. Die pressures and forces could be reduced by flattening in two steps, with an intermediate anneal.

For *pinching* the center, follow the same sequence:

(a) $e_c = 0.18$ or 18%
(b) $\epsilon = 0.2$
(c) $\sigma_f = 190 \times 0.2^{0.3} = 118{,}000$ psi
(d) Consider the geometry of the operation; compared with Fig. 4.4, $L = 0.1$ in, $h = 0.205$ in, and $w = 0.25$ in (assuming that little material goes into the width):

$$h/L = 0.205/0.1 \simeq 2$$

deformation must be inhomogeneous, and from Fig. 4.5

$$Q_i = 1.5$$

$$p_i = 1.5 \times 118 = 177{,}000 \text{ psi}$$

(e) $P_i = 0.25 \times 0.1 \times 177{,}000 = 4{,}400$ lb (or 2.2 tonf)

Indentation pressure is independent of friction, therefore, no lubricant is necessary except for wear prevention and surface quality.

4.2 A blanking punch is made of D2 tool steel. For best performance and favorable flow line orientation, a billet of $d_o = 2$ in and $h_o = 2$ in is hot upset at 1000°C to a height of $h_1 = 0.5$ in on a hydraulic press at $v = 600$ in/min. No lubricant is used. Calculate, for the end of the press stroke, the press force.

(a) The new diameter d_1 is found from constancy of volume:

$$V = (2^2\pi/4)\,2 = 6.28 \text{ in}^3$$

$$A_1 = 6.28/0.5 = 12.56 \text{ in}^2; \ d_1 = 4 \text{ in}$$

(b) $e_c = (2 - 0.5)/2 = 0.75$ (or 75%)

(c) $\epsilon = \ln (2/0.5) = 1.4$

(d)   In hot-working, strain rate is important. At the end of the stroke

$$\dot{\epsilon} = v/h_1 = 600/(60 \times 0.5) = 20/s$$

(e)   From Table 4.2, after conversion

$$\sigma_f = C\dot{\epsilon}^m = 30 \times 20^{0.12} = 30 \times 1.43 = 43{,}000 \text{ psi}$$

(f)   In axial upsetting only friction need be considered. From Fig. 4.7, for $d/h = 4/0.5 = 8$ and sticking friction

$$Q_a \simeq 2.8$$

$$p_a = 2.8 \times 43{,}000 = 120{,}000 \text{ psi}$$

(g) $P_a = A_1 \, p_a = 12.56 \times 120{,}000 = 1{,}500{,}000$ lb or 750 tonf.

4.3 An 1100 Al can (container) of 50 mm OD and 48 mm ID is to be produced by the back extrusion (Fig. 4.19d) of $d_0 = 50$ mm diameter annealed slugs. Lanolin is used as a lubricant. Calculate the container pressure during steady-state extrusion.

From Sec. 4.5.5 (note that the extrusion is a hollow tube) and Eq. (4.19):

(a) $A_0 = 50^2 \pi/4 = 1960 \text{ mm}^2$

   $A_1 = (50^2\pi/4) - (48^2\pi/4) = 154 \text{ mm}^2$

   $R_e = A_0/A_1 = 12.7$

(b) $\epsilon = \ln R_e = 2.54$

(c)   We are to calculate the pressure for the steady-state condition, therefore σ_{fm} is needed. According to Sec. 4.3.1, a σ_f vs. ϵ curve could be plotted, or we can integrate the area under the stress-strain curve within the limits of $\epsilon = 0$ and $\epsilon = 2.54$:

$$\sigma_{fm}\epsilon = \int_0^\epsilon K\epsilon^n \, d\epsilon = K \left[\frac{\epsilon^{n+1}}{n+1} \right]_0^\epsilon$$

From Table 4.3, $K = 140$ MPa; $n = 0.25$. Then

$$\sigma_{fm}\epsilon = 140 \, \frac{2.54^{1.25}}{1.25} = 140 \, \frac{3.2}{1.25} = 358$$

$$\sigma_{fm} = 358/2.54 = 141 \text{ MPa}$$

(d) With the given geometry and lubricant, wall friction can be ignored. From Eq. (4.20)

$p_e = 141[0.8 + (1.2 \times 2.54)] = 141 \times 3.85 = 540$ MPa

(e) This pressure acts on the base of the container

$P_e = 1960 \times 10^{-6} \times 540 = 1.12$ MN ($= 125$ tonf)

(f) Check whether punch pressure would give a higher value:

$p_i = 141 \times 3 = 423$ MPa

$P_i = (1960 - 154)\, 10^{-6} \times 423 = 0.76$ MN

thus the value calculated in (e) is safe.

4.4 A 5% Al-bronze part is made by cold-upsetting to $e_c = 50\%$. Calculate the flow stress.

Table 4.2 does not contain flow stress data for this material. The UTS $= 400$ MPa, and elongation is 65%. From a comparison with other materials, this should be equivalent to a fairly high n value, say 0.4. Since $n = \epsilon_u$, the UTS should be found at

$$\epsilon_u = \ln \frac{A_0}{A_u} = 0.4$$

$$\frac{A_0}{A_u} = \exp 0.4 = 1.49$$

and the starting area $A_0 = 1.49$ for a necked area $A_u = 1$; thus, at the point of necking, the engineering compressive strain (the area reduction) is

$$e_c = \frac{1.49 - 1}{1.49} = 0.33$$

The force at necking

$$P = \text{UTS} \times A_0 = \sigma_f A_u$$

and the true flow stress at necking, after 33% reduction in cross section, is

$$\sigma_f = \frac{\text{UTS} \times A_0}{A_u} = 400 \times 1.49 = 596 \text{ MPa}$$

Further deformation to 50% reduction in area (which would be equivalent to 50% reduction in height, since the tensile and compressive stress-strain curves are presumed to be identical) would require still higher stresses, say $\sigma_f = 700$ MPa.

Problems

4.1 Repeat the calculations for Example 4.3 but assume that extrusion is carried out at 300°C on a fast press with a ram speed of $v = 0.75$ m/s.

4.2 It is proposed that the end of a $d_0 = 0.25$ in, 1015 steel bar be upset over a length of $h_0 = 0.2$ in to form a flat head of $h_1 = 0.03$ height (thickness). To assess feasibility, calculate the upsetting stresses and forces, (a) first assuming that a good lubricant reduces friction to $\mu = 0.1$, and then (b) for rough, unlubricated dies (as would apply if the lubricant had broken down or the lubricant supply had failed). (c) If the operation is feasible, suggest a suitable die material.

4.3 A bolt of the indicated geometry is to be produced for high-temperature service. A material similar to H13 is proposed, and it is planned to forward-extrude the shaft and, in a separate operation, back-extrude the head, both at 1000°C, in a hydraulic press with a ram speed of $v = 0.5$ m/s. Determine if the proposition is feasible as far as tool loading is concerned, by calculating (a) the size of the starting billet, (b) maximum forward-extrusion pressure, assuming that a graphitic lubricant of $\tau_i = 70$ MPa shear strength is applied to the container, (c) the backward extrusion pressure (remember to check both container and punch pressures).

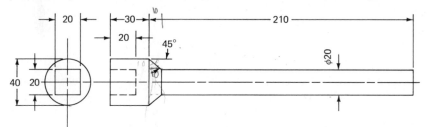

All dimensions in millimeters

4.4 A medium-carbon steel (1045) is to be cold-extruded. Determine, from the equilibrium diagram (a) what phases one should expect, (b) in what proportion in this material. From a consideration of the properties of two-phase structures, suggest (c) the optimum metallurgical condition for this application.

4.5 A bolt head is produced by cold-heading (upsetting) the head on an annealed bar. (a) Make a sketch of the longitudinal cross section of the bolt, and indicate the grain-size variation one should expect if the bolt is annealed after cold-heading. (*b*) Point out the weakest cross section and indicate (*c*) a method of production that would avoid this weakness.

4.6 A screw thread is cold-rolled on a 1015 low-carbon steel bolt of 0.25 in diameter. Observation shows that the contact zone $L = 0.05$ in between die and screw blank. The average strain hardening during rolling corresponds to a strain of $\epsilon = 0.4$. Calculate (*a*) the applicable flow stress and (*b*) the average interface pressure in the contact zone. Determine (*c*) if there is any danger of internal fracture (make sketch to illustrate the point).

4.7 A small, shallow U channel of 5% Sn bronze is cold-rolled. The shape is shallow enough to regard it as a $w = 10$ mm wide, $h = 2$ mm thick strip of rectangular cross section. A 30 percent reduction in height is taken in a single pass, on a mill with 150-mm diameter rolls, rotating at $v = 0.8$ m/s speed, with a mineral-oil lubricant. Calculate (*a*) the roll force and (*b*) the power requirements.

4.8 A $d_0 = 1$ in and $h_0 = 2$ in billet of a free-machining (leaded) brass is to be compressed to an $h_1 = 0.4$ in height in a hydraulic press (ram velocity $v = 600$ in/min) at 800°C between unlubricated anvils. (*a*) For the end of the press stroke, calculate interface pressure and press force. (*b*) What increase in stresses would occur if the workpiece were to cool to 600°C? (*c*) If a very good lubricant were applied to reduce forces, would this introduce other problems?

4.9 Take a common (flathead) nail, measure the diameter and the thickness (height) of the head. Calculate (*a*) the head volume and (*b*) the length of wire that had been upset. From these data, (*c*) would you expect buckling during free heading? If the answer to (*c*) is yes, how would the head be formed?

4.10 It is proposed to make a tube-shaped part of 0.15% C steel by cold-piercing (Fig. 4.13) a $d_0 = 20$ mm billet with a $d_p = 10$ mm diameter punch. The length of the part (the depth of the hole) is 50 mm; for constructional reasons, the punch is 60 mm long. (*a*) Calculate the punch pressure, (*b*) suggest a suitable punch material. (*c*) Check the punch for bending and compression. (*d*) Specify a lubricant suitable for this task.

4.11 A component is produced by forward extruding a $d_0 = 20$ mm diameter billet through a $d_1 = 14$ mm die of $\alpha = 45°$ half angle. Part of the billet remains unextruded to serve as the head. Many components are found to have center-burst defects. What could be done to get out of trouble (the material cannot be changed). Make a sketch (to scale) to justify your answer.

4.12 Commercial purity (1100 Al) aluminum is routinely rolled, in several passes but without annealing, to a total reduction of over 98 percent. (*a*)

Find the uniform strain (ϵ_u), the total elongation (e_f) and the reduction in area (q) for this material. (*b*) Compare these to the rolling reduction obtainable, and explain the reasons for the difference.

4.13 A copper (99.94% Cu) bar is to be cold-rolled into a shape, which must have a minimum UTS of 60,000 psi. If the finished cross section $A_n = 0.05$ in^2, what bar diameter should one start with to achieve this? (Hint: A look at Figs. 2.10 and 2.11 will show that the true stress σ and σ_{eng} must be very similar in a strain-hardened material.)

4.14 A flat, annealed 70/30 brass wire is to be drawn from a 10 mm × 2 mm cross section to a fluted shape of an average thickness $h = 1$ mm and an unchanged width of $w = 10$ mm. The die half angle is $\alpha = 7°$ and the lubricant is an emulsion. Calculate (*a*) the relevant flow stress; (*b*) the drawing force. (*c*) Check if the process is feasible. (*d*) Suggest a way of achieving the required end result.

5

SHEET-METALWORKING PROCESSES

Because of the low cost of mass-produced sheet of high quality, *sheet metalworking* has gained an outstanding position among manufacturing processes. Originally, sheet was rolled and supplied in limited sizes; however, with the appearance of continuous tandem rolling mills, the sheet is now really produced in a *wide strip* form. Coils may be cut up for easier handling in the customer's facilities; however, there is an increasing trend to ship entire coils (sometimes slit into narrower widths) which are then fed into the press lines of the manufacturer.

5.1 THE METAL

All wrought alloys (Sec. 4.2.2) are suitable for sheet-metalworking processes. The critical properties are, however, somewhat different from those for bulk deformation.

Foremost is the requirement that the sheet should survive deformation without localized necking (which, apart from its weakening effect, would be objectionable on parts that are visible in service) and fracture (which would destroy the mechanical integrity of the part). Therefore, in addition to reduction of area q, the factors that delay necking (high n value, Sec. 2.1.5) or prevent localized contraction of the neck (high m value, Sec. 2.1.9, and phase transformations, Sec. 2.2.8) become important. Depending on the deformation mode imposed, textures (Sec. 2.1.3) may contribute desirable or undesirable features, as will be discussed in Sec. 5.5.2.

Many sheet-metalworking operations involve a relatively modest tensile strain (say, under 10 percent). This creates a peculiar problem of its own when the sheet material shows yield-point elongation or serrated yielding (Sec. 2.2.2). As stretching begins, yielding is localized into a thin, highly visible band (Lüders line). On continued stretching (Fig. 2.17), families of these lines criss-cross the surface until at the end of yield-point elongation (typically, after 5–10 percent strain) the whole surface is uniformly deformed. Stretcher-strain marks (or, as they are called in the shop, *worms*) are harmless, but they are objectionable on exposed surfaces.

The material used in largest quantities, *mild steel* (low-carbon steel), suffers from yield-point elongation. It presents problems even in feeding a press with annealed material from a coil, because the sheet unrolls with sharp localized *kinks,* and *roller leveling* just makes the kinks more closely spaced. It is usual, therefore, to give a small, 0.5 to 1 percent rolling reduction (*temper pass*) to the annealed material. This initiates yielding in an extremely closely spaced form, not visible to the eye, and suppresses localized yielding during subsequent press working. This remedy is permanent in aluminum-killed steel, but strain aging (Sec. 2.2.2) brings back the yield-point elongation in rimmed steels within 1 to 3 weeks. Rimmed steel solidifies with the evolution of gases (Sec. 2.3.3) which drive all contaminants toward the center of the cast ingot. Therefore the surface of the ingot (and of the rolled sheet) is very pure, ductile, and clean. Coupled with a lower cost, these features make rimmed steel a very attractive starting material. If it cannot be used within a week or so of temper rolling, the Lüders lines can again be suppressed by roller-leveling just before pressing.

5.2 SHEARING

Irrespective of the size of the part to be produced, the first step involves *cutting* the sheet or strip into appropriate shapes by the process of *shearing.* This process of separating adjacent parts of a sheet through controlled fracture cannot be described as either purely plastic deformation or as machining. The sheet is placed between two edges of the shearing tools—in the instance of *blanking,* a *punch* and a *die* (Fig. 5.1). On penetration of the tool edges the sheet is first plastically deformed, then cracks are generated at a slight angle to the cutting direction. When these cracks meet, the shearing action is complete even though the cutting edges had moved only partly through the thickness of the sheet (Fig. 5.1*a*). The cut edge is not perfectly perpendicular to the sheet surface. It also exhibits some roughness (evidence of crack penetration), a smooth burnished zone (formed when the part is pushed through the die) and a rounded edge (formed by the initial plastic deformation). Nevertheless, the finish thus produced is acceptable for a great many applications.

The quality of the cut surface is greatly influenced by the *clearance*

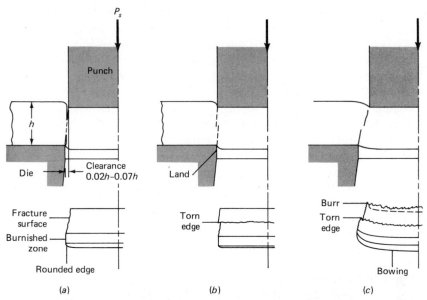

FIG. 5.1 The process of shearing and the appearance of the cut surface with (*a*) optimum clearance, (*b*) insufficient clearance, and (*c*) excessive clearance (exaggerated).

between the two shearing edges. With a very tight clearance, the cracks, originating from the tool edges, miss each other and the cut is then completed by a *secondary tearing* process, producing a jagged edge approximately midway in the sheet thickness (Fig. 5.1*b*). Excessive clearance allows extensive plastic deformation, and when the crack finally forms, it may still miss the opposing crack; separation is delayed and a long fin (*burr*) is pulled out at the upper edge (Fig. 5.1*c*). The jagged edge of the burr with its sharp roots acts as a stress concentrator which initiates fracture during subsequent forming or in the service of the part. Therefore, proper choice of the clearance is a vital aspect of the process. From experience, the clearance is taken between 2 and 7 percent of the sheet thickness (the smaller clearance goes with a more ductile material).

Shearing is practiced in a number of processes. When shearing is conducted between rotary blades, the process is referred to as *slitting;* cutting along a straight line is simply shearing. A contoured part (whether it be circular or more complex in shape) is cut between a punch and die in a press, and the process is called blanking. The economy of the process greatly depends on the proper physical arrangement of the parts to minimize scrap losses. Essentially, the same process is also used to remove unwanted parts of a sheet, but then one refers to *punching* a hole, of circular or any other shape.

A number of special processes are available for producing very clean

cut edges of a surface finish sufficiently smooth to allow immediate use of
the parts, e.g., as gears in lightly loaded machinery. The principle is generally that of imposing a compressive stress on the *shear zone* by means of
a specially shaped *blankholder,* so that crack initiation is delayed and the
whole thickness is plastically sheared (Fig. 5.2*a*). Alternatively, one may
shear with a *negative clearance,* so that the part is actually pushed (extruded) through the cutting die (Fig. 5.2*b*), or a conventionally blanked part
may be *finish-shaved* with a tight clearance (Fig. 5.2*c*). In all instances,
a *counterpunch* prevents bowing of the part.

The size of the press required to perform the cut is readily calculated
for conventional shearing (Fig. 5.1). Since the process is partly plastic
deformation and partly shear, and deformation is concentrated in a very
narrow zone where strain hardening takes place, a good approximation is
obtained from

$$P_s = 0.7\ \text{UTS} \cdot h \cdot l \tag{5.1}$$

where h is the sheet thickness and l is the length of cut. When the
shearing edges are parallel, the entire length of the contour must be taken.
The cutting force can be reduced by having one of the shearing edges at
an angle to the other (at a *shear*), thus only the actual length instanta-
neously sheared needs to be considered. The ultimate tensile strength of
most materials is known (as in Tables 4.2 and 4.3). If only the K and n values are available, the UTS may be approximated by substituting
$\text{UTS} \simeq K \cdot n^n$ (this is really the true stress at the point of necking and is thus
higher than the UTS, but using it will just increase the margin of safety of
calculation).

FIG. 5.2 Methods of shearing parts with a finished edge: (*a*) precision blanking, (*b*) negative
clearance blanking, and (*c*) shaving.

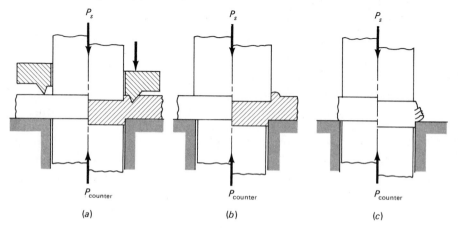

Blanking is a high-productivity process, limited only by the rate of feeding material into the press and by the number of strokes of the press. Several parts may be blanked or punched simultaneously, and complex geometries can be created in *compound dies* (in which several cutting edges work simultaneously, Fig. 5.3) or in *progressive dies* (in which several punching and blanking operations are sequentially performed with die elements fastened to the same die plates, Fig. 5.4). Holes of standard sizes and shapes can be cut on general-purpose *punch presses*. Numerically controlled presses allow rapid location of the sheet and selection of the punch, and thus permit low-cost production of small and medium quantities.

The punch and die are made of tool steel in mass-production dies (Figs. 5.3 and 5.4). The *scrap skeleton* (or, in punching, the part) would bind on the punch (or in the die), and must be stripped with a fixed or spring-supported *stripper plate* (Fig. 5.3) or with a plastic (usually polyurethane) *foam pad* (Fig. 5.4). For smaller quantities (say, a few hundred pieces) the die cost can be lowered if a greater scrap loss is tolerable. The die is simply a steel plate cut to size, and the cutting action occurs by

FIG. 5.3 Principal elements of a compound die for making washers.

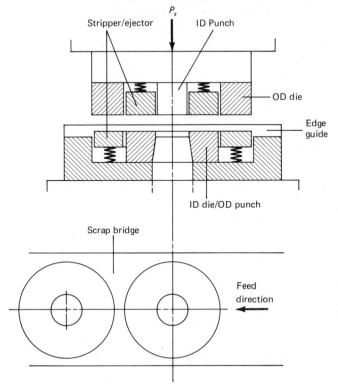

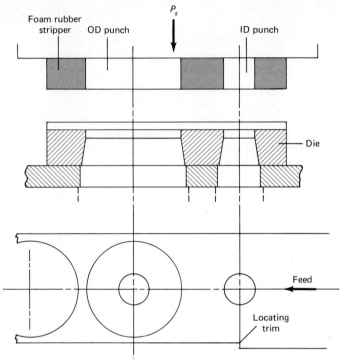

FIG. 5.4 Principal elements of a progressive die for making washers (die elements for cutting the locating trim are not shown).

FIG. 5.5 Blanking with a rubber foam die.

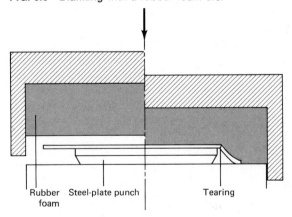

pressing the sheet around this die with a *rubber cushion* (Fig. 5.5). The overhanging part of the sheet is bent down and clamped against the base plate by the cushion, and tearing occurs around the edges of the die plate.

5.3 BENDING

Many parts are further shaped by the relatively simple process of *bending* in one or several places. Characteristic of this process is the stretching (tensile elongation) imposed on the outer surface and compression of the inner surface (Fig. 5.6). There is only one line (the *neutral line*) which retains its original length. For a given sheet thickness, the tensile strain increases with a decreasing forming radius. The first consequence of stretching the outer surface is that it reveals the grain size of the material, and while a coarse-grained appearance (*orange peel*) may be aesthetically undesirable, it is not really a bending defect since it can be remedied by choosing a finer-grain material. The minimum permissible radius (or, more generally, *radius-to-thickness ratio*) is given by one of two failure conditions.

Localized necking and thus a structural weakening of the bent part occurs when elongation in the outer fiber, e_t, exceeds the uniform elongation of the material, e_u, in the tensile test (Fig. 5.6):

$$e_t = \frac{1}{2R_b/h + 1} \leq e_u \tag{5.2}$$

Because the strain is redistributed during bending to adjacent zones, a somewhat higher strain is usually permissible. For materials that obey the power law of strain hardening, Eq. (2.7), $\epsilon_u = n$ and the engineering uniform strain e_u may be calculated from

$$e_u = (\exp n) - 1 \tag{5.2a}$$

FIG. 5.6 Characteristic dimensions and stresses during bending.

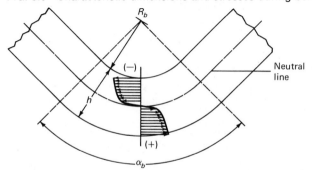

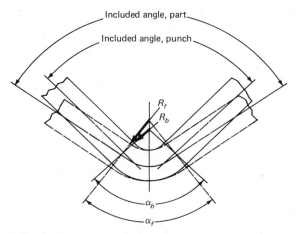

FIG. 5.7 Springback after bending (broken lines).

The second and absolute limit is given by the actual fracture of the material. This is more directly related to the reduction in area q measured in the tensile test (Eq. 2.9 and Tables 4.2 and 4.3). The *minimum permissible bending radius* may be estimated from the following formulas[1]

$$R_b = h \left(\frac{1}{2q} - 1 \right) \qquad \text{for } q < 0.2 \tag{5.3}$$

$$R_b = h \frac{(1 - q)^2}{2q - q^2} \qquad \text{for } q > 0.2 \tag{5.4}$$

Since fibered structures possess directional properties (Sec. 2.2.4), it is often more favorable to bend with the bend line across the rolling direction than parallel to it.

The stress state is extremely complex in bending. The complete tensile and compressive stress-strain curves of the material are generated on the tensile and compressive sides of the bend, respectively. This means that close to the neutral plane the stresses are elastic and are released when the forming-tool pressure is removed. Consequently, the bent material springs back and both the angle of the part and the bend radius increase (Fig. 5.7). The elastic zone is more extensive with a relatively gentle bend (large R_b/h ratio), and elastic stresses are larger for a material with a high yield stress $\sigma_{0.2}$ and low elastic modulus E; therefore *springback* also increases according to the approximate formula[2]

---

[1] J. Datsko and C. T. Yang, *Trans. ASME, Ser. B (J. Eng. Ind.)*, 82:309–314, 1960.

[2] F. J. Gardiner, *Trans. ASME*, 79:1–9, 1957.

$$\frac{R_b}{R_f} = 4\left(\frac{R_b\sigma_{0.2}}{hE}\right)^3 - 3\left(\frac{R_b\sigma_{0.2}}{hE}\right) + 1 \tag{5.5}$$

where R_b is the radius of the bending die and R_f is the radius obtained after pressure is released.

Since the length of the neutral line does not change, the angle after springback, α_f, can be obtained (in radians) from

$$\alpha_f\left(R_f + \frac{h}{2}\right) = \alpha_b\left(R_b + \frac{h}{2}\right) \tag{5.6}$$

If springback is known and the material is of uniform quality and thickness, springback can be compensated for by overbending (Fig. 5.8a and b). Alternatively, the elastic zone can be eliminated at the end of the stroke by applying a pressure sufficient to deform the entire sheet thickness compressively (Fig. 5.8c). Special techniques have also been developed that maintain a compressive stress in the bend zone during the entire forming process (e.g., Fig. 5.8d). The latter method imposes a hydrostatic pressure on the bend zone and permits bending beyond the normal limits given by Eqs. (5.3) and (5.4).

A very simple estimate of the *bending force* may be obtained from

$$P_b = \frac{lh^2 \text{ UTS}}{w_b} \tag{5.7}$$

where l is the length of the bend and w_b is the width of the die opening (Fig. 5.8a), and UTS is as before (see discussion to Eq. 5.1).

The equipment used for bending depends on the size (mostly length)

FIG. 5.8 Methods of obtaining a desired angle of bend (in this instance, 90°): (a) and (b) overbending, (c) plastic deformation at the end of the stroke, and (d) bend zone compressively deformed during bending. (Part d after V. Cupka, T. Nakagawa, and H. Tyamoto, *CIRP,* 22:73–74, 1973.)

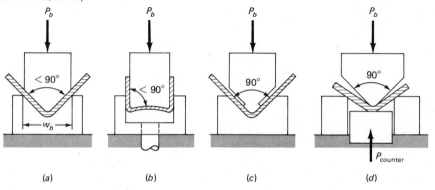

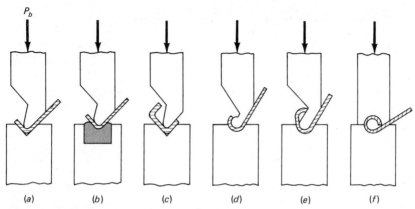

FIG. 5.9 Press-brake forming of (*a*) a 90° angle, (*b*) the same but with a polyurethane female die, (*c*) a U channel, and (*d–f*) a bead.

of the bent part.   Short lengths can be bent at high rates in mechanical presses in one or more dies, depending on shape complexity.   Longer lengths call for special presses with very long beds (*press brakes*).   In the latter, simple tooling forms complex shapes by repeatedly bending a long sheet (Fig. 5.9).   The *female die* half may be replaced by a slab of *polyurethane foam,* further reducing tooling costs.

Continuous production and very high production rates become possible in *roll forming;* bending is now done in stages, between contoured, driven rolls placed in tandem (in a line).   Thus, sections that replace hot-rolled or extruded sections are formed, as well as corrugated sheet, and tube for subsequent welding (Fig. 5.10a).   The strip can be bent easily into

FIG. 5.10 Bending with rollers: (*a*) longitudinal rolling of tube and U-sections and (*b*) pyramidal rolls for plate and section bending.

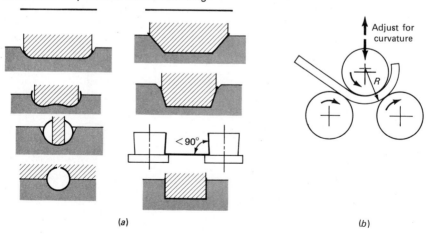

complex shapes, if necessary, with the use of idling rollers pressing from the sides.

An adjustable but uniform curvature may be imparted to a sheet, plate, or section by passing it through three rolls arranged in a *pyramidal* fashion (Fig. 5.10*b*). This is an important preparation step for making large welded-plate structures.

Besides bending of sheet metal, *bending of sections and tubes* is an important manufacturing activity. The problem in *free bending* (e.g., through pyramidal benders) is usually that of distortion and buckling of more complex shapes, and better results are obtained when the profile or tube is *wrapped* around a *form block.* Bending is now accomplished gradually by passing a *wiper roll* or *wiper surface,* hinged at the center of the radius of curvature, around the section or tube. To prevent the collapse of tubes when bending over tight radii, the inside can be supported with sand, a low-melting-point metal, or more practically by a mandrel that is made up of individual sections (rather like a string of pearls) or by a fixed mandrel over which the tube is drawn as it is bent around a shaped, rotating die.

An important form of bending is *flanging.* Flanging of the outer edge of a blank (Fig. 5.11*a*) is similar to a shallow deep-drawing operation and

FIG. 5.11 Deformations in flanging (a) a rim, (b) a hole (showing cracks caused by an excessive tensile strain), and (c) a tube; and (d) in necking a tube.

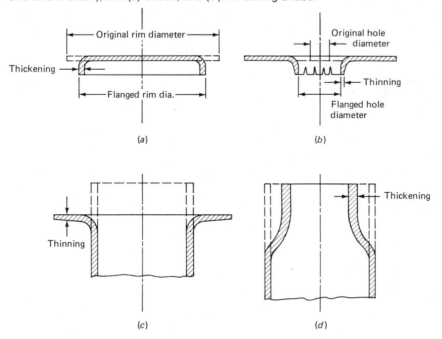

sets no great demand on ductility. In contrast, flanging of a hole (Fig. 5.11*b*) imposes severe tensile strains on the edge of the hole and is one of the quickest ways of finding material defects. The same applies to *expanding* or flanging the ends of a tube (Fig. 5.11*c*). In contrast, the *necking* of a tube (Fig. 5.11*d*) imposes compressive stresses and the reduction taken in a single operation is limited only by the collapse of the tube.

5.4 SPINNING

A special variety of bending is *spinning* a rotationally symmetrical part. The starting blank is held against a *male die* (*form*) which in turn is rotated by some mechanism such as a lathe spindle. Specially shaped tools are then pressed against the blank so that it is gradually laid up against the surface of the form (Fig. 5.12*a*). The process is called *power spinning* when the tools are held and moved in specially driven slides, and *shear spinning* when the wall thickness of the product is also reduced (Fig. 5.12*b*). The maximum reduction obtainable is limited by the ductility of the material and correlates well with the reduction of area in the tensile test. Very large, thick-walled shapes can be spun hot.

5.5 SHEET FORMING

Enormous numbers of sheet-metal products are formed into more or less deep, containerlike components in a great variety of shapes. They can be

FIG. 5.12 Spinning: (*a*) parabolic shape spun by hand spinning, and (*b*) reducing the wall of a container by shear spinning.

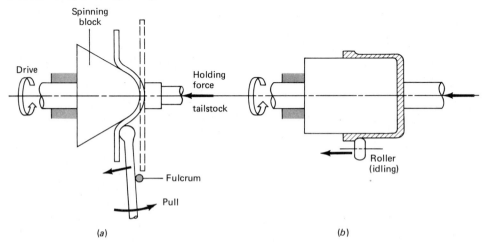

(*a*)   (*b*)

produced by two essentially different techniques, stretch forming and deep drawing, or their combinations.

5.5.1 Stretch Forming

In pure *stretching,* the sheet is completely clamped on its circumference and a form (male die) is pushed against it (Fig. 5.13). The shape is developed entirely by stretching the sheet material; therefore the limitation is the appearance of a neck. When the geometry of the part is such as to produce essentially uniaxial tension, this coincides with the necking limit (i.e., the uniform elongation ϵ_u) in the uniaxial tensile test. When the shape is such that *transverse strains* are also imposed, the appearance of a neck is delayed and it becomes possible to reduce thickness further. The relationship is conveniently expressed in experimentally determined *forming limit diagrams* (Fig. 5.14). The forming limit line moves higher with a material of greater formability. The tension-tension quadrant (the right-hand side of Fig. 5.14) encompasses all strain combinations for stretching.

The technique sketched in Fig. 5.13*a* is too slow for mass production. Instead, a *matched pair* of male and female dies are made, and clamping is accomplished with an independently movable ring (blankholder, Fig. 5.13*b*). For very shallow parts, as produced by *embossing,* the dies provide enough restraint (Fig. 5.13*c*). Matched die pairs give a better definition and allow recessed shapes.

Just as in bending, orange peel is objectionable and grain size must be kept to specified maximum values. Stretcher-strain marks (Sec. 5.1) are a cause for rejection if the part will be visible in service. Another possible problem is initial *wrinkling* (*puckering*) of the sheet between punch and blankholder, which may be difficult to smooth out even between mating male and female dies.

FIG. 5.13 Stretching processes: (*a*) stretch forming, (*b*) stretch drawing, and (*c*) embossing.

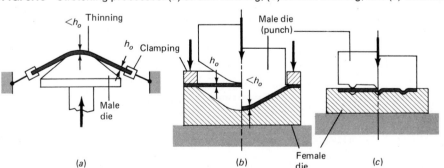

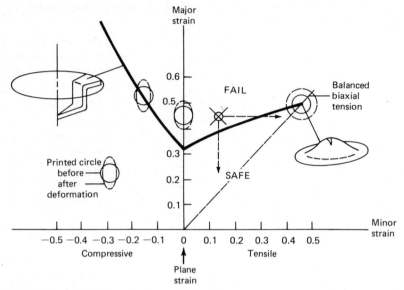

FIG. 5.14 A forming limit diagram typical of low-carbon steel. (Left-hand side typically after G. M. Goodwin, *SAE* paper 680092, 1968; right-hand side after S. P. Keeler, *SAE* paper 650535, 1965.)

5.5.2 Deep Drawing

In the simplest case of *deep drawing,* a circular blank of diameter d_0 is converted into a flat-bottomed cup by drawing it through a *draw die* with the aid of a punch of diameter d_p (Fig. 5.15*a*). Both the die and punch must have well-rounded edges, otherwise the blank might be sheared. The finished cup must be *stripped* from the punch—for example, by machining a slight recess (a ledge) into the underside of the draw die. After the cup

FIG. 5.15 Deep drawing (*a*) without and (*b*) with blank holder.

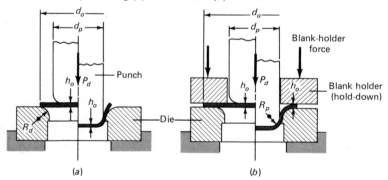

has been pushed through, the top edge springs out because of springback and gets caught in the ledge on the return stroke of the punch.

The difference between stretching and deep drawing is substantial; in the former, the blank is clamped and depth attained at the expense of the sheet thickness; in the latter, the blank is allowed (and even encouraged) to draw into the die, and thickness remains nominally the same.

The circumference of the blank is reduced while it is forced to conform to the smaller diameter of the punch. The resulting circumferential compressive stresses cause the blank to thicken and also to wrinkle, unless it is sufficiently resistant to buckling. Therefore, *drawing without a blankholder* is possible only when the *drawing ratio* $d_0/d_p < 1.2$, or at a greater drawing ratio when the thickness of the blank is sufficient (Fig. 5.16). This depends also on the die profile, which determines the rate of circumferential compression.

Beyond the indicated limits, deep drawing must be conducted with a *hold-down* or *blankholder* (Fig. 5.15b). The blankholder must exert sufficient pressure to prevent wrinkling (Fig. 5.17b), but excessive pressure would restrict free movement of the material in the draw ring and thus cause excessive *thinning* and finally fracture in the partly formed cup wall (Fig. 5.17d). As a first approximation, the *blankholder pressure* may be taken as 1.5 percent of the yield stress $\sigma_{0.2}$ of the material, to produce a sound drawing (Fig. 5.17c). The maximum diameter of the circle that can be drawn under ideal conditions is expressed as the *limiting drawing ratio* (LDR)

$$\text{LDR} = \frac{d_{0(\text{max})}}{d_p} \qquad (5.8)$$

and is a function of the maximum stress that the partly drawn cup wall can

FIG. 5.16 Limiting drawing ratio in drawing low-carbon steel without a blank holder, with different die geometries. (After G. S. A. Shawki, *Werkstattstechnik, 53*(1):12–16, 1963.)

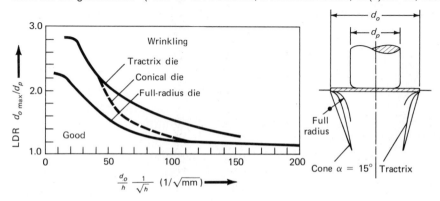

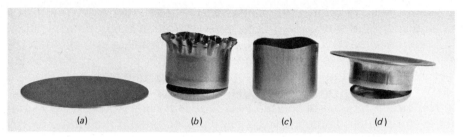

FIG. 5.17 Deep drawing of low-carbon steel cups from (a) a round blank, with (b) insufficient, (c) optimum, and (d) excessive blank-holder pressure; note in (c) the typical earing due to planar anisotropy.

take. The axial tensile force applied to the cup wall provides all the force required to pull the sheet in, and is composed of the forces required to bend and unbend the sheet around the draw radius, compress the sheet circumferentially, and overcome friction between blankholder and die surfaces and around the draw radius. Therefore the LDR is also a function of the draw radius and of the lubrication practice. Comparable values can be obtained only by agreeing on a standard die geometry and lubricant (very often a polymeric film such as PTFE).

   A very approximate estimate of the *drawing forces* may be obtained from the formula

$$P_{\mathrm{d}} = \pi d_{\mathrm{p}} h \, \mathrm{UTS} \left(\frac{d_0}{d_{\mathrm{p}}} - 0.7 \right) \tag{5.9}$$

where h is again the sheet thickness and UTS the ultimate tensile strength as in Eq. (5.1).

   The draw force does, of course, increase with increasing strain-hardening rate, but this also strengthens the walls of the cup. Therefore, in contrast to stretching, the strain-hardening exponent n has only a small effect on the limiting draw ratio. Of much more powerful influence is the normal anisotropy of the material (Sec. 2.1.3) as expressed by the r value (Fig. 5.18). Face-centered cubic materials with their isotropic behavior ($r \simeq 1$) give a limiting drawing ratio of around 2.1 to 2.2 (which corresponds to a cup of approximately $0.9d_{\mathrm{p}}$ depth). For the deepest possible draw, it is preferable to have a material that fits easily into the draw die, that is, one that reduces its width by virtue of the slip mechanism while maintaining most of its thickness (as in Fig. 2.6b). This translates into a high r value and is most clearly evidenced by titanium. On the other hand, a sheet with a low r ratio thins out (Fig. 2.6a) and might fail prematurely, as exemplified by zinc. Body-centered cubic materials, particularly α-iron, can develop a favorable structure through appropriate primary processing controls, and aluminum-killed low-carbon steel can give r values as high

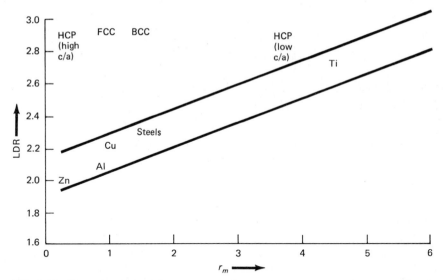

FIG. 5.18 Deep-drawing performance as a function of the *r* value.

as 2. This allows somewhat deeper draws than aluminum, which suffers also from a low (and sometimes even negative) strain-rate sensitivity. A high strain-rate sensitivity (high *m* value) is helpful in that it strengthens an incipient neck and thus delays strain localization and fracture (Sec. 2.1.9). The maximum forming limit is again a material property and can be plotted on the forming limit diagram by extending it to encompass a tensile major strain and a compressive minor strain (Fig. 5.14), corresponding to the deformation of the flange during drawing.

A material with planar anisotropy (Sec. 2.1.3) shows different properties in the rolling, transverse, and 45° directions ($r_0 \neq r_{90} \neq r_{45}$). These result in varying heights of the drawn cup; the *ears* reflect the crystal symmetry and come in pairs (4, 6, or 8; Fig. 2.17c).

Cylindrical products of greater depth but essentially unchanged wall thickness are formed in *redrawing* operations (Fig. 5.19a), while vessels with thin walls but heavy bases are produced by *ironing* (Fig. 5.19b). Ironing forces may be approximated by analogy to wire drawing (Sec. 4.7.2). A basic phenomenon, not mentioned hitherto, is that a cold-worked material exhibits greater ductility when the deformation direction is reversed (*strain softening*); this is exploited in the *reverse drawing* of cups (Fig. 5.19c).

There is, of course, wide opportunity (but often combined with greater difficulty) to change the basic shape of the drawn part, e.g., to a square or rectangular one (in which case earing in the corners is helpful). A punch with a curved or hemispherical end imposes a different, combined deformation state, to be discussed next.

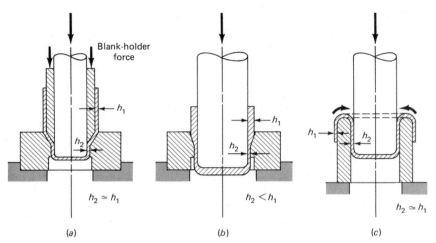

Blank-holder force

$h_2 \simeq h_1$

$h_2 < h_1$

$h_2 \simeq h_1$

(a)

(b)

(c)

FIG. 5.19 Further deformation of a cup-shaped part by (a) redrawing, (b) ironing, and (c) reverse redrawing.

5.5.3 Combined Stretching-Drawing

In many practical applications, most notably in the production of automotive body and chassis parts, the drawing process is neither pure stretching nor pure drawing. The sheet is not entirely clamped (therefore it is not pure stretching) neither is it allowed to draw in entirely freely (thus it is not pure drawing). Instead, the complex shapes are developed by controlling the draw-in of the sheet, retarding it where necessary with *draw beads* formed on the die and blankholder surfaces (Fig. 5.20). Very often, forming is taken close to the limits allowed by the material, and fracture could easily occur in the absence of tight controls. In the last few years, there has been a remarkably swift acceptance of formability concepts for production control purposes.

The simplest case is that of predominantly stretch forming, to which a forming limit diagram (Fig. 5.14) applies. When a part is found to fail during stretching, a *grid* of small circles is printed or etched on trial sheets. During drawing these circles distort into ellipses, the major axis of which gives the major strain and the minor axis the minor strain. When these strains are transferred onto the forming limit diagram, various remedial actions may be taken. The minor strain may be increased (horizontal arrow in Fig. 5.14) by increasing the restraint of the sheet in that direction (e.g., with draw beads), or the major strain may be reduced (vertical arrow in Fig. 5.14) by reducing the depth of the draw, or localized thinning in a deep part of the drawing can be reduced by increasing friction on that part of the male tool.

The maximum permissible pure draw is directly given by the experi-

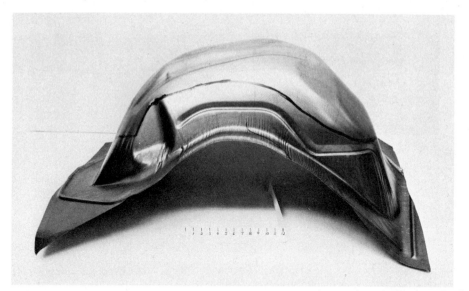

FIG. 5.20 A severely formed part (an automotive wheelhouse) made in a single operation by a combination of stretching and drawing. Note the etched circle grid, the drawbeads, and the line, marking the boundary between the stretched nose and the drawn-in sides. (Courtesy of A. S. Kasper, Chrysler Corporation, Detroit.)

mentally determined LDR, Eq. (5.8). Increased punch friction (a rough punch) helps on marginal draws.

The most difficult analytical task is encountered in *combined stretching-drawing* operations. For these, combined *stretch-draw limit curves* are constructed using the LDR and the stretching limits as the end points (Fig. 5.21); the task then remains of identifying the relative contribution of the two mechanisms in the production of a given part. When the tooling is already in existence, analysis follows the principles indicated for stretching. The really important application is, of course, in predicting success or failure before the expensive tools are built, so that modifications can be made in time. This is a rapidly developing field which can already boast some successes and should, ultimately, allow fitting of the process to the material (or vice versa) before a production die is finalized.

Because these processes incorporate both stretching and drawing, all material properties mentioned hitherto become important. Thus, in addition to the LDR governed by normal anisotropy (the r value), factors that delay necking or allow greater deformation after necking (the strain-hardening coefficient n and the strain-rate-sensitivity exponent m) are also beneficial. Furthermore, any transformations that result in strengthening (such as the formation of martensite during deformation) will also aid in obtaining deeper draws.

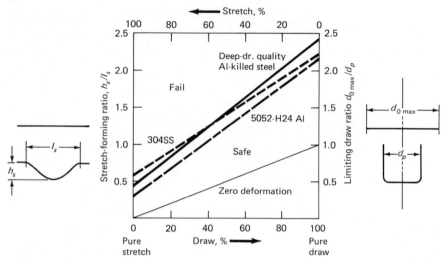

FIG. 5.21 Typical forming line diagram. (After A. S. Kasper, *Metal Progress,* *99*(5):57–60, 1971. Copyright 1971, American Society for Metals.)

5.5.4 Press Forming

In industrial usage, the term *press forming* serves to describe all sheet-metalworking operations performed on power presses with the use of mostly permanent (steel) dies. It incorporates all steps required to complete a part of any complexity from the sheet, whether it be blanking, punching, bending, drawing, stretching, ironing, redrawing, embossing, flanging, trimming, and so forth.

Each operation can, of course, be performed separately, in individual, single-operation dies and presses. Die costs still add up and labor and handling costs can be high. Nevertheless, this is the only option when the part is very large, or total production quantities are insufficient to justify more complex dies.

Compound dies perform two or more operations in a single stage and assure the greatest accuracy of the product, but are limited to relatively simple processes such as blanking, punching, flanging, perhaps combined with bending or a single draw (or, in special processes, also multiple draw).

Many parts are of a geometry that cannot be directly formed, either because the depth-to-diameter ratio is too large or because the shape has steps, conical portions, etc., requiring several successive draws. A compound die is often inadequate, and progressive or transfer dies are needed.

Progressive dies contain, within one *die set,* all the die elements needed to complete the part from the sheet. Coiled stock is fed to the die;

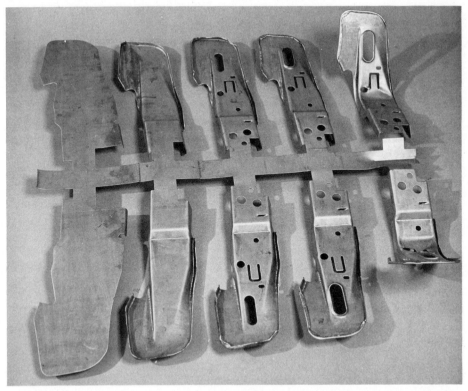

FIG. 5.22 A typical example of progressive die work: Forming of two seat-frame parts at a time, by a sequence of blanking, flanging, piercing, flattening of flange, and in the final stage, cutting off and bending. (General Seating Products Div., Lear-Siegler Industries Ltd., Kitchener, Ontario.)

the blank is only partially cut so as to remain attached with *connecting tabs* to the remnant of the strip, and this skeleton is used to move the part through the forming stages, with the final separation reserved for the last stage (Fig. 5.22). A finished part is obtained for every press stroke. Die costs are very high but labor costs are low.

Transfer dies are constructed on the same principle, but the blank is cut out first and then mechanically moved through successive stages, sometimes arranged in a circle.

Presses for both progressive and transfer dies have to be large enough to accommodate all die stages on the press bed and to provide the force for all operations simultaneously. Very high die costs are compensated by low labor costs in mass production at high production rates, since one finished part is obtained for each press stroke.

A great variety of shapes can be produced at high rates in special-purpose machines, some of which are the so-called *four-slide machines,* originally developed for complex wire-bending operations, but now increasingly used also for sheet metalworking.

5.5.5 Special Operations

There are a great number of specialized processes designed to assure greater depths of draw, more complex shapes, lower die cost, or a combination of any of these features. It is, for example, possible to draw over a male punch (often made of a resin or a Zn alloy) with a *rubber cushion,* thus eliminating the need for the more expensive steel die. Similar aims are accomplished but with better control of the process when, instead of the rubber cushion, a *liquid,* contained by a rubber diaphragm, is used.

A single die (male or female) is used and the press dispensed with in the various *high-energy-rate forming* (HERF) processes. The energy required for deformation is derived from an explosive, a magnetic field, or the pressure shock created in water by the sudden evaporation of a wire. The pressure application is very sudden but the rate at which the material deforms is usually not much higher than in a fast mechanical press. Of the many possible applications, *drawing-in* of necks and *internal expansion* of tubular and containerlike parts is frequently encountered. The latter serves as an alternative to expansion with a rubber plug or hydraulic fluid.

5.6 SHEET-METALWORKING EQUIPMENT

Apart from special-purpose equipment, most press forming makes use of hydraulic and, mostly, mechanically driven presses. Suitable clutches permit operation in single strokes, initiated by the operator, or continuously at rates of 30–300 strokes per minute.

The principle of construction is similar to presses used in bulk deformation (Table 4.6) but special features and, for the same tonnage, much larger beds make them more adaptable to the working of sheet metal. Smaller presses often have *inclinable press* frames which facilitate removal of the stamped part by gravity. Larger presses may have two or even three independently adjustable rams, one moving inside the other. Such *double-* and *triple-acting presses* provide built-in facilities for holding or clamping, and for ejecting, and allow more complex operations. Spring, air, or hydraulically powered *cushions* can also serve for blankholders or add flexibility to the operation. Mechanical feeding and part removal speeds up production.

Tool materials are chosen mostly on the basis of the expected size of the production run. Blanking tools are subjected to severe wear and are made from the various cold-working die steels (Table 4.7). Bending and drawing dies are made from the same materials, although cast iron and even hard zinc alloys or plastics are suitable for short production runs or the softer workpiece materials. In contrast to bulk deformation, die materials seldom limit sheet-metalworking processes. The problem is, more likely, that of finding an economical die material and die-making method.

5.7 SUMMARY

Well over one-half of the total metal production ends up in the form of sheet-metal parts. The variety of products is immense, from aircraft skins and automobile bodies to appliance shells, from construction girders and truck frames to furniture legs, from supertankers and bathtubs to beer cans, and from wheel rims and fan blades to watch gears. Combined with joining processes, the scope of sheet metalworking is very broad, but some fundamental limitations must be understood. Some of these limitations may be circumvented, but usually at extra expense. For lowest-cost production, the design of the part and of the process must recognize that:

  1 Shearing (blanking, punching) does not result in a perfectly smooth and perpendicular cut, but acceptable quality can be obtained with the proper die clearance, provided that bending of the sheet is prevented. This sets a limit on the closeness of two neighboring cuts.

  2 An increased hydrostatic pressure changes the shearing process to resemble extrusion with a resulting smooth "cut" edge.

  3 Bending, hole flanging, spinning, and stretch forming are limited by the onset of necking or by fracture; the former is related to uniform elongation (and thus the n and m values), the latter to resistance to triaxial tension (and thus to reduction in area q in the tensile test). Changing uniaxial to biaxial tension, and increased friction on the punch, delay localization of the neck; nevertheless, minor (and functionally insignificant) adjustments to the shape of the part (typically, more generous radii) often provide the most economical relief from production problems. If a complex shape is genuinely required, it can be produced in a sequence of operations.

  4 Deep drawing in the first stage is limited primarily by the r value. Subsequent redrawing or ironing permit the production of parts of large depth-to-diameter ratio, thin wall combined with thick bottom, zero corner radius and tapered or stepped shape.

  5 Combination of several processes is possible, and the variety of shapes is then almost unlimited. Parts may have varying cross sections (e.g., a neck or bulge on a cup-shaped part) and transverse features (e.g., holes pierced into the side of vessels), as discussed in Chap. 11.

6 The scope of processing can be further expanded if conventional limitations are relaxed. A good example is the multicompartment dinner tray, in which wrinkling and folding is not only permitted but even encouraged, to provide the necessary stiffness while also facilitating deep draws that would far exceed the stretchability of the Al sheet.

Further Reading

DETAILED PROCESS DESCRIPTION:
  Metals Handbook, 8th ed., American Society for Metals, Metals Park, Ohio, vol. 4., *Forming,* 1969.

TEXTS:
  EARY, D. F., and E. A. READ: *Techniques of Pressworking Sheet Metal,* 2d ed., Prentice-Hall, Englewood Cliffs, N.J., 1974.
  See also Chap. 4 references to: Schey, Cook, Kalpakjian, Lissaman and Martin, Parkins, Sachs, Backofen, Johnson and Mellor, Geleji, Grainger, Schey, Wilson.

SPECIALIZED BOOKS:
  BROWER, D. F., and P. WILDI: The Magnetic Pulse Metal-Forming Technique, in *Techniques of Metals Research,* R. F. Bunshah (ed.), Interscience, New York, 1968, vol. 1, pt. 3, pp. 1539–1566.
  BRUNO, E. J. (ed.): *High-Velocity Forming of Metals,* revised ed., Society of Manufacturing Engineers, Dearborn, Mich., 1968.
  DAVIES, R. S.: Explosive Working of Metals, in *Techniques of Metals Research,* R. F. Bunshah (ed.), Interscience, New York, 1968, vol. 1, pt. 3, pp. 1589–1607.
  DAVIES, R. S., and E. R. AUSTIN: *Developments in High Speed Metal Forming,* Industrial Press, New York, 1970.
  RINEHART, J. S., and J. PEARSON: *Explosive Working of Metals,* Pergamon, New York, 1964.
  SACHS, G.: *Principles and Methods of Sheet Metal Fabricating,* 2d ed., Reinhold, New York, 1966.
  SPRINGBORN, R. K. (ed.): *Cold Bending and Forming Tube and Other Sections,* Society of Manufacturing Engineers, Dearborn, Mich., 1966.
  WILLIS, J.: *Deep Drawing,* Butterworths, London, 1954.

Problems

5.1 Circular blanks of diameter $d_0 = 10$ in are to be cut from $h = 0.125$ in thick, annealed 5052 aluminum alloy. What press force is needed?

5.2 Circular blanks (slugs) of 1100 Al, diameter $d = 25$ mm and thickness $h = 3$ mm, are to be mass-produced as the starting material for toothpaste-tube extrusion. The available presses are of 500 kN capacity and can take maximum 220-mm-wide strip. Economy of material utilization increases with increasing numbers of rows cut from one strip width. Calculate (a) the force required for blanking a single slug, and (b) the maximum number of slugs that can be blanked simultaneously with the available press. (c) Suggest the optimum layout for the slugs if the web (remaining material between cuts and at edges) is approximately h, but seldom less than 1 mm (0.40 in).

5.3 Derive Eq. (5.2).

5.4 An annealed cartridge brass (70/30 brass) strip of 25 mm width and thickness $h = 1$ mm is bent into a sharp 90° angle with the bend running across the width (thus the bend length $l = 25$ mm). The die opening is $w_b = 20$ mm. Determine: (a) the minimum permissible bending radius R_b, (b) the angle of the bending punch, taking into account springback, and (c) the bending force P_b.

5.5 It is proposed that, as a cost-saving measure, the part of Prob. 5.4 be made of a low-carbon (say, 1008) steel. What changes in the process should one anticipate (recalculate all values and compare).

5.6 The blanks of Prob. 5.1 are to be drawn with diameter $d_p = 5$ in punches into flat-bottomed cups. Assuming that the drawability of annealed 5052 is at least as good as that of the H24 temper (strain hardened and then partially annealed to give a strength typical of half-hard material), determine from Fig. 5.21 if this is feasible. If the answer is yes, calculate (a) the blankholder force, (b) the drawing force, and (c) assuming that the average wall thickness of the part remains unchanged and the punch nose radius is $R_p = 0.5$ in, the height of the cup to be expected.

5.7 A dish of diameter $d = 1000$ mm and depth $h = 250$ mm is to be produced of 5052 Al, in the shape of a spherical segment. Because of the danger of puckering in deep drawing and because of the small quantity required, it is proposed that the dish be made by pure stretch forming. Determine whether this is feasible.

5.8 It is found from experience that the neutral line in bending is displaced toward the compressive side. This displacement increases with decreasing R_b/h ratios, and the neutral line is at approximately $0.4h$ when $R_b/h = 1$. Recalculate Eq. (5.2) for this case.

5.9 In order to reduce weight, the part of Prob. 5.4 is to be made of 6061-T6 aluminum alloy (YS = 275 MPa, UTS = 310 MPa, elongation = 12 percent, reduction of area = 30 percent, and $E = 70 \times 10^3$ MPa). Recalculate all values and compare with those of Prob. 5.4 to see if any changes are needed.

5.10 Deep-drawing quality, aluminum-killed steel of $r_m \simeq 1.7$ has an LDR of 2.4 (Fig. 5.21). Calculate, for a sheet of $h = 2$ mm thickness, and a punch of $d_p = 100$ mm diameter and $R_p = 5$ mm nose radius, (a) the maximum blank diameter $d_{0(max)}$, (b) the depth of cup, assuming constant wall thickness of 2 mm, and (c) the height-to-diameter ratio. Compare the h/d ratio with that given in the text for $r_m = 1$; note the effect of LDR on height-to-diameter ratio.

5.11 A 2 m × 1 m × 2 mm thick 304 stainless-steel sheet is to be cut lengthwise into strips. (a) Calculate the shearing force for parallel shear blades, then (b) estimate the force if the blades are inclined at a 2° angle (assume that cutting always takes place simultaneously over the entire projected contact length).

6

MANUFACTURING PROCESSES FOR POLYMERS

The term *polymer* refers to any substance in which several (often many thousands) molecules or building units (*mers*) are joined into larger, more complex molecules. Customarily the term is applied to *organic substances*, commonly called *plastics*, but *inorganic* polymers, commonly called *glasses*, exhibit many of the same properties and share manufacturing processes. Therefore, both groups will be discussed here.

6.1 ORGANIC POLYMERS

The manufacturing properties of plastics are very much a function of their *molecular structure*. Two broad categories can be distinguished: *thermoplastic* and *thermosetting polymers*.

6.1.1 Thermoplastic Polymers

Thermoplastic polymers could be compared to a bowl of spaghetti: individual covalently bonded molecules may twist and turn; nevertheless, they are essentially linear in shape and are held to each other by relatively weak van der Waals forces. Their response to mechanical stress is most important from the manufacturing point of view.

Polymer rheology

At some relatively high temperature, the molecules—even if kinked and curved—slide readily relative to each other and the polymer behaves like a *viscous fluid* (Fig. 6.1*a*). Viscous flow is readily modeled by comparison to a *dashpot* (Fig. 6.1*e*) in which a piston is displaced against the resistance of an oil (a *Newtonian fluid*) of viscosity η. The shear stress τ required to move the piston is proportional to the *shear strain rate* $\dot{\gamma}$:

$$\tau = \frac{\eta\gamma}{t} = \eta\dot{\gamma} \tag{6.1}$$

It should be noted that this expression is the same as Eq. (2.11),

$$\sigma = C\dot{\epsilon}^m \tag{2.11}$$

except that σ is normal (tensile) stress and $\dot{\epsilon}$ is normal (tensile) strain rate and, of course, the strain-rate sensitivity index m for a perfectly viscous, Newtonian substance is unity. Polymer melts often show non-Newtonian characteristics but always have a high m value.

On cooling, one of two events may take place. First, if the molecule is of a simple, straight shape, with chemical regularity (order) along the chain, and conditions are favorable, substantial *ordering* of the molecules may occur. While there is no change in the structure of the molecule, it may kink and fold several times and become intertwined and closely packed with its neighbors. Such crystallization results in a sudden contraction of volume, just as it does in metals (at T_m and the broken line in Fig. 6.1*a*). Once this ordering is attained, the material behaves like an *elastic solid* and any further volume change is a result solely of the diminishing thermal excursions of molecules; thus the volume shrinkage is much more gradual.

Alternatively, and especially if the molecule is of a more complex shape, has side branches, and has no chemical regularity, cooling may not lead to crystallization. Instead, the structure remains *amorphous* and the volume shrinks, essentially because thermal vibrations are reduced (and sometimes also because further structural bonds are established). Below a certain temperature, often described as the *fictive point* or the *glass-transition temperature* T_g, the mobility of the molecules is so much reduced that no further bonds are established, sliding of the molecules relative to each other becomes virtually impossible, and the substance again behaves like an elastic solid, just as if it had crystallized (solid line in Fig. 6.1*a*). Well below T_g, its behavior can be modeled by a single *spring:* when a tensile stress σ is imposed, the tensile strain e_t is proportional to a *spring constant,* now called Young's modulus E (Fig. 6.1*c*), as given by Eq. (2.3),

$$e_t = \frac{\sigma}{E} \tag{2.3}$$

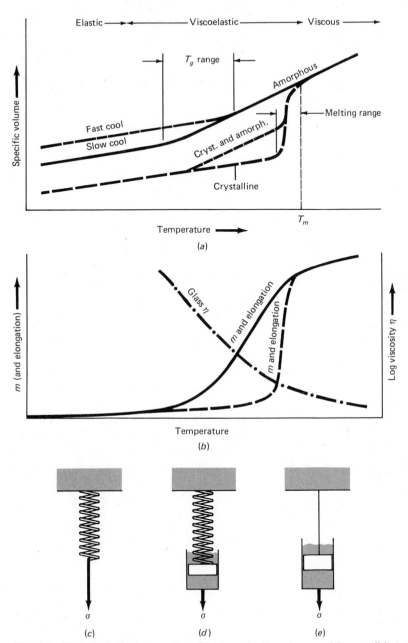

FIG. 6.1 Manufacturing properties of thermoplastic polymers: (a) reversible transitions from melt to solid; (b) changes in mechanical properties; and (c) elastic, (d) viscoelastic, and (e) viscous behavior.

The relationship will be the same for a shear strain, except that now the shear stress τ and *shear modulus G* should be used:

$$\gamma = \frac{\tau}{G} \tag{6.2}$$

Above a critical strain, the spring snaps and fracture occurs.

From below the glass-transition temperature to above the melting temperature, thermoplastics of a high molecular weight exhibit a *viscoelastic* behavior. When stretched in this range (sometimes described as the *leathery* or *rubbery* range), some of the deformation is elastic while some of it is permanent, because molecules become displaced relative to each other by viscous flow. Viscoelastic behavior can be described by various (and often very complex) models, the simplest of which is a *Maxwell element,* a spring and dashpot in series (Fig. 6.1*d*). The total strain is now the sum of elastic (instantaneous) and viscous (strain-rate sensitive and therefore time-dependent) strains:

$$\epsilon_{\text{total}} = \frac{\sigma}{E} + \frac{\sigma t}{C} \tag{6.3a}$$

$$\gamma_{\text{total}} = \frac{\tau}{G} + \frac{\tau t}{\eta} \tag{6.3b}$$

Another basic model is the *Voigt element,* a spring parallel to a dashpot, and real polymers are often modeled by a Maxwell element and one or more Voigt elements in series. Obviously, when such a polymer is deformed in the viscoelastic temperature range and then the forming stress removed, the deformation corresponding to the spring component is reversed (with a delay due to the Voigt element), and the formed product would have a highly variable and poor shape. Therefore thermoplastics must be cooled well below T_g before the forming stress is removed.

Viscoelastic behavior reflects thermally activated processes. Just as in metals (Sec. 2.1.9), this leads to a time-temperature equivalence. Particularly at slow strain rates, such as prevail in creep, an increase in temperature is equivalent to a decrease in strain rate or increase in time.

From the manufacturing point of view, the viscous and viscoelastic behaviors are of utmost importance. Viscous flow is synonymous with high strain-rate sensitivity or, in terms of Eq. (2.11), high strain-rate sensitivity index *m*. As shown in Fig. 6.1*b*, strain-rate sensitivity increases on heating above the glass-transition temperature. When the material is stretched, a neck appears after only 2–15 percent uniform elongation, but, because of strain-rate effects (Sec. 2.1.9) and alignment of molecules, the neck material becomes stronger, and the neck, instead of localizing and further reducing, spreads out along the entire stretched length (Fig. 2.9c). Successive necks may form, and total elongations of several hundred or thousand percent may be attained before fracture occurs.

When stretched or otherwise deformed, the molecules become aligned and the polymer acquires directional properties (a texture, Sec. 2.1.3) with the maximum strength in the direction of alignment. Therefore drawing is an important step in making textile fibers and thin sheet (film).

Polymer structures

Polymers with this behavior may be produced by the *addition* of short elements or monomers (Fig. 6.2). If the resulting long-chain molecule (*linear polymer*) is uniform and "straight," such as in high-density polyethylene and in PTFE, crystallization occurs readily, although some of the polymer always remains amorphous. Solidification takes place over a temperature range and the presence of amorphous material results in a glass transition (Fig. 6.1a). In contrast, a low-density, branched polyethylene remains almost entirely amorphous, as do polymers with less symmetrical molecules, such as PVC and polystyrene. Some guidance on softening may be gained from the deflection temperatures in Table 6.1.

Linear polymers may also be made by *condensation* mechanisms: two species may react to form a polymer and a by-product (e.g., hexamethylene diamine and adipic acid form nylon 6,6 and water, or ethylene glycol and adipic acid form a polyester resin and water, Fig. 6.2).

Copolymers are the polymeric analogs of alloys. Two or more species of repeat units, randomly arranged, make up the structure of random copolymers, while long sequences of each species alternate in block copolymers.

The concept of homologous temperature (Sec. 2.1.1) is useful also for polymers. To a first approximation, $T_g \simeq 0.5\ T_m$, although the exact ratio depends much on composition and structure. The value may range from 0.25 to 0.9, with 0.66 a good average for homopolymers, while higher values are typical of random copolymers and lower values of block copolymers.

Some polymers are formulated with *plasticizers* (essentially nonvolatile solvents) which dissolve in the polymer and effectively depress the glass-transition temperature T_g by promoting sliding movement of polymer chains. Polymers such as PVC thus remain flexible at lower temperatures.

The important point about thermoplastic polymers is that the softening and hardening cycles are *reversible.* It is possible to form a polymer into a semifabricated shape (pellet, bar, tube, sheet, or film), which is then cooled and shipped to the secondary manufacturer who reheats and forms it into the final shape. On cooling, it acquires sufficient strength

ADDITION POLYMERS

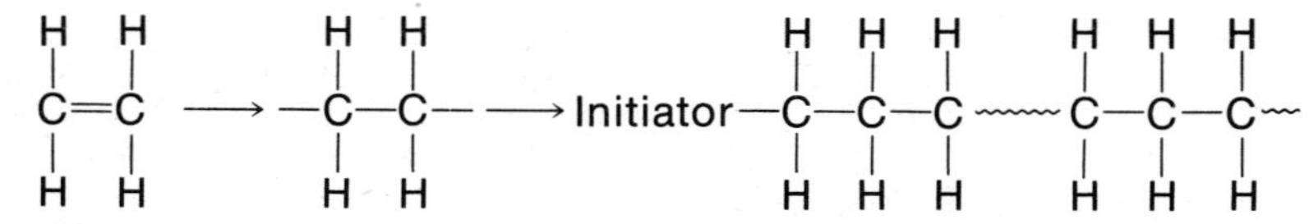

Ethylene monomer

→ Initiator → Polymer: PE (Polyethylene)

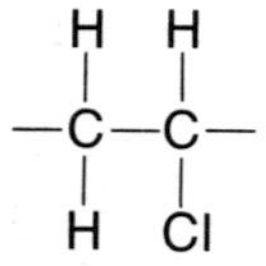

(repeat unit)

PVC (Polyvinyl-chloride)

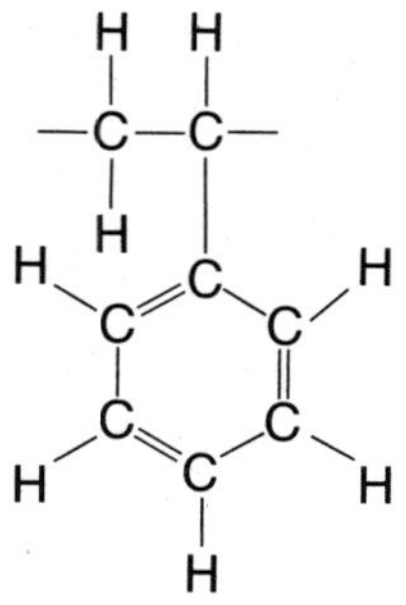

Polystyrene

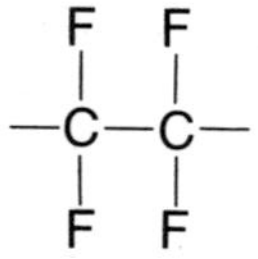

PTFE (Polytetra-fluoroethylene)

CONDENSATION POLYMERS

Linear polymers

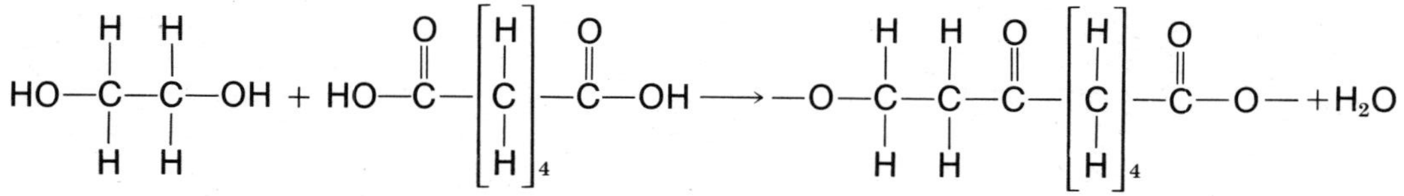

Ethylene glycol + Adipic acid → Polyester + Water

Network polymers

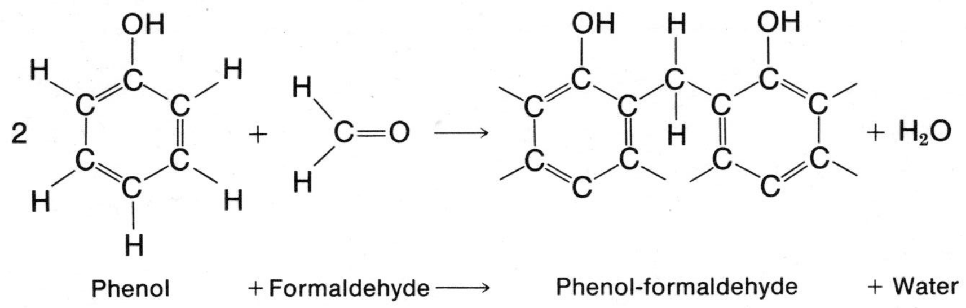
Phenol
+ Formaldehyde ⟶
Phenol-formaldehyde
+ Water

ELASTOMERS

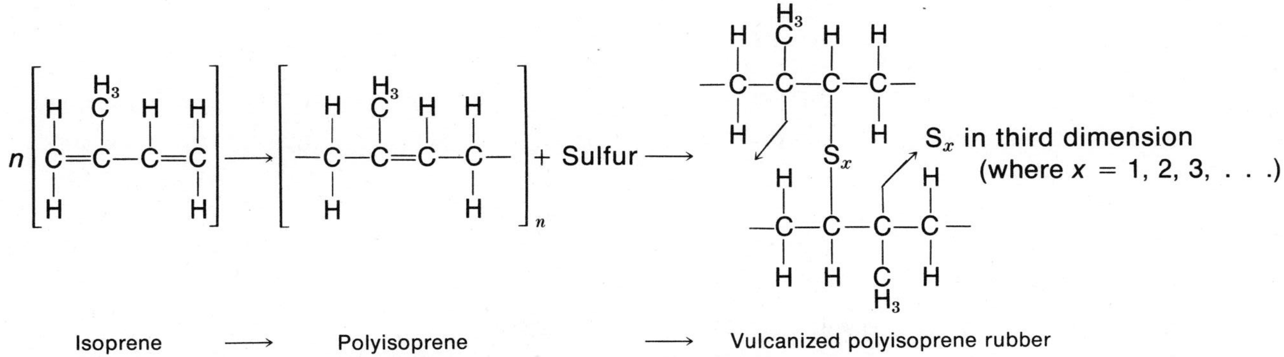
+ Sulfur ⟶
S$_x$ in third dimension
(where x = 1, 2, 3, . . .)
Isoprene ⟶
Polyisoprene ⟶
Vulcanized polyisoprene rubber

FIG. 6.2 Formation and structure of some common polymers.

TABLE 6.1 MANUFACTURING PROPERTIES OF SELECTED THERMOPLASTIC POLYMERS*

Type	Deflection temperature, °C		Compression molding		Injection molding		Shrink-age, %	Tensile properties		Machina-bility§
	1850 kPa†	450 kPa†	Tempera-ture, °C	Pres-sure, MPa†	Tempera-ture, °C	Pres-sure, MPa‡		Strength, MPa‡	Elonga-tion, %	
ABS	95	100	150–230	7–55	190–270	50–170	0.4–0.9	30–55	5–70	E–G
Acetal copolymer	110	155	170–205	7–35	195–250	70–140	~2	60	40–75	E
With 25% glass	160	165			190–250	70–140	0.2–0.6	130	3	F–G
Acrylic (PMMA)	80	85	150–220	15–70	160–260	70–140	0.2–0.8	50–80	2–10	E–G
Cellulose acetate	50	55	125–220	1–35	165–255	55–230	0.3–0.8	20–65	6–70	E
Fluorinated ethyl-enepropylene		65	320–400	7–15	330–405	35–140	3–6	20	300	E
Nylon 6	75	190			270–330	<70	0.8–1.5	85	60–300	E
Polycarbonate	130	135	250–330	7–15	250–345	70–140	0.5–0.7	55–70	100–130	E
Polyethylene, Low density	30	35	135–180	1–6	150–315	55–200	1.5–5	4–15	90–800	G
High density	45	60	150–230	3–6	150–315	70–140	1.5–5	20–45	20–1000	E
Polyimide	140		350	25–35				120	10	E
Polypropylene	50	95	170–230	3–7	205–290	70–140	1–2.5	30–40	200–700	G
Polystyrene	95	100	120–205	7–70	160–260	70–200	0.1–0.6	35–85	1–2	F–G
PVC, rigid	60	62	140–205	5–15	150–210	70–300	0.1–0.5	35–65	2–40	E
Flexible			140–180	3–15	160–195	50–170	1–5	10–25	200–450	
Styrene-butadiene block copolymer	<0	<0	120–160	1–20	150–220	100–200	0.1–0.5	4–20	300–1000	P

* Compiled from *Modern Plastics Encyclopedia*, McGraw-Hill, New York, 1974.

† Divide by 7 to get psi.

‡ Divide numbers by 7 to get 1000 psi.

§ Approximate rating: E = excellent; G = good; F = fair; P = poor.

to provide the contemplated service functions. Clean scrap can be re-cycled by adding it to the virgin polymer, at least to a limited extent.

6.1.2 Thermosetting Polymers

The second important class of polymers is formed of units with a molecu-lar configuration that allows extensive cross-linking or the growth of the polymer molecule in a *spatial network*. Thus, for example, the reaction of phenol with formaldehyde yields a polymer (a phenol-formaldehyde resin) while rejecting water (Fig. 6.2). The reaction is accelerated by the applica-tion of heat (or other source of energy, e.g., light or radiation) and the polymer acquires its strength more rapidly at an elevated temperature (Fig. 6.3). Prolonged heating either causes cross-linking, thus strengthening the resin, or leaves it unaffected. Many *thermosetting polymers* form spa-tial networks and have, therefore, only limited ductility. When heated to excessively high temperatures, the polymer is destroyed; it burns up.

Thus, thermosetting polymers acquire their strength through an irre-versible polymerization and cross-linking mechanism, and can be shaped only by keeping the reacting species separate until brought together in the mold, or by taking polymerization only part way, so that the relatively short, partially polymerized molecules can be moved relative to each other. Heat is applied to complete polymerization and cross-linking only after bringing the partially cured polymer into the required shape (Table 6.2). Once cured (polymerized and cross-linked), a thermosetting resin cannot be reused.

The essential difference between thermoplastic and thermosetting polymers, then, is that the thermosetting polymer is heated to stabilize its

FIG. 6.3 The development of strength in a thermosetting polymer as a function of time and temperature.

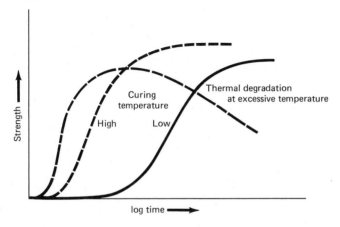

TABLE 6.2 MANUFACTURING PROPERTIES OF SELECTED THERMOSETTING POLYMERS*

Type	Deflection temperature, °C		Compression molding		Injection molding		Shrinkage, %	Tensile properties		Machinability§
	1850 kPa†	450 kPa†	Temperature, °C	Pressure, MPa‡	Temperature, °C	Pressure, MPa‡		Strength, MPa‡	Elongation, %	
Epoxy, casting	45–290						0.1–1	30–90	3–6	G
Molding, glass-filled		120–260	150–160	2–35			0.1–0.5	70–140	4	F
Melamine-formaldehyde	150		150–165	15–35			1.1–1.2			F
Phenol-formaldehyde	115		130–160	15–30			1–1.2	50–60	1–1.5	F–G
Wood-flour-filled	150		145–195	15–35	165–205	70–140	0.4–0.9	35–65	0.4–0.8	F–G
Polyester, glass-filled	>200		140–180	3–15			0.1–0.2	20–70	0.5	G
Polyurethane elastomer			140–180	1–35			0.9–3	30–60	100–650	F–E
Urea-formaldehyde, cellulose-filled	125–145		135–175	15–55	145–160	100–140	0.6–1.4	40–90	0.5–1	F–G

* Compiled from *Modern Plastics Encyclopedia*, McGraw-Hill, New York, 1974.

† Divide by 7 to get psi.

‡ Divide numbers by 7 to get 1000 psi.

§ Approximate rating: E = excellent; G = good; F = fair; P = poor.

shape, while the thermoplastic polymer is heated to make it moldable and must be cooled to fix its shape.

6.1.3 Elastomers

A third class of polymers exhibits a highly elastic, rubbery behavior. In these, highly kinked molecules are tied to each other by occasional covalent-bond *cross links* (Fig. 6.2). Thus, when a deforming force is imposed, the kinked molecules not only unwind, but can also move relative to each other as far as the cross links permit. When the load is removed, they take up their original shape. Most *elastomers* are manufactured into the requisite shape by cross-linking after molding. Thus, sulfur is used to *vulcanize* rubber at elevated temperatures. Some of the newer elastomers have a dual behavior: they acquire thermoplastic properties on heating but behave as elastomers after cooling (e.g., styrene-butadiene block copolymer in Table 6.1).

Thorough *mixing* of ingredients is vital for all classes of polymers and special equipment (some resembling dough mixers, others rolling mills) has been developed for the purpose.

6.1.4 Molding Processes

Molding processes correspond to casting and bulk-deformation processes in metals. Many of them may be applied to both thermoplastic and thermosetting resins (Fig. 6.4).

Compression molding is essentially a forging process, performed in a heated die that forms a premeasured quantity of the polymer (Fig. 6.4a).

FIG. 6.4 Polymer processes related to forging: (a) compression molding and (b) transfer molding.

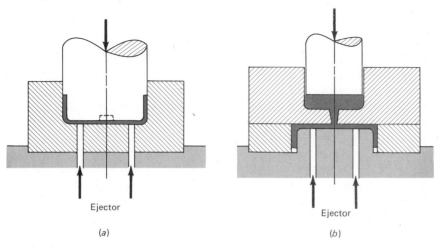

Ejector

(a)   Ejector

(b)

In a true closed (trapped) die, variations in polymer quantity are reflected in variations of part thickness; when a small flash is allowed to extrude (usually along the punch perimeter), closer tolerances may be held. The process is applied mostly to powdered polymers and, as in other powder compaction, high temperatures and pressures (Tables 6.1 and 6.2) must be developed. Thermosetting polymers reside in the die until fully cured, for a cycle time of the order of 1 to 5 minutes.

Cycle times are much shorter in *cold molding.* A powder or fibers (often of refractory materials) are mixed with a binder and compacted in a cold die. These procedures are followed by *curing* in a separate oven.

Another variant of the forging process, *cold forging,* is applied to thermoplastics just above the glass-transition temperature. Deformation and set are instantaneous, and production rates can be as high as in metalworking, but the viscoelastic behavior of the polymer (Sec. 6.1.1) results in shape changes after forming.

Transfer molding utilizes an extrusion-molding principle (Fig.6.4*b*). An excess quantity of the polymer is loaded into the *transfer pot* from which it is pushed through an *orifice* (sprue) into the mold cavity by the action of a punch at approximately 15,000 psi (100 MPa) pressure. The die is of closed construction and multiple cavities may be used, thus reasonable production rates are achieved on both thermosetting and thermoplastic compounds. The polymer acquires uniform temperature and properties in the die chamber prior to transfer; further heating by shear through the orifice reduces its viscosity so that it can fill intricate mold details and flow around delicate inserts.

Injection molding is perhaps the most widespread technique, utilized for thermoplastic and, more recently, thermosetting resins as well. The process (Fig. 6.5) resembles the hot-chamber die casting of metals: the die, split to allow removal of the solidified product, is kept shut with an appropriate press force (Tables 6.1 and 6.2) and ejectors are provided for removing the molded component. Fine (0.02 to 0.08 mm by 5 mm) *vents* assure that no air is trapped. Multiple cavities are readily accommodated, but the principles of assuring uniform flow through properly designed runners and gates must be followed. The difference between metals and plastics lies in the supply of the polymer, which is usually fed in a solid form, as pellets or powder, through a *hopper* to a cylinder or *barrel,* the die-end of which is surrounded with *heaters* that gradually bring the polymer to the required temperature. For thermoplastic resins, preheating must take the polymer above the melting point (170–320°C), while the mold is held at a lower (typically 90°C) temperature to assure cooling below the glass-transition temperature prior to removal of the part from the mold; typically, two to six cycles can be achieved per minute. For thermosetting resins, the barrel is preheated just sufficiently (to 70–120°C) to allow molding while the mold is heated to 170–200°C to effect curing in a

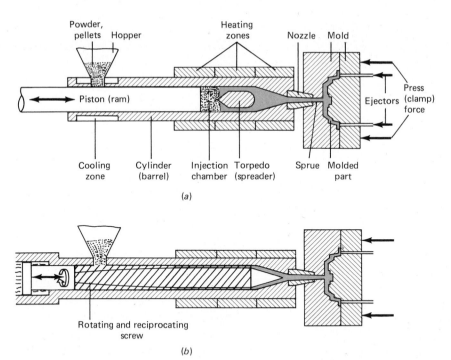

FIG. 6.5 Main operational features of injection-molding machines with (a) plunger and (b) reciprocating screw injection. (After *Petrothene Polyolefins: A Processing Guide,* U.S. Industrial Chemicals Co., New York.)

reasonable length of time. The pressure may be exerted by a *reciprocating,* hydraulically driven *plunger* capable of developing pressures of 10,000 to 25,000 psi (70 to 180 MPa, Fig. 6.5a). Alternatively, and much more frequently, transport of the polymer and consolidation into a continuous mass is accomplished with a *rotating screw* (Fig. 6.5b). Intense shearing of the polymer assures homogeneity and also contributes to heating. In the *reciprocating screw machine* the screw is supported by a hydraulic ram that is pushed back when the pressure in front of the screw builds up. This results in a pressure drop, which allows the hydraulic ram to push the screw forward and thus inject the plastic into the mold.

Production rates are greatly increased with *multistation, rotary turntable machines* on which loading, injection, and stripping (and, if appropriate, placing of inserts) take place simultaneously.

Liquid polymers can be *cast* into any desired shape. Thermoplastics, primarily nylon, are cast above the melting temperature by the same techniques as metals (Sec. 3.2.2). Thermosetting polymers are poured as liquids made up of the reacting components (as in the case of epoxies, phenolics, polyesters, and polyurethane) or of a monomer-polymer-catalyst system (acrylics). Curing is effected at room or elevated temperature.

Gases are either removed from the liquid by vacuum or are kept in solution through the application of pressure during curing. Molds may be made of metal, glass, and rigid or flexible polymers; the latter can be peeled off the casting. An important castable group are *plastisols*. When polymer particles (usually PVC) are dispersed in a plasticizer, the resulting paste can be poured into any mold. When heated, the polymer dissolves in the plasticizer and a flexible, permanently plasticized part is obtained. *Slush-casting* yields thin-wall products such as toys and gloves.

Large hollow parts may be made of polyethylene and other powders by *rotational molding*. The thin-walled metal mold is heated while rotating around two mutually perpendicular axes. The polymer powder melts and coats the inner surface of the mold, whereupon the mold is cooled and the part removed. The technique is suitable also for plastisols.

In all molding processes the polymer reproduces the surface finish of the mold, and, once the often expensive die is made, a smooth surface can be obtained at no further expense. Dimensional variations, however, can be rather large unless sufficient time is allowed for the polymer to become rigid prior to removal from the mold. Minimum radius is under 1 mm (around 1.5 mm for cold molding) and drafts are quite small, between 0.2 and 1° (but 3–5° in cold molding).

Extrusion molding is used mostly for the production of sheet, tube, and bar in thermoplastic materials. The principle is like that shown in Fig. 6.5b, but an *extrusion die* takes the place of the mold and the screw does not reciprocate. The screw is designed to give a gradually decreasing clearance toward the die orifice, assuring consolidation of the polymer particles. The highest pressure is obtained just in front of the orifice, and thus a sound product is extruded; best quality and highest production rates are achieved with screw designs that assure thorough mixing by breaking up the polymer stream. Because the thermoplastic polymer still exhibits a substantial resistance to fracture on emergence from the extrusion die (the m value is still high), it is often further thinned by *continuous stretching*. As discussed in Sec. 6.1.1, this also improves the mechanical properties because some alignment of the molecules is achieved. Extruded tube, blown into a larger diameter, is an important source of thin, biaxially stretched sheet (blown film).

6.1.5 Thermoforming

These processes take the products of extrusion molding and further form them into the required final shape after reheating above the glass-transition temperature. The processes are similar to sheet metalworking, although they fall mostly into the family of stretch forming, and relatively little deep drawing is practiced. Stretch forming (Fig. 6.6) is particularly favorable because the high strain-rate sensitivity (m value) of the thermo-

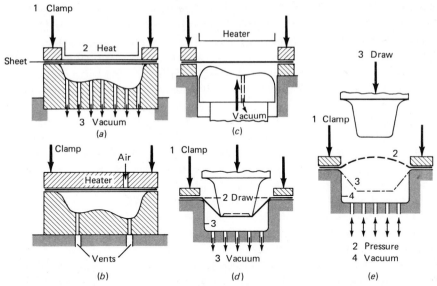

FIG. 6.6 Typical thermoforming methods: (*a*) vacuum, (*b*) pressure, (*c*) drape-vacuum, (*d*) plug-assist, and (*e*) pressure-bubble plug-assist forming.

plastic polymers allows large elongation without fracture (Sec. 6.1.1), although the surface of the formed part will be somewhat uneven.

In principle, *thermoforming* employs a *clamp* that grips the sheet around its circumference, a *heater* to bring the polymer above the glass-transition temperature (usually around 130–190°F or 55–90°C), and a die which may be male or female. Conformance to the die shape may be achieved by mechanical means or by air pressure. Since the die is cooler, the polymer is chilled (and stiffened) by die contact and portions of the workpiece that first touch the die remain thicker. Deformation is then limited to the freely deforming portions of the workpiece and could lead to fracture. Much of process design is aimed at controlling the wall-thickness distribution.

In the simplest (straight) techniques the polymer is heated above T_g; *vacuum* is then sufficient to draw the sheet into intricate recesses of the female die (Fig. 6.6a). Alternatively, hot *air* may be applied at *pressure* (Fig. 6.6b) to drive the sheet into the female die cavity (provided, of course, that venting holes are furnished at the underside). A more complex shape can be developed by *drape forming* (similar to the stretch forming of metals): reentrant shapes can be formed by providing the punch with holes through which a vacuum is drawn, and thus the polymer is pulled into the recessed parts (Fig. 6.6c). The corners of such parts will be thin.

Corners can be strengthened by deforming the sheet with a punch (now called a *plug*) prior to applying vacuum or air pressure (*plug-assist forming,* Fig. 6.6*d*). Further wall-thickness control is achieved if the clamped sheet is first free-formed into a dome by pressure (*pressure bubble*) or vacuum; this preform is then deformed by the plug, and the final shape is achieved by making the sheet conform to the plug or die configuration with the application of pressure or vacuum. One possible combination of these process elements is shown in Fig. 6.6*e*. Some polymers (typically, ABS) are suitable for forming by sheet-metalworking techniques.

Blow-forming is the application of the same principle to hollow preforms (*parisons*). The tube emerging from the blow-molding machine is immediately cut and thus sealed at one end and is then expanded (blown), by compressed air, into bottles and other containers (Fig. 6.7). The pinch-off seam is eliminated, and more accurate and complex neck shapes are produced if the parison is made to the optimum shape and wall-thickness distribution by injection molding.

Success in all thermoforming hinges on exploiting the large elongations attainable with high *m* values; therefore, both temperatures and strain rates must be carefully controlled.

It should be mentioned that the sheet used in thermoforming is produced by extrusion, casting, or, for PVC, by *calendering*. The latter is akin to the rolling of metals, except that a mass of plastic is passed through, typically, three roll gaps formed by four rolls in a single stand.

FIG. 6.7 Bottle-blowing sequence: Mold arrives at extrusion die (*a*) and pinches off by closing (*b*); bottle is blown (*c*) at next station; bottle is cooled and removed at third and fourth stations (not shown).

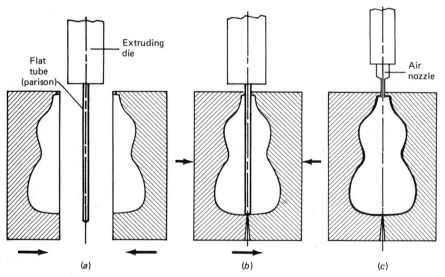

6.1.6 Filled and Laminated Polymers

The cost of polymeric components can be reduced and their strength (or other properties) improved by embedding a *filler* or a *strengthening fiber* in the polymer.

Wood flour is a low-cost filler (e.g., in phenol-formaldehyde). Short, strong fibers (cellulose, asbestos, graphite, carbon black, or glass) impart greater strength while still permitting standard molding procedures (Tables 6.1 and 6.2) and result in a composite of uniform properties in all directions.

Long fibers and sheets (of paper, cloth, glass cloth, metal, or graphite) impart great strength to the polymer. *Laminated sheets* and structural sections may be formed by pressing, casting, extrusion, roll coating, or by laying up liquid resins on the fiber skeleton. They broaden the application of polymers enormously.

A very useful filler is gas. *Foamed plastics* have become important in many applications, because *discrete porosity* improves buoyancy and elasticity, and acoustical, heat, and electrical insulating properties, while *interconnected porosity* makes the structure absorbent. The weight and cost savings are important bonuses. Porosity is induced with *foaming agents* which may be *chemical* (substances that decompose to release a gas such as nitrogen), or *physical* (liquids that evaporate or gases that expand). They can be added to both thermosetting and thermoplastic compounds; in the former they act during heating to a higher temperature and in the latter during either heating or cooling. A great variety of techniques are used, many in standard plastic-processing equipment. *Structural foams* are made with an integral solid skin, formed by the collapse of bubbles against a solid mold.

6.1.7 Plastic-Processing Equipment

For general guidance, production characteristics of various types of *plastic-processing equipment* are given in Table 6.3. Smaller units than shown are available for laboratory and production purposes and larger production units can also be obtained. The size of the machine is dictated by the size of the part to be produced and by the total clamping or forming force to be developed. The calculation is based on the molding pressure (from Tables 6.1 and 6.2), exerted over the entire projected area (including parts as well as gates, runners, etc.). It is usual to allow a minimum 20 percent safety factor.

6.1.8 Process Limitations and Design Aspects

Many of the principles discussed for casting (Secs. 3.1.2, 3.2.2, and 3.4), forging (Secs. 4.4.2 and 4.4.3) and sheet metalworking (Secs. 5.2, 5.3, and

TABLE 6.3 CHARACTERISTICS OF PLASTICS PROCESSING EQUIPMENT*

	Extruder		Press (clamp)					
Equipment	Capacity kg/h	Barrel dia., m	Shot size, kg	Size, MPa	Bed, m	Daylight,† m	No. of parisons	No. of stations (rotary table)
Extruders								
Single screw	1–3000	0.01–0.3						
Twin screw	20–1000	0.01–0.2						
Injection molders								
Plunger type			0.01–4	100–20,000	0.07 × 0.12–1 × 1	0.15–0.9		
Reciprocating screw			0.01–25	100–40,000	0.07 × 0.15–2 × 2.5	0.15–1.8		(1–14) (1–6)
Blow molders								
Continuous extrusion	20–200			10–200	0.07 × 0.07–2.5 × 2.5	0.15–0.75	1–8	(1–2)
Rotary table	20–350			10–500	0.35 × 0.35	0.25–0.5	1–4	(2–9)
Accumulator	5–800		2–50	50–3000	0.3 × 0.4–3 × 3	0.5–5.5	1–8	
Injection	20–200		0.1–8	200–2500	0.3 × 0.5–0.6 × 1	0.1–0.75	1–12	2–4
Reciprocating screw	20–350		0.3–2.5	250–20,000	0.25 × 0.25–1 × 1.5	0.25–1	1–12	1–4
Compression molders				250–5000	0.15 × 0.15–2.5 × 3.6	2–3.5		
Transfer molders								
Clamp					0.15 × 0.15	0.1–1		
Transfer				10–2000	0.1 × 1.5	0.15–0.5		

* Data extracted from *Modern Plastics Encyclopedia*, 1974–1975, McGraw-Hill, New York, 1975.

† Press stroke is usually 60–80% of daylight (maximum open height).

5.5) are also applicable to the processing of plastics, provided that the characteristic behavior of polymers is taken into account.

Thus, extreme care is needed in allowing or encouraging the removal of gases produced by reactions or entrapped during compaction of particulate starting materials. This sets a practical limit of several inches (100–200 mm) to the thickness attainable without gross porosity. The low heat conductivity of plastics also limits heating rates and thus the economical thickness of thermoformed parts, typically to below 0.25 in (6 mm).

The minimum attainable wall thickness (Chap. 11) is limited by the difficulty of removing very thin parts from the mold, and also by the high pressures required to fill at a high width-to-thickness ratio. Large wall-thickness variations are just as undesirable as in metal castings.

Tight corners can be filled but generous radii increase die life and prevent stress concentrations in service. Sufficient draft is important, particularly on ribs and bosses. Through-holes are limited only by the strength of the *core pin* and are usually held below a length-to-diameter ratio of 8. Blind holes are produced by freely extending core pins and are limited to a depth-to-diameter ratio of 4 for $d \geq 0.06$ in (1.5 mm) and to a ratio of 1 for smaller holes. Threaded holes of 0.20 in (5 mm) diameter and over can be molded directly. Smaller holes are best drilled, and smaller threads of reasonable strength are best provided by *metal inserts.*

The use of molded-in inserts greatly expands the scope of application for plastics and very often eliminates subsequent assembly, although at some expense. Binding posts, electric terminals, anchor plates, nuts, and other metallic components are molded into polymers by the millions. Some precautions are necessary, however. The shape of the metal part must assure mechanical interlocking (for example, by heavy knurling), since there is no adhesion between metals and plastics (at least not without special surface preparation). The thermal expansion of plastics is usually much larger than that of metals; this helps to shrink the plastic onto the insert but could also cause cracking of a brittle plastic. The wall thickness around the insert therefore must be made large enough to sustain the secondary tensile stresses.

6.2 GLASSES

While it may seem strange to classify glasses among polymers, they are in fact spatial networks not unlike their cross-linked organic counterparts.

6.2.1 The Structure of Glass

Glasses are based on a *glass-forming oxide* such as SiO_2 in which the oxygen atoms surround the smaller silicon atom, and the free valences of the oxygen join in a rather loose, three-dimensional, covalently bonded network (Fig. 6.8a). Just as in other polymers, very slow cooling (or holding at elevated temperatures) allows crystallization (in which case SiO_2 is called quartz or cristobalite), while faster cooling supresses crystallization (as in Fig. 6.1a). The viscous-flow behavior of the melt is retained (glasses are not viscoelastic), but the viscosity gradually increases, causing a stiffening of the amorphous polymer structure. When the viscosity rises to very high values (by convention, above 10^{13} poises or 10^{12} N·s/m^2), the glass (now called fused silica) becomes an elastic-brittle solid.

A few other oxides (such as B_2O_3 or P_2O_5) can form similar networks, and yet others (such as Al_2O_3) enter into the SiO_2 network. Many other oxides (e.g., Na_2O, CaO, PbO) depolymerize the network by breaking up oxygen-to-oxygen bonds; their oxygen attaches itself to a free bond, while the metal atom, in the ionic state, is distributed randomly (Fig. 6.8b). *Depolymerization* lowers the bond strength, thus also the melting point and the viscosity at a given temperature, making the glass more suitable for manufacturing purposes. The most common glass contains soda (Na_2O) and lime (CaO), and is generally used for windows, tumblers, and other mass-produced glassware. Its relatively high thermal expansion (Table 6.4) does not allow it to withstand thermal shocks; therefore glasses of lower expansion (such as borosilicates and aluminosilicates) are used for chemical and cooking applications. The presence of sodium makes glass susceptible to corrosion by water. In the fully depolymerized form (when one Na_2O molecule combines with one SiO_2 molecule) it becomes water-

FIG. 6.8 Simplified, two-dimensional representation of (a) fully polymerized and (b) partially depolymerized glass (fourth bonds are outside the plane of the illustration).

◯ Oxygen

○ Silicon

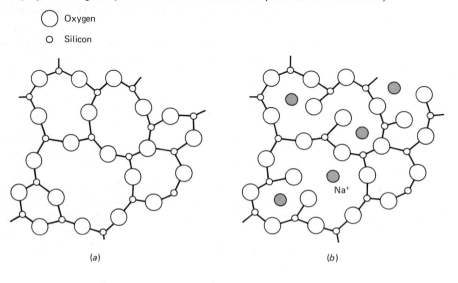

(a)   (b)

TABLE 6.4 MANUFACTURING PROPERTIES OF SOME GLASSES*

Property	Corning Glass Works code number and type				
	7940 fused silica†	7740 boro-silicate	1720 alumino-silicate	0080 soda-lime-silica	8871 potash-lead
Composition, weight %					
SiO_2	99.9	81	62	73	42
B_2O_3		13	5		
Al_2O_3		2	17	1	
Na_2O		4	1	17	2
K_2O					6
Li_2O					1
CaO			8	5	
MgO			7	4	
PbO					49
Viscosity, poise‡ at °C					
$10^{14.5}$ (strain point)	956	510	667	473	350
10^{13} (annealing point)	1084	560	712	514	385
$10^{7.6}$ (softening point)	1580	821	915	695	525
10^4 (working point)		1252	1202	1005	785
Coefficient of linear expansion $\times 10^{-7}/°C$	5.5	33	42	92	102
Typical uses	High temperature, aerospace windows	Chemical, baking ware	Ignition tube	Container, sheet, plate	Art glass, optics, capaci-tors

* Data compiled from *Properties of Glasses and Glass-Ceramics,* Corning Glass Works, Corning, New York, 1973.

† Produced by vapor deposition.

‡ Multiply poise by 0.1 to get N·s/m².

soluble; such water glass is useful as a binder for sand molds (Sec. 3.2.1) or grinding wheels (Sec. 8.8.3) because it regains some strength when the water is driven off.

6.2.2 Properties and Manufacture

In the amorphous state glasses exhibit Newtonian viscosity; that is, they have a strain-rate sensitivity index m of unity. In line with the importance of this discussed in Sec. 6.1.1, they can sustain very large extension without the localization of a neck. Therefore, they have long been formed by a variety of techniques involving stretching, similar to thermoforming (Sec. 6.1.5).

If the temperature is high enough to lower viscosity to around $10^3 - 10^4$ poises ($10^2 - 10^3$ N·s/m^2), a low air pressure is sufficient for blowing. Free form blowing (*hand-blowing*) is now limited to artistic and laboratory work. For mass-production, a preform (parison) is formed in a preliminary step, with a plunger preforming the cavity while often also shaping such elements as the neck of a bottle. Cast-iron or steel molds, split if necessary to permit release of the finished part, are then used to control the blown shape, and large quantities of bottles are thus produced. To maintain a good surface finish, the cast-iron mold may be coated with a wax-sawdust mix that is then converted into carbon; prior to blowing, the carbon layer is slightly dampened, and the steam generated during blowing separates the *paste-mold* from the glass, assuring a smooth finish. Light bulbs are produced on rotary, multistage blowing units at the rate of 2000 per minute.

At temperatures that give viscosities of 500–7000 poises (50–700 N·s/m^2) the glass can be formed by *pressing* into cast-iron or steel molds, resembling closed (trapped) forging dies. Sometimes a flash is allowed to form (usually in the direction of punch movement), and the excess glass is broken off after the glass surface has been scored. The edge is finished by playing a hot pencil flame on it.

Sheet glass is really a semifinished product; it is made by *drawing* the viscous glass vertically up (or horizontally out) from the furnace through rolls. High-quality plate glass is produced at higher temperatures by casting (followed by grinding) or, more recently, by *floating* the sheet over a bath of tin in a nonoxidizing atmosphere.

Glass fibers may be made by extrusion through orifices of very fine diameter (a few micrometers) at high speeds and in great numbers (up to 400 filaments simultaneously). Short fibers (*glass wool*) are *spun* by flinging them out from hundreds of holes in a rotating head; they are often immediately *matted* to form insulating blankets.

Glass that becomes highly viscous without touching a die has an extremely smooth surface (*natural fire finish*). Glass exposed to a humid atmosphere soon develops minute surface cracks that make it susceptible to tensile stresses. Residual stresses from uncontrolled cooling can

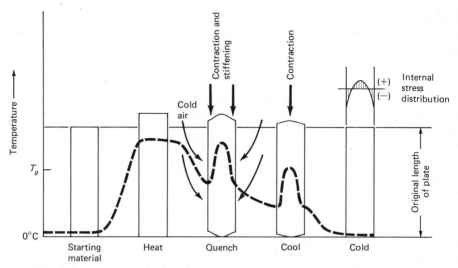

FIG. 6.9 The sequence of operations in toughening a glass and the resultant internal (residual) stress distribution.

cause spontaneous disintegration; therefore annealing is a crucial step of processing. Glass deforms under its own weight at the softening point (Table 6.4); internal stresses are relaxed in 15 min at the annealing point, in 4 h at the strain point. If a glass is suddenly cooled from above T_g with a blast of air, the surface stiffens while the center is still contracting; thus large compressive residual stresses (Fig. 6.9) are induced in the surface which then allow the glass to take substantial tensile stresses in service. Such *toughened glass* finds wide application in eyeglasses, car windows, and various industrial products. Bulletproof glass is further strengthened by *bonding* two or more layers of toughened glass with a thin sheet of a polymer (plasticized polyvinyl butyral). Regular glass similarly bonded is used for automobile windshields.

Glass properties can be readily modified to meet service requirements. Small traces of metallic ions impart a through-the-thickness color (the iron oxide content of white domestic glass is kept below 0.05 percent to avoid coloration). Prolonged heating at elevated temperatures causes crystallization (*devitrification*) which makes the glass opaque and more dense, and endows it with properties similar to sintered ceramics, but with a greater consistency and uniformity. Such crystallized glasses are widely used as ovenware.

6.3 SUMMARY

Polymers have been indispensable to humans for millenia. Animal and vegetable fibers were the only organic polymers available until fairly recently, but the corrosion-resistant, transparent, amorphous inorganic glasses have served first as containers, then as windows, mirrors, and lenses, and are still used in vast quantities. Manufactured organic polymers, at first cellulose acetate and later Bakelite, were initially regarded with the suspicion accorded to substitute materials but have changed our technological outlook in the last 40 years. From containers to packaging, from household utensils to toys, from electrical insulators to fabrics, they have moved from victory to victory against established materials, because understanding of properties and processes, stemming from basic research, has been successfully transplanted into practice.

From the manufacturing point of view, there are two major classes: thermosetting and thermoplastic substances.

1 Thermosetting resins are always organic. Once formed, their structure is fixed (in the large majority of instances); therefore shaping must take place while they are in the prepolymer state. They are then, usually, subject to the laws of viscous flow and can be treated as a melt or liquid.

2 Thermoplastic substances, including organic polymers and glasses, may be crystalline (and opaque) with a definite melting point T_m or amorphous (and transparent) with a gradual approach during cooling to brittleness at the glass transition temperature T_g. Organic polymers above T_m and glasses above T_g exhibit viscous flow characteristics and are amenable to solidification processing including sheet and shape casting, die casting (injection molding), and melt forging (pressing). Because of their high strain-rate sensitivity (high m value), they are also amenable to stretch forming (blowing). Organic thermoplastics show viscoelastic characteristics above T_g (or above T_m if crystalline), but their strain-rate sensitivity is high enough to permit extensive stretch forming (thermoforming) in addition to bulk deformation (compression and transfer molding).

3 Elastomers have a T_g well below room temperature. They are mostly related to thermosetting resins in that cross-linking must take place during or immediately after shaping, although thermoplastic varieties are now available.

Further Reading

DETAILED PROCESS DESCRIPTIONS:
SHAND, E. B.: *Glass Engineering Handbook,* 2d ed., McGraw-Hill, 1958.
Modern Plastics Encyclopedia, McGraw-Hill, New York (annual).
Plastics Engineering Handbook, 3d ed., Society of the Plastics Industry, Reinhold, New York, 1960.

TEXTS:

ALFREY, T., and E. F. GURNEE: *Organic Polymers,* Prentice-Hall, Englewood Cliffs, N.J., 1967.

BILLMEYER, F. W., JR.: *Textbook of Polymer Science,* 2d ed., Wiley-Interscience, New York, 1971.

KAUFMAN, M.: *Giant Molecules, the Technology of Plastics, Fibers and Rubber,* Doubleday Science, New York, 1968.

KINGERY, W. D.: *Introduction to Ceramics,* Wiley, New York, 1960.

MANDELKERN. L.: *An Introduction to Macromolecules,* Springer, New York, 1972.

MORTON, F. H.: *Elements of Ceramics,* 2d ed., Addison-Wesley, Reading, Mass., 1974.

O'DRISCOLL, K. F.: *Nature and Chemistry of High Polymers,* Reinhold, New York, 1964.

RODRIGUEZ, F.: *Principles of Polymer Systems,* McGraw-Hill, New York, 1970.

ROSEN, S. L.: *Fundamental Principles of Polymeric Materials for Practicing Engineers,* Barnes and Noble, New York, 1971.

TRELOAR, L. G. G.: *Introduction to Polymer Science,* Wykeham, London, 1974.

VAN VLACK, L. H.: *Physical Ceramics for Engineers,* Addison-Wesley, Reading, Mass., 1964.

WILLIAMS, H. L.: *Polymer Engineering,* Elsevier, Amsterdam, 1975.

SPECIALIZED BOOKS:

BERNHARDT, E. C. (ed.): *Processing of Thermoplastic Materials,* Reinhold, New York, 1959.

LENK, R. S.: *Plastics Rheology,* Wiley-Interscience, New York, 1968.

NIELSEN, L. E.: *Mechanical Properties of Polymers and Composites,* 2 vols., Dekker, New York, 1974.

SCHILDKNECHT, C. E. (ed.): *Polymer Processes,* Interscience, New York, 1956.

TOBOLSKY, A. V.: *Properties and Structure of Polymers,* Wiley, New York, 1960.

Problems

6.1 The behavior of strain-rate sensitive materials is sometimes expressed by the equation

$$\dot{\epsilon} = b\,\sigma^n \qquad \text{or} \qquad \dot{\gamma} = B\,\tau^n$$

where b and B are material constants and n is a strain-rate sensitivity exponent (not to be confused with our n value, the strain-hardening exponent). What is the value of n for (*a*) a Newtonian fluid and (*b*) an ideal rigid-plastic, non-strain-rate-sensitive material? (*c*) What is the relation of this n to m in Eq. (2.11)?

6.2 The part shown in Example 11.1 is to be made of a wood-flour-filled phenol-formaldehyde resin. (*a*) From data in Tables 6.2 and 11.5, choose an appropriate process. (*b*) Taking average values from Table 6.2, specify the process conditions. (c) Design, in principle, the die, showing the main die elements. (d) Determine the size of equipment needed.

6.3 Collect three different plastic bottles and containers (include, if available, a bottle with a handle). Inspect the outer surfaces for evidences of manufacturing techniques. After sectioning, inspect the inner surfaces and gage the wall-thickness variations. From your conclusions, describe the most likely manufacturing process for each container.

6.4 Some special glassware is made to resist fracture by a chemical toughening process: lithium-rich glass is formed to shape, then immersed in molten NaCl. In a surface layer, Li atoms are replaced, one for one, by Na atoms. Why should this process work?

6.5 In the process of toughening a glass plate (Fig. 6.9), the plate is heated to, say, 150°C above the annealing point and then cooled with a blast of air. For simplicity, let us assume that a thin surface layer cools by 300°C without any change in the bulk temperature. The surface tensile stresses generated by the sudden contraction will cause fracture if they exceed the tensile strength of the glass, assumed here to be 70 MPa. Young's modulus $E = 7 \times 10^4$ MPa. Using simple analysis, (a) derive a formula for the magnitude of tensile stresses generated during quenching, and (b) with the data given in Table 6.4, indicate which glasses may fracture. (For a slightly more sophisticated analysis, Poisson's ratio is needed and can be taken as 0.2 for all glasses).

6.6 A bottle is blown, as in Fig. 6.7, from a PVC tube of 30-mm OD and 28-mm ID. What will the finished wall thickness be at the waist (diameter 70 mm) and at the bulge (diameter 110 mm)?

7

PARTICULATE PROCESSING

As indicated in Fig. 1.3, manufactured components or articles of consumption may be directly produced by bringing a powder of the starting material into the desired end shape. The material may be metal, ceramic, or, indeed, a polymer. The term powder is rather restrictive, because many *particulate materials* of fairly large particle size and of irregular shape may also be processed. This is especially true of organic polymers (Chap. 6), for which particulate processing is so generally practiced that it is frequently not even recognized as such. The most ancient applications of this technique are to ceramics, and we have already come across them when discussing the making of molds for some expendable-mold casting processes (Sec. 3.2.1).

Irrespective of the nature of the particulate material, it is first *blended* to assure an even distribution of constituents. Following this, the two essential steps in processing are *compacting* (that is, imparting the desired configuration while also bringing individual particles into close proximity), and then *sintering* of the *green compact* (that is, establishing strong bonds between adjacent particles). Handling is, in general, easier when the green compact has adequate strength. This is provided by mechanical interlocking of irregularly shaped particulates; however, only adhesion (Sec. 2.4) or a *binder* can assure green strength with spherical powders.

7.1 COMPACTION

The method chosen for shaping depends on the nature of the particulate and on the intended density of the compacted material.

7.1.1 Dry Particulate

A mold containing the negative of the component shape may be filled under *gravity,* giving a compact of low density and very low strength. Only with very careful handling can it be converted into a porous, sintered product (more likely, the mold itself will be heated to at least initiate sintering). *Vibration* of the mold during filling aids arrangement of the particulate and results in densities as high as two-thirds of the *theoretical density* (the density of the solid material, as in Table 1.3), especially if the particles are of favorable shape and contain both larger- and smaller-size fractions so that interstices between the larger particles are filled by smaller ones.

A further increase in density is obtained by applying pressure to the powder. If the part shape is fairly simple and a die can be made of steel, high applied pressures are permissible, and, if the particulate can deform plastically, densities in excess of 90 percent of theoretical can be achieved. The effectiveness of *pressing* with a punch is limited, however, because particulate material does not transmit pressures as a continuous solid would. The pressure tapers off rapidly, giving compacts of higher density close to the punch and of diminishing density farther away (Fig. 7.1*a*). The situation obviously improves with a *floating container* (Fig. 7.1*b*) or with several punches—e.g., two *counteracting punches* compacting from the two ends of the die cavity (Fig. 7.1*c*). The logical extension of this concept is *isostatic pressing,* in which a deformable mold (reusable rubber or single-use metal) contains the particulate while hydrostatic (omnidirectional) pressure is applied by means of a hydraulic fluid inside a pressure vessel (Fig. 7.1*d*). Improved consolidation is obtained when individual particles are moved relative to each other, as in the *extrusion* or *rolling of powders.*

A substantial barrier to attaining high green compact densities is the friction of individual particles relative to each other and also to the die wall. A thin film of lubricant, chosen so as not to interfere with subsequent sintering, can greatly improve the density of the compact.

7.1.2 Plastic Forming

When the proportion of the lubricant or other liquid is high enough to allow the relative displacement of individual particles within a liquid matrix, the mixture acquires rheological properties suitable for processing by plastic-deformation techniques. The most familiar example is clay, a layer-lattice compound of hydrated alumina and silica, in which individual platelets of the ceramic adsorb sufficient water to allow movement under pressure. The plasticized mixture can be formed by forging (pressing) techniques into shapes of reasonable complexity, or it may be extruded into tubes, bars, and sections of great variety. The extrusion process may

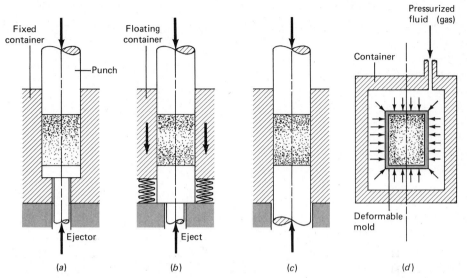

FIG. 7.1   Methods of compacting particulate matter and the resulting density distribution: (*a*) single punch, fixed container; (*b*) single punch, floating container; (*c*) counteracting punches; and (*d*) hydrostatic (isostatic) compaction.

be made continuous and excellent mixing of the constituents assured by the screw-extrusion technique (Sec. 6.1.4).   Hollow shapes may be formed by *jiggering* (a mechanized form of throwing on the wheel): the plastic mass is pressed against a rotating mold with appropriately shaped, profiled *templates* (in this respect the process resembles spinning, Sec. 5.4).

7.1.3   Slurries

An even larger proportion of liquid phase imparts viscous flow properties to the mixture, so that it can be treated by techniques familiar from casting.   The mold is very often a porous material so as to absorb excess liquid and allow firming up of the *slurries* (*slips*).   When the slip is allowed to sit in the mold until a sufficiently thick wall of firm material has built up, and the excess is then drained off, a hollow component is obtained that stands a much better chance of surviving sintering.

7.2   SINTERING

Depending on the liquid content and wall thickness of the green compact, a shorter or longer drying period is required to drive off excess liquid that would vaporize at the sintering temperatures and could result in a disintegration of the compact.

The compact contains particles of the material in close proximity to each other. The energy of the system will decrease by reducing the total surface area—in other words, by joining adjacent particles (Fig. 7.2). This, however, requires diffusion, which in turn can occur only at elevated temperatures (particularly above 0.5 T_m, Sec. 2.1.8). To begin with, inter-atomic bonds are established between adjacent surfaces, and necks form (Fig. 7.2b), with a marked increase in strength (Fig. 7.2a). More massive diffusion is needed to reduce the size of *pores* (Fig. 7.2b); of course, this is accompanied by a *shrinkage* of volume (and increasing density, Fig. 7.2a), and, if the material is ductile, there is also a noticeable increase in ductility. Fatigue properties are likely to be inferior as long as there is any porosity left.

The whole process is accelerated when one of the constituents melts and envelopes the higher-melting constituent (*liquid-phase sintering*).

FIG. 7.2   The process of sintering: (a) changes in properties and (b) development of micro-structure.

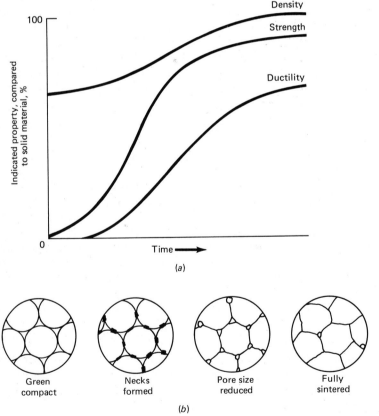

7.3 HOT COMPACTION

All particulate processing sequences involve shaping and sintering; advantages are often gained by combining the two into one simultaneous operation. Sufficient pressure is applied, at the sintering temperature, to bring the particles together and thus accelerate sintering. If possible, individual grains are deformed to assure greater compliance and also to move grains relative to each other, so that surface films that may slow down or prevent diffusion are also broken, and porosity is completely eliminated. Since a particulate body has no strength whatsoever prior to pressing, the applied pressures must be all compressive, and preferably omnidirectional.

Hot pressing in heated *graphite or ceramic dies* is feasible but suffers from the difficulties of transmitting the pressure uniformly to all parts of the compact. Therefore, *hot isostatic pressing* has also been developed. In this the powder is encased in a deformable metal *can* or *shroud* which is then placed inside a heated, pressurized chamber where a glasslike liquid or an inert gas transmits the pressure (up to 15,000 psi or 100 MPa) to all parts of the compact. Hot-rolling, extrusion, and forging of particulate materials is also feasible, although provision must be made to prevent undesirable reactions with the surrounding atmosphere. Therefore the particulate is sometimes encased in a can made of a metal that resists the high temperatures, the process is conducted in a protective atmosphere, or carbon is added to produce a reducing environment.

7.4 CERAMICS

Ceramics are compounds of metallic and nonmetallic elements. In addition to a vast variety of naturally occurring silicates and oxides, the definition includes manufactured materials, sometimes of similar composition but of greater purity, and at other times carbides, nitrides, and other compounds not found in nature. Because of similar application, the element carbon, in both the graphitic and diamond form, is sometimes included. Most ceramics are hard but also brittle, and therefore easily *crushed* to a suitable particle size.

7.4.1 Structural Ceramics

The oldest and still widely used constructional material is clay, irreversibly transformed into brick and tile by a *firing* process. Depending on the composition of the starting material (clay) and the temperature of firing, the finished product may be porous, yet very durable and resistant to water (*earthenware*), or, at increasing firing temperatures, of reduced porosity and increased hardness (*stoneware*). Selected clays mixed with feldspar and flint (SiO_2) fired at high temperature become partially vitrified, resulting in *porcelain*. After a first firing, the body may be coated with a slurry of a glass which forms a *glaze* on second firing.

Apart from their everyday use, clay products find extensive industrial application for their great compressive strength, abrasion resistance, resistance to chemicals, and for their great durability in general, but often also for their electrical- and heat-insulating qualities.

7.4.2 Industrial Ceramics

In addition to clay-type products, a great variety of other ceramics have found increasing use, especially since the rapid development of the electronics industry created a demand for ceramics of very high purity and exceptional electrical and magnetic properties. The raw materials for these are either very carefully processed and purified natural ceramics or artificial compounds.

Abrasives form a special class of industrial ceramics. Selected for their great hardness, abrasives were the first tools ever used for shaping other ceramics (stones, Table 1.1). Abrasive machining retained its significance through the ages partly because it is very often the only process available to shape a particularly hard material, and partly because it imparts a surface finish and accuracy often unmatched by other processes. A more detailed discussion of abrasives is given in Sec. 8.8.3; however, we note here that many abrasives are bonded by particulate technology, either by solid-phase sintering or, as in the production of grinding wheels, by liquid-phase sintering. In the latter instance, individual particles of the abrasives are bonded by a *vitreous* (glassy), continuous matrix.

7.5 POWDER-METALLURGY PRODUCTS

In principle, all metals can be processed in the particulate form (Fig. 1.3).

7.5.1 The Powder

The starting material could be an extremely fine powder produced by *deposition* from the vapor phase (Zn), *precipitation* from a chemical solution (Cu, Ni), *reduction* of an oxide (Fe, Cu, Mo, W) or vapor of a compound (Ni, Fe), or *electrolysis* (Fe, Cu, Be). The metal may be reduced from its ores in a liquid form, refined, and then worked by some previously discussed technique in the solid form. The solid is then *comminuted* into small particles, for example, by pounding until it disintegrates into small flakes, by machining fine chips, or even by chopping up fine wire. Alternatively, the metal is remelted and poured through an orifice, and the issuing stream is broken up (*atomized*) with powerful jets of water, air, or an inert gas into small (2–150 μm diameter) particles. In special processes, the melt is broken up as it is formed, for example, by *spinning* an *electrode*. Some alloys may be produced by mixing powders of the constituent metals in the correct proportion, others are first alloyed by melting and then broken up into particles.

The *shape* of the powder, the *distribution of particle sizes,* and *surface conditions* have a powerful effect on subsequent consolidation and sintering. Some metals (e.g., iron) are likely to be oxidized, but the oxide is readily reduced by a suitable atmosphere during sintering. Others, such as titanium, dissolve their own oxide and are thus reasonably suitable for powder processing. Still others are covered with a thin but very tenacious and persistent oxide film that greatly impairs the properties of the finished part, and these materials (typically those containing chromium and, in general, the high-temperature superalloys) must be treated by special techniques to keep oxygen content very low. Any contaminant that segregates on the surface is bound to create not only consolidation and sintering problems but will also greatly detract from the service properties of the material. Any remainders of a surface film automatically occupy the worst possible position on grain boundaries and act as crack initiators (Sec. 2.3.2).

7.5.2 Pressing of Metal Powders

Consolidation of metal powders follows the various routes described in Sec. 7.1. A small quantity of a lubricant, usually a stearate, is added to aid the densification of powders such as copper and iron that will be sintered with free access of the sintering atmosphere to the powder body. Iron powders are often mixed with varying quantities of carbon, copper, or nickel to impart better mechanical properties, aid sintering, and often also provide built-in lubrication for bearing applications.

For pressing in a rigid die, an accurately measured quantity of the powder is fed into the die and is compacted with pressures ranging up to 120,000 psi (800 MPa). Dies are usually of the true closed-die (trapped) configuration (Fig. 7.1). A punch penetrating from one side is suitable for only thin (low-height) parts (Fig. 7.1a), although an action similar to opposed-punch pressing is obtained with a floating die (Fig. 7.1b). Greatest control is achieved with two punches penetrating from opposite sides, and special presses equipped with lower and upper rams are commonly used for thicker parts. In parts of greatly varying thicknesses (measured in the direction of punch movement), uniform density necessitates *multiple punches* guided within each other (Fig. 7.3). Clearances between moving parts must be kept extremely small (below 25 μm) to prevent entry

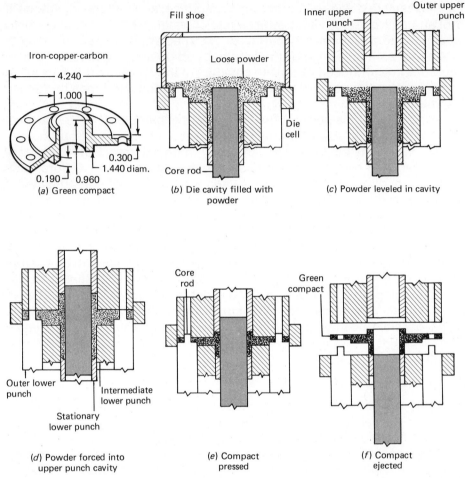

FIG. 7.3 An example of a complex powder pressing operation (dimensions in inches). (By permission, from *Metals Handbook,* vol. 4, copyright 1969, American Society for Metals.)

of powder. The dies are built of high-strength tool steel or, for larger production runs and severely abrasive conditions, of cemented tungsten carbide. Initial compaction may be accomplished by isostatic pressing or, less frequently, by slip casting.

7.5.3 Sintering

The green compact is sintered in an *atmosphere* chosen to provide a nonoxidizing, reducing, or, for steel, sometimes also a carburizing environment. With proper allowance for shrinkage, tolerances can be held to a 0.008-in (0.2-mm) range in the axial direction on a 1-in (25-mm) dimension.

The surface finish will be rougher than that of the compacting die because *porosity* is still significant, 4–10 percent depending on powder characteristics, compacting pressure, and sintering temperature and time. Particularly for bearing applications, density is often kept even lower so as to allow *infiltration* with a lubricating oil, a lubricating polymer (such as PTFE), or a metal (such as Pb or Sn). The infiltrant is drawn into the *sintered skeleton* by capillary forces after it is heated above the melting point of the infiltrant. Copper infiltrated into iron increases strength and improves machinability.

Cold restriking (coining or sizing) of the sintered compact increases its density and improves dimensional tolerance to 0.001 in (0.025 mm).

Further densification and strength improvement can be achieved by *resintering* the repressed compact (Fig. 7.4).

Instead of the traditional pressing, sintering, repressing, and resintering sequence, the green compact may be preheated to the forging temperature and directly *hot forged* to close tolerances at full theoretical density. Such parts can possess the same properties (including toughness) as conventionally forged pieces and can be given shapes otherwise too complex to attain. For example, bevel gears can be forged to finish dimensions requiring only minimum surface finishing.

Compaction by rolling, followed by sintering and perhaps rerolling, is used both for manufacturing sheet and for *cladding* a solid base metal.

FIG. 7.4 The effect of re-pressing at 100,000 psi (700 MPa) and resintering an electrolytic powder-iron compact (sintering at 2050F or 1120°C for 1 h). (By permission, from *Metals Handbook,* vol. 4, copyright 1969, American Society for Metals.)

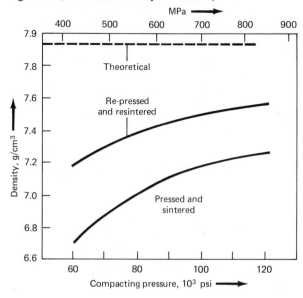

A common feature of powder-metallurgy processes is that only as much material is used as is needed for the finished part. Even though the starting material may be more expensive, the savings in the intermediate processing steps often more than compensate for this, particularly on parts of complex shape.

In all metal-powder processing, great care must be taken to avoid *ignition* or *explosion of powders* that, because of their very large surface area-to-volume ratio, could oxidize at explosive rates.

7.6 COMPOSITE MATERIALS

Particulate processing is especially suitable for combining the desirable properties and cost advantages of dissimilar materials.

A *ceramic composite, concrete,* is used in vast quantities. The low-cost ceramic filler is silica in the form of the gravel and sand *aggregate.* Strength is derived from an irreversible hydration reaction of the bonding agent, *cement,* which leads to the formation of a *gel* that binds the agglomerate particles together and imparts to the structure great compressive strength and reasonable tensile strength. Tensile and bending properties may be improved by creating a further composite with *steel reinforcing bars,* often *prestressed.*

Composites of great manufacturing importance are the tool and die materials called *sintered carbides,* containing a carbide (most frequently WC) in a ductile matrix (usually cobalt). The compact is produced by liquid-phase sintering into the desired form. Once manufactured, it is too hard to be further shaped by any other technique but grinding. The hardness increases, but ductility decreases, with increasing carbide content, and at least 3–6 percent cobalt is necessary to assure a reasonable toughness. Die components subjected to bending stresses often contain up to 10 percent and sometimes even 30 percent cobalt. Further improvements, at least for cutting-tool purposes, are obtained by replacing some of the WC with TiC, and the matrix can be further strengthened by using superalloys rather than pure cobalt or nickel. Some typical applications are noted in Table 4.7 and in Sec. 8.3.1.

In a broader sense, *enamels* (glass films) applied to a metal substrate also form ceramic-metal composites, as does graphite mixed with copper for electric-motor brushes.

In general, the unique feature of particulate processing is its ability to create composites of dissimilar materials that could not be obtained by any other technique. We have already mentioned *porous compacts* (air-solid composites), used extensively as filters and sound- and vibration-damping elements, and oil-filled compacts, found by the dozen in every household as *prelubricated* ("permanently" lubricated) *bearings. Fiber reinforcement* presents no problem; *layered structures* (with a high-density, high-strength load-bearing portion of the part intimately joined to a porous portion) are feasible. Particulate processing extends the range of engineering materials by combining the desirable properties of pure metals with the grain-growth-inhibiting, strengthening properties of finely divided, well-distributed ceramic inclusions. Examples are sintered aluminum powder (SAP) and nickel reinforced by a thoria dispersion (TD nickel), used for high-temperature turbine parts. It is often the only process that can be applied for making parts of very high melting point materials (such as tungsten bars for drawing into lamp wire) or of very great hardness (including the making of polycrystalline diamond for tool bits).

7.7 EQUIPMENT

Except for the artistic applications of clay, particulate processing best lends itself to mass production.

The equipment used in the consolidation of ceramics has many common features with plastic-processing equipment (Sec. 6.1.7), while powder metallurgy relies mostly on presses derived from metalworking (Sec. 4.6). The tooling (dies) is made according to the principles discussed in Chap. 4. Molds and dies for slip casting are designed according to the principles discussed in Chap. 3.

An indispensable part of particulate processing is the *furnace,* either for sintering or for preheating prior to or during hot compaction. Atmosphere control, essential in powder metallurgy, often demands large *atmosphere generators.*

7.8 PROCESS LIMITATIONS AND DESIGN ASPECTS

Particulate processing technology is subject to limitations that can be readily deduced from comparison with other techniques previously discussed. The limitations have essentially two main sources: first, the particulate material must be able to fill the mold or die cavity, and second, the completed compact must be of a shape that can be released from the mold or die.

Slip casting is the most versatile, and any shapes (including hollow and undercut ones) can be formed provided that they can be released from the mold. It is possible to *join* several separately molded pieces and assure the virtual disappearance of the joint during sintering. One needs to think only of the complex yet low-cost mass-produced figurines made in porcelain and other ceramics to realize the potential of the process. How-

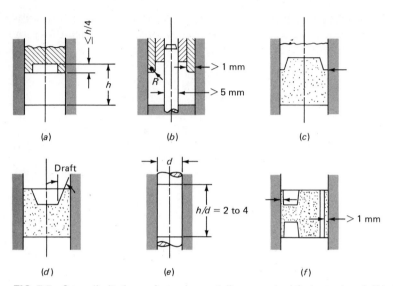

FIG. 7.5 Some limitations of powder-metallurgy parts: (a) stepped end, (b) sleeve thickness, (c) sharp-nosed punch, (d) draft on punch, (e) length-to-diameter ratio, and (f) minimum wall thickness.

ever, diffusion must be assured if this technique is to be used for powder metals. The next-greatest freedom is afforded by flexible isostatic compacting molds which permit undercuts or reverse tapers, but not transverse holes.

The limitations posed by rigid dies are best understood if the die action is contemplated. A single punch cannot assure uniform density if the part is of varying thickness (actually, varying axial height); therefore steps are limited to one-quarter of height (Fig. 7.5a). Much larger steps are allowable with a multiple-sleeve die; however, it must be remembered that a very thin sleeve is impracticable (Fig. 7.5b), and that the sleeve should be radiused to prevent excessive wear. Knife-edge punches wear excessively and should be changed to present a flat face (Fig. 7.5c). On withdrawal, a deeply penetrating punch would damage the compact and should be tapered (Fig. 7.5d). Holes can be made with parallel walls but should be of minimum 0.2 in (4–5 mm) diameter to prevent premature core rod failure. The maximum depth-to-diameter ratio is practically limited to 2–4 (Fig. 7.5e). Even under pressure, the powder cannot fill very thin sections (Fig. 7.5f).

7.9 ATOMIC PROCESSING

The smallest particulate is the atom (or, for a compound, the molecule). Components may be produced through controlled *deposition of atoms* on a surface: one speaks of *plating* and *coating* when the deposit is to stay in place (as in the chrome plating of a car bumper) and of *forming* when the deposit, *stripped* from the form (variously called *matrix, mandrel, die,* etc.), serves as a component. Our interest is in the latter techniques.

7.9.1 Electroforming

To achieve electrochemical deposition of metals or alloys, a plate or slab (the *anode*) of the metal is immersed into an aqueous solution of a salt of the same metal (the *electrolyte*), and is connected to the positive terminal of a low-voltage, high-current dc power supply. An *electrically conductive mold,* called matrix, of the desired shape is immersed at some distance from the anode and is connected to the negative terminal (and thus becomes the *cathode*). Metal atoms are removed as positive ions from the anode, transported through the electrolyte, and deposited on the cathode as neutral atoms.

It takes 96,500 C (coulomb = ampere·second) to remove 1 mole of monovalent metal (*Faraday constant*). The *metal transfer rate* is then

$$W_e = \frac{j}{96,500} \frac{M}{Z} \eta \tag{7.1}$$

where W_e is g/s·m², j the current density (A/m²), M the gram atomic weight (g/mole), Z the valence (charge/ion) and η the efficiency (typically around 0.9).

Electrolyte composition, temperature, and circulation and current density need careful control. Once a deposit of sufficient thickness is obtained (and this may take hours or days), it is stripped from the matrix. Permanent matrixes may be made of metal or glass (or rigid plastic), the surface of which has been *metallized* (e.g., by a chemical-deposition technique). Adhesion is minimized with a thin coating of a parting compound, and a slight taper is allowed to facilitate stripping. Expendable matrixes are made of a metal that can be chemically dissolved, or of a metal, wax, or plastic that can be melted out. Since the finished part is not stripped, great freedom in shape complexity is gained (comparable to investment casting, Sec. 3.2.1).

The atom-by-atom deposit reproduces the matrix surface with the greatest accuracy, and this, together with the attainable shape complexity, defines the economical application range of the process to finished parts (such as waveguides, venturi tubes, reflectors, and typing wheels) and dies (for the stamping of high-fidelity records and for plastic molding in general). Internal stresses can be severe, and there is much art in producing sound parts.

7.9.2 Vapor Deposition

If a metal (or any other substance) is heated in vacuum above its boiling point and the vapor is then condensed on a cooler surface, a coating is formed by *physical vapor deposition.* More importantly, a compound of the substance can be vaporized and reacted with another gas or vapor to form a deposit on a (generally heated) substrate; such *chemical vapor deposition* is faster and is suitable for forming structural parts such as pyrolytic carbon, boron nitride, and refractory metals (W, Mo), usually in the form of tubes. Nickel is reduced from its carbonyl commercially, but the process is not competitive for the direct manufacture of parts.

The major application of vapor deposition remains surface coating and diffusion.

7.10 SUMMARY

Particulate matter, ranging in size from atoms to coarse powder, has been consolidated into usable products from the earliest times. Ceramics provided the first sanitary containers and permanent structures, and are still the material of choice when durability, chemical inertness, and aesthetic appeal are important. Bricks, tiles, bathroom fixtures, dinnerware, grinding wheels, ferrite cores, and ceramic permanent magnets share their processing history with tungsten carbide tools, iron-alloy gears, porous bearings, and metal filters. Common points of importance are:

1 Coherence and strength are developed through sintering, which results in volume shrinkage. Control of shrinkage may set limits on attainable strength, or, if the material is ductile, may require a subsequent sizing operation.

2 Strength develops (even in liquid-phase sintering) through adhesion, and contaminants that are not dissolved by the system reduce strength (particularly in tension), limit ductility, and impair fatigue resistance. Control of surface and process conditions is, therefore, of prime importance.

3 Shape complexity is highly variable, depending on the molding (shaping) method. Distortion, internal stresses, and cracking are more common than with other processes, mostly because the materials themselves are more brittle, and design must recognize this limitation.

Further Reading

DETAILED PROCESS DESCRIPTIONS:
    HAUSNER, H. H.: *Handbook of Powder Metallurgy,* Chemical Publishing, New York, 1973.

SHALER, A. J.: Powder Metallurgy Techniques, in *Techniques of Metals Research,* R. F. Bunshah (ed.), Interscience, New York, 1968, pp. 1663–1737. Also several articles on thin-film, chemical, and electrolytic deposition techniques, pp. 1191–1405.

Metals Handbook, 8th ed., American Society for Metals, Metals Park, Ohio, vol. 4, *Forming,* 1969, pp. 449–464 (Powder Metallurgy).

TEXTS:

HAUSNER, H. H.: *Powder Metal Processes,* Plenum, New York, 1967.

HIRSCHORN, J. S.: *Introduction to Powder Metallurgy,* American Powder Metallurgy Institute, New York, 1969.

JONES, W. D.: *Fundamental Principles of Powder Metallurgy,* Arnold, London, 1960.

WAYE, B. E.: *Introduction to Technical Ceramics,* Maclaren, London, 1967.

See also Chap. 6 references: Kingery, Norton, Van Vlack.

SPECIALIZED BOOKS:

HAUSNER, H. H. (ed.): *Modern Developments in Powder Metallurgy,* Plenum, New York, 1966.

POWELL, C. F., J. H. OXLEY, and J. M. BLOCHER, JR. (eds.): *Vapor Deposition,* Wiley, New York, 1966.

SPIRO, P.: *Electroforming,* Robert Draper, Teddington, England, 1968.

Problems

7.1 Perfectly spherical nickel powder of 0.1 mm particle diameter is compacted by vibration. (*a*) What percentage of the theoretical density can be achieved? (*b*) Will this increase or decrease if the particle diameter is uniformly increased to 0.2 mm?

7.2 A compacted body is sintered into a 30-mm-diameter, 50-mm-tall cylinder which, upon weighing, is found to have a mass of 290 g. Calculate (*a*) the apparent density, (*b*) the percentage of theoretical density, and (*c*) the void volume (porosity) in percent.

7.3 If the cylinder of Problem 7.2 were sintered until full theoretical density was obtained, what would be the dimensions of the cylinder?

7.4 An electrical insulator block of 10 × 20 × 150 mm dimensions is slip cast. Immediately after removal from the mold it weighs 48 g. After drying, the weight is 35 g, and the length has shrunk to 130 mm, with proportional shrinkage in the thickness and width directions. Calculate (*a*) the weight loss, percent; (*b*) the coefficient of linear shrinkage; (*c*) the dry dimensions and volume.

7.5 Bathtubs are often made of enameled iron sheet. The bathtub is normally at room temperature, but is subjected to sudden heating by hot water. Should the thermal expansion of the enamel be greater or lower than that of iron?

7.6 It is proposed that the part shown in Example 11.1 be made by powder metallurgy (with the contour shown in broken lines). (a) Analyze the part shape and suggest a cold-pressing die configuration; (b) calculate the press size if the cold-compaction pressure is the maximum indicated in Sec. 7.5.2. (c) If, for a higher density and greater dimensional accuracy, the part is to be repressed and resintered, can the same die be used?

7.7 Is it physically possible to make the part of Problem 3.3 by powder metallurgy techniques? If it is, is it likely to be technically and economically attractive?

7.8 A cylinder of $d_0/h_0 = 1$ is compacted by cold pressing an atomized steel powder. Full density and high strength are to be obtained by hot-upsetting the cylinder to ¼ its original height. (a) What diameter should one expect, approximately, after upsetting? (b) Should one anticipate cracking in upsetting? If yes, where and why? (Illustrate with a sketch.) (c) If cracking is a danger, how could it be prevented?

7.9 Nickel powder is sometimes consolidated by cold rolling into a thin strip. After trimming the edges of the green strip, it is sintered and cold rolled again. During this second cold-rolling operation, edge cracking may occur. (a) What feature of the rolling process is responsible for cracking? (b) What feature of the sintered product contributes to cracking? (c) What could be done to eliminate or reduce cracking?

8

MACHINING

In the processes discussed so far, the shape of the workpiece was obtained by molding or plastic deformation of the material. If necessary, the shape may also be carved out of a solid body, and the tolerances and surface finish of a previously formed workpiece may be improved by removing the excess material. Machining removes material which has already been paid for; therefore it can be economically undesirable. For this reason, modern developments aim at reduction or, if at all possible, elimination of machining, especially in mass production. Nevertheless, machining is capable of creating geometric configurations, tolerances, and surface finishes often unobtainable by any other technique (Chap. 11). For these reasons, machining has lost some important markets, yet at the same time it is also developing and growing.

If absolutely essential, a machining process can be found for any engineering material, even if it may be only grinding or polishing. Nevertheless, economy demands that a workpiece should be machinable to a reasonable degree. Before the concept of machinability can be explored, it is necessary to identify a basic process, that of metal cutting. *Machining* is a generic term, applied to all metal removal, while *metal-cutting* refers to processes in which the excess workpiece material is removed by a harder tool, through a controlled-fracture process.

8.1 THE METAL-CUTTING PROCESS

The variety and range of metal-cutting processes is very large, nevertheless, it is possible to idealize the process of chip removal by considering *orthogonal cutting* (Fig. 8.1). In this, a rectangular workpiece is machined by a *tool* with a *face* at a *rake-angle* γ_c, measured from the normal of the surface to be machined. Contact with the machined surface is relieved at the back or *flank* of the tool by a *clearance angle θ*. The tool is set to cut a depth (an *undeformed chip thickness*) of h_c, and, when the tool is pushed forward, intense shearing takes place at an angle ϕ, resulting in the formation of a chip of thickness h_2. The *chip thickness h_2* and the *cutting ratio h_c/h_2* are determined by the *shear angle ϕ*. The magnitude of this shear angle is of fundamental importance: for any given undeformed chip thickness, a small angle means a long *shear plane* and, therefore, a high cutting force and energy.

8.1.1 Forces in Cutting

The magnitude of the shear angle ϕ depends on the relative magnitudes of forces acting on the tool face (Fig. 8.1*b*). Externally, on the tool holder, two forces can be observed and measured: the *cutting force P_c* is exerted parallel to the surface of the workpiece, while the thrust force P_t is applied perpendicular to the workpiece surface. Alternatively, the *resultant force P_R* may also be regarded as being composed of two forces acting on the tool itself: the normal force P_n perpendicular to the tool face and the friction force F along the face. Even more important: the shear force F_s in the plane of shear and the compressive force P_h that exerts a hydrostatic pressure on the material being sheared are developed in the material itself.

The magnitudes of P_c and P_t are easily measured with a dynamometer, and from these and the rake angle γ_c, the forces P_n and F are readily calculated:

$$P_n = P_c \cos \gamma_c - P_t \sin \gamma_c \tag{8.1a}$$

$$F = P_c \sin \gamma_c + P_t \cos \gamma_c \tag{8.1b}$$

FIG. 8.1   The orthogonal cutting process: (*a*) geometry and (*b*) forces.

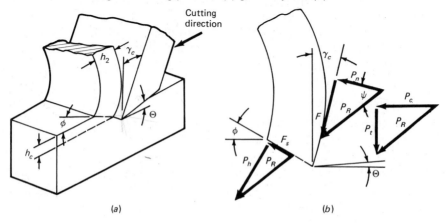

(*a*)   (*b*)

Since the chip thickness ratio h_c/h_2 is obtainable by direct measurement, the shear angle ϕ can be determined from

$$\tan \phi = \frac{(h_c/h_2) \cos \gamma_c}{1 - (h_c/h_2) \sin \gamma_c} \tag{8.2}$$

and the forces F_s and P_h acting on the material can be directly calculated.

A simple way of interpreting the results would be in terms of a model that postulates sliding of the chip along the rake face of the tool. The magnitude of the friction force F would then determine a *friction angle* ψ for which

$$\tan \psi = \frac{F}{P_n} \ (= \mu) \tag{8.3}$$

and, if some assumption is made about the relationship of forces, the shear angle can also be predicted. Thus, from the premise that the material will choose to shear at an angle that minimizes the cutting force, it can be shown[1] that

$$2 \phi = 90° - \psi + \gamma_c \tag{8.4a}$$

Another approach[2] leads to a qualitatively similar result,

$$\phi = 45° - \psi + \gamma_c \tag{8.4b}$$

Thus, the shear angle decreases and the shear force (and with it the *work of cutting*) increases with a decreasing rake angle (by approximately 1.5 percent for each degree change) and an increasing friction angle. One may conclude that favorable conditions (in terms of *energy consumption*) could be secured by using large *positive rake angles* and minimizing friction along the die face.

8.1.2 Chip Formation

While these conclusions are qualitatively correct, the model is oversimplified in many ways.

First of all, in practice, the shear plane broadens into a *shear zone* (usually denoted as *primary shear zone*, Fig. 8.2), the thickness of which increases in a strain-hardening material (with increasing n) and also in a

[1] H. Ernst and M. E. Merchant, in *Surface Treatment of Metals,* American Society of Metals, New York, 1941, pp. 299–378.

[2] E. H. Lee and B. W. Shaffer, *Trans. ASME (J. App. Mech.),* 73:405–413, 1951.

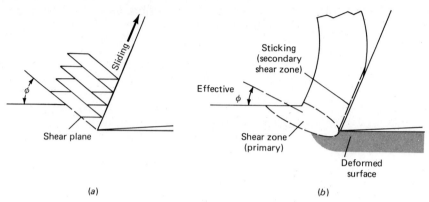

FIG. 8.2 The process of cutting: (a) idealized and (b) more realistic mode.

strain-rate-sensitive material of high *m* value (unless heating from localized shear reduces the flow stress more than it would rise because of a high *m*). The work zone, over which shearing must be effected, thus becomes longer and energy consumption higher; furthermore, the *new surface* suffers plastic deformation.

Second, the interface pressure on the tool face is high; therefore, just as in plastic deformation processes, the workpiece material adjacent to the tool surface may remain immobile (sticking friction, Sec. 4.3.2), rendering the concept of the coefficient of friction meaningless. The chip material must then flow (in the so-called *secondary shear zone*) over a stationary material zone attached to the tool face, and the properties of the workpiece material in promoting this shearing become extremely important.

Finally, a low energy consumption is only one of the desirable conditions. It is equally or more important that wear of the tool be minimized and that the surface produced should be of a controlled, uniform roughness, free of damage. This makes the behavior of the workpiece material relative to the tool of great importance.

With ductile materials such as low-carbon steel, several domains of behavior are often distinguishable.[1] At very low speeds (below 5 fpm or 0.02 m/s), *chip formation* is somewhat *discontinuous* and chunks of metal are lifted out from the surface, leading to a badly damaged, scalloped surface appearance (Fig. 8.3a). At moderately low speeds, say 20 fpm (0.1 m/s), chip formation is *continuous,* the shear zone is well defined and the chip slides up on the tool face (Fig. 8.3b). Under these conditions, a cutting fluid finds access to both the rake and flank faces and can act as a lubricant, and the general relationships shown in Eq. (8.4) are valid. With

[1] M. C. Shaw, in *Machinability,* Spec. Rep. 94, The Iron and Steel Institute, London, 1967, pp. 1–9.

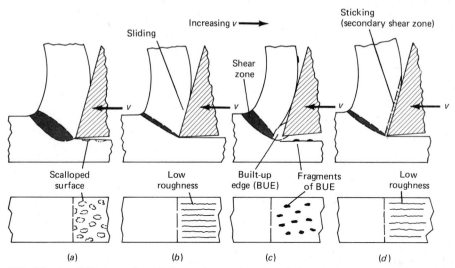

FIG. 8.3 (a–d) Changes in chip formation and surface finish with increasing cutting speed.

increasing speed (say 60 fpm or 0.3 m/s) and increasing heat generation, friction rises until sliding at the tool face is arrested; the system then seeks to minimize the energy of cutting by shearing along a nose of stationary material attached to the tool face. This so-called *built-up edge* acts like an extension of the tool (Fig. 8.3c): shear occurs on the face of this built-up edge, hence the *effective rake angle* becomes quite large and the energy consumption drops. However, a penalty is paid in a poor surface finish, because the built-up edge not only fails to give a well-defined cut surface but also becomes periodically unstable, leaving occasional large lumps of metal and damaging cracks behind. At yet higher speeds, particularly above 180 fpm (1 m/s), the built-up edge gradually disappears (and the energy requirement increases), the chip makes full contact with the tool face, and movement of the chip occurs by shearing close to the tool face, in the secondary shear zone mentioned earlier (Fig. 8.3d). The work of deformation is converted into heat, and since much of this heat is concentrated in the secondary shear zone, with a *temperature rise* of several hundred degrees C (Fig. 8.4), the strength of the shear zone and the energy consumption drop, but conditions favorable for diffusion between the tool face and the workpiece material are set up. If alloying elements of a steel or carbide tool are allowed to diffuse into the workpiece, very rapid wear (crater wear) can develop.

Under special conditions, the chip may be continuous, yet show a *periodic change* in thickness. This is typical of titanium, which is hard and has a low thermal conductivity. The large cutting energy causes severe localized heating and a marked drop of flow stress in the shear zone. Since

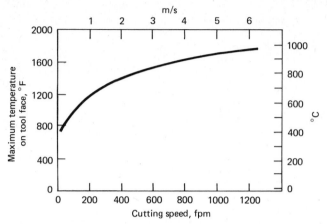

FIG. 8.4 Maximum temperatures attained on the tool face in the cutting of steel. (From *Manufacturing Planning and Estimating Handbook,* American Society of Tool and Manufacturing Engineers, Dearborn, Mich., 1963.)

the heat is not conducted away, the initial shear zone continues to deform and thin out during the passage of the chip along the tool face, while a new shear zone begins to build up ahead of the tool edge.

Further complications that defy analysis are the deformation caused by the pushing, *ploughing action* of the tool edge (which can never be of zero radius) and the friction on the flank of the tool, both of which are additional to the force (or energy) of chip formation.

8.1.3 Interactions between Workpiece and Tool

Any valid consideration of the metal-cutting process must extend to the entire system composed of the workpiece, tool, coolant, and machine tool, and the mechanics and chemistry of their interactions.

Cutting fluids

Lubricant function can be inferred only when speeds are low and the lubricant can penetrate into the zone between chip and tool face; at higher speeds, only the *cooling effect* remains, except that the flank (back) face of the tool that rubs on the already cut surface may still need lubrication for reduced friction and wear. *Cutting fluids* are designed, whenever possible, to perform both functions; therefore, the majority are of the soluble oil variety, which is really an emulsion of the lubricating phase in water, the best heat-transfer medium readily available. Only at low cutting speeds is it profitable to use neat oils (Table 8.1).

Coolants fulfill an additional, and sometimes extremely important,

TABLE 4.1 TYPICAL CUTTING FLUIDS

Process	Tool	Workpiece material						
		Mg alloys	Al alloys	Cu alloys	Steel	Stainless steel, Ni alloys	Cast iron	Ti alloys
Turning	HSS	Dry MO CH	MO-EM CH FA-MO	MO-EM CL-EM FO-MO	EP-EM CH	CL-MO CL-EM	EP-EM EP-MO CH	EP-EM EP-CH
	WC		CH FA-MO Dry	MO-EM FO-MO CH	Dry EP-EM EP-CH	Dry CL-MO CL-EM	Dry EP-EM CH	EP-MO EP-EM
Milling	HSS			MO-EM FO-MO CH	EP-MO EP-EM	CL-MO CL-EM	EP-EM CH	EP-EM EP-CH
	WC				Dry EP-MO EP-EM	Dry CL-MO CL-EM	Dry EP-EM CH	EP-MO EP-EM
Drilling	HSS		MO-EM FO-EM		EP-CH EP-MO EP-EM	CL-MO	EP-EM EP-CH	EP-MO
	WC		EP-EM FA-MO		EP-CH EP-EM	CL-MO CL-EM	Dry EP-EM EP-CH	
Reaming Broaching Tapping	HSS				EP-MO EP-EM	CL-MO CL-EM	Dry EP-EM EP-CH	
Grinding		MO CH		CH FO-MO	EP-CH	CL-EM CL-CH	EP-CH EP-EM	EP-CH EP-EM

* Data extracted in part from R. K. Springborn (ed.): *Cutting and Grinding Fluids*, American Society of Tool and Manufacturing Engineers, Dearborn, Mich., 1967.

Lubricant descriptions as in Table 4.4, with the addition of CH = water-based chemical solutions and surface-active compounds.

function. They flush away chips from the cutting zone and prevent clogging or binding of the tool.

Surface finish

Machining is very often chosen as the means of producing a prescribed surface finish. The surface bears witness to all variables that entered into producing it. The tool and process geometries define the basic or *ideal surface roughness;* superimposed on this are the results of chip formation and friction between tool and workpiece materials. A well-lubricated low-speed operation, or high-speed cutting with no built-up edge, gives the best finishes.

Tool wear

After energy consumption and surface finish, wear is the most important consideration. Since the tool is exposed to extremely high pressures and temperatures, which are sometimes further aggravated by shock loading, tool wear can be rapid and the tool may even be totally destroyed. Thus, the economy of the process is controlled very largely by *tool life.*

As might be expected, *tool wear* can take several forms (Fig. 8.5): *rounding* of the cutting edge, *crater wear,* and *flank wear* are gradual processes, while *chipping* and *fracture* occur suddenly. Gradual wear is produced either by adhesive wear (often accelerated by diffusion of elements that assure the hardness of the tool) or by abrasive wear (especially when the workpiece material contains hard particles). Chipping and fracture are encountered in discontinuous cutting processes and also when the workpiece contains hard spots or inclusions, or when the tool is overloaded in a deep, low-speed cut.

Gradual wear must obviously be a function of the rubbing distance and of the temperature generated in cutting; therefore, it often follows a power function first observed by Taylor[1]

$$vt^n = C \tag{8.5}$$

where v is the *cutting speed* (in fpm), t is the tool life (in minutes), n is the *Taylor exponent* (not to be confused with the strain-hardening index or n value), and C is the cutting speed (fpm) for 1 min tool life.

Both n and C are characteristics of the tool and workpiece material combination. Since the equation plots as a straight line on a log-log scale, n (the slope) and C (the intercept at $t = 1$ min) are readily determined if a few cutting tests are performed at various speeds, and the time

---

[1] F. W. Taylor, *Trans. ASME,* 28:31–350, 1906.

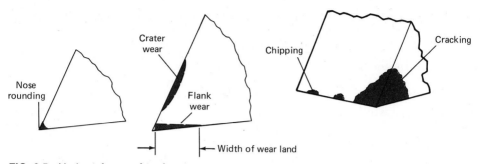

FIG. 8.5 Various forms of tool wear.

required to produce a given wear (depth of crater, or width of flank wear land) is measured. As a very rough guide, the following values may be used: HSS, $n \simeq 0.12$; WC, $0.25 < n < 0.3$; and ceramics $0.5 < n < 0.7$ (except Al alloys: $n = 0.4$ for both HSS and WC).

To obtain an approximate value for C, multiply v_s (from Figs. 8.28 and 8.29) by 1.75 for HSS and 3.5 for WC (see Example 8.3 at end of chapter).

Tool life is also affected by other factors such as feed, but to a much lesser extent; therefore, to a first approximation, it is sufficient to consider speed alone.

An important and often overlooked consequence of tool wear is the *damage* done to the *machined surface.* A dull tool deforms the surface to a greater depth and may even tear it (thus impairing the mechanical properties and, particularly, the fatigue resistance of the part). Also, higher localized temperatures are generated which are extremely dangerous in heat-treatable steel because untempered martensite (of no ductility) may be produced.

Machine tool rigidity

The behavior of the system is greatly affected by the characteristics of the tool holder and the machine tool itself; *Elastic deflections* must be minimized to retain a well-defined cutting geometry, and *vibrations* must be prevented, not only because the resultant *tool chatter* results in an objectionable surface appearance and waviness, but also because vibrations can cause—or at least contribute significantly to—tool wear and tool breakage.

8.2 THE METAL

It is now obvious that no element of the machining system should be considered in isolation. Thus, it is not possible to rank materials according to their machinability, disregarding the mechanical and temperature conditions set up by the process itself. Neither can the interactions with the tool material, the coolant, and the machine tool be ignored. Therefore, *relative ranking* is always very approximate, and the order of merit may reverse when cutting processes different from those used to establish machinability ratings are considered.

8.2.1 Machinability

Machinability, like castability, is a very broad technological property: it expresses the ease and economy with which a material may be cut under average conditions. There are neither universally accepted tests nor specific properties that could be singled out as a measure of machinability.

The most frequently quoted *machinability index* is an average rating stated in comparison with a *reference material:* for steels, a free-machining Bessemer steel B1112, very similar to the present AISI 1212 steel; for copper-base alloys, leaded (free-machining) brass, and for aluminum alloys, 7075-T6 Al. The system can be very misleading because it implies quantitative relationships that are not actually there.

A more quantitative measure is tool life to total failure (by chipping or cracking) under specified conditions. The results are given as the *maximum cutting speed* for a given tool life in minutes or seconds.

Another measure of machinability is tool wear. This could be related to the gradual wear of the flank face. It is given as the change in the machined part dimension due to wear per unit time for a given cutting speed and feed, or as the time required for a given standard flank-land wear to develop. In other cases, the depth of crater wear on the rake face is given.

Another quantitative measure is the surface finish (Sec. 11.1.2) produced at standardized cutting speeds and feeds.

8.2.2 Machinable Materials

Since machinability is such a many-faced property, it is influenced by a number of material properties. Most importantly, good machinability means cutting at a minimum energy with minimum tool wear. This means that:

1 A material of low ductility is desired, in which chip separation occurs after minimum sliding. This is exactly the opposite of what one looks for in plastic deformation (Sec. 4.2); thus, desirable properties now include a low strain-hardening index n and a low resistance to void formation, thus low reduction in area q.

2 To minimize cutting energy, the shear strength or—what is more practically measured—the strength (UTS) of the material should be low.

3 A very strong metallurgical bond between the tool and workpiece, usually expressed as adhesion (Sec. 2.4), is undesirable when it also promotes diffusion and weakening of the tool material by depletion of alloying elements. When diffusion does not take place, high adhesion helps to stabilize the secondary shear zone.

4 Very hard and sharp compounds [such as some oxides (Al_2O_3), all carbides, silicon, and many intermetallic compounds] embedded in the workpiece material act as cutting tools themselves and accelerate tool wear. Second phases that are soft or softened at the high temperatures reached in the shear zone are beneficial because they promote localized shear.

Once the composition and metallurgical condition of the tool and workpiece are fixed, the only material variable remaining is that of the workpiece temperature. A higher temperature lowers the shear strength of the material (thus making possible the machining of some very difficult materials), but it also increases adhesion and accelerates diffusion, and it can shorten tool life. If the latter consideration predominates, every effort is made to keep the work zone cool with large quantities of cutting fluid. This may be difficult when the workpiece material has a low thermal conductivity, which not only allows a very steep temperature gradient in the cutting zone, but also prevents conduction of heat to the surface where it could be removed by the cooling fluid.

It is obvious from the foregoing discussion that some of the most ductile materials favored for plastic deformation are the most difficult to machine. This is true of pure aluminum, copper, and iron and their ductile alloys. Even more difficult are the ductile but also high-strength materials such as stainless steels. Two-phase materials are often desirable because ductility is impaired by the presence of platelike or, in general, sharp second-phase particles, especially if they are also brittle and of low strength.

Some materials may be brought into a more machinable condition through control of their strength and ductility. A prime example of this is the series of carbon steels ranging from pure iron to hypereutectoid steels (Fig. 8.6), which can be obtained in three different forms:

1 In the fully annealed condition, in which strength increases with the increasing amounts of carbide present in the lamellar pearlitic form; at the same time, ductility decreases.

2 Heat-treated to bring the carbide into a spheroidal form; strength drops while ductility rises (consider Fig. 2.19).

3 Finally, there is the option of cold-working the steel, thus raising its strength and depressing its ductility.

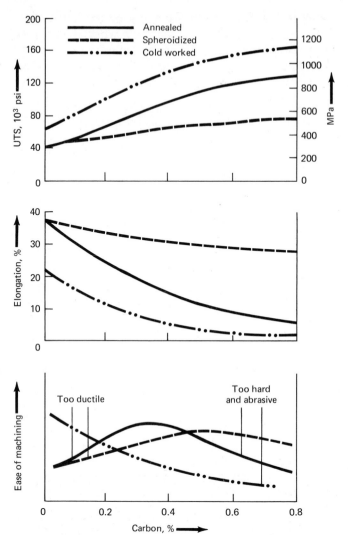

FIG. 8.6 The ease of machining carbon steels as a function of their metallurgical condition.

On this basis, one can readily choose the optimum treatment that assures the best machinability for a given carbon content. At low carbon levels (typically below 0.2% C), the annealed material is much too ductile, and the cold-worked material with its low ductility offers the best machinability. At intermediate carbon levels (typically up to 0.45% C) the strength of the cold-worked material would give rise to excessive cutting forces and the lamellar pearlite with its lower ductility and moderate strength is preferable. At yet higher carbon levels, the large quantities of carbide present in the lamellar pearlite act as a cutting tool and cause premature abrasive wear of the cutting tool proper. Thus, the spheroidal condition with its relatively harmless globular carbides and lower strength is preferable even though the ductility is higher.

Similar considerations can be applied to other materials; thus, ductile, soft materials are more machinable in the strain-hardened condition, while very hard materials, especially those containing hard second phases, are best machined in an annealed (or solution-treated and overaged) condition that also assures a globular shape of the hard second-phase particles. Sometimes this is not possible, for example, in the case of the hypereutectic aluminum-silicon alloys used for engine blocks; only the hardest cutting tools such as diamond can then be used, or machining must be limited to abrasive processes. Heat-treatable materials of relatively lower strength (e.g., precipitation-hardening aluminum alloys) are best machined in the fully heat-treated (aged, Sec. 4.2.2) condition.

8.2.3 Free-Machining Alloys

Until now we have considered the ease of generating a chip. Obviously, the chip must also be removed, preferably without special devices. This is best accomplished by assuring that it breaks into short, easily disposable segments—in other words, that a discontinuous chip (Fig. 8.7) is formed. Materials conforming to this requirement are called *free-machining.*

Some materials are free-machining by nature. Thus, gray cast iron with its large graphite flakes produces short chips without extra alloying elements or process control. There is, however, a penalty to pay: the machined surface is rough, because graphite particles break out. Refining

FIG. 8.7 Typical (a) continuous and (b) free-cutting chips.

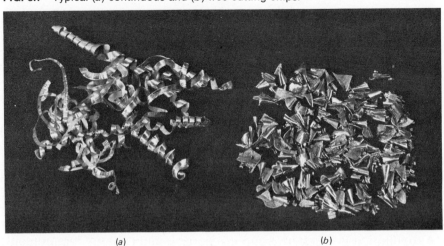

(a)   (b)

the graphite plate size improves the finish without impairing the free-machining properties.

Ductile materials are not free-machining because the chip comes off as a continuous fiber (Fig. 8.7) which at best can be made to *curl* but not to break. Free-machining properties can be imparted by alloying with an element that creates local stress concentrations and planes of weakness, thus assuring breakup of the chip, obviously at some sacrifice in mechanical properties. For many applications, where the absolute greatest strength and ductility is not important, the economy of machining justifies such methods.

One of the most universal free-machining additives is lead (and the related element bismuth), which is insoluble in iron, copper, and aluminum and their alloys. Finely dispersed, lead helps to break the chip, and is also believed to ease machining by providing built-in lubrication and thus easy shear on the tool face. Graphite in iron, in the globular form, acts in a similar manner. Sulfur added to steel combines with manganese to form manganese sulfide, which, if of the appropriate globular shape, imparts free-machining properties. It is possible that sulfur (and the related tellurium) also aids machining by forming a reaction film of low shear strength on the tool face, although no solid proof for this seems to exist. Phosphorus in low-carbon steels hardens the ferrite, making it more machinable. A very important discovery of the last 10 years is that steels deoxidized with calcium form a calcium aluminosilicate which appears to improve machinability by affecting the shear strength in the secondary shear zone without impairing either workability or mechanical properties for service purposes.

8.2.4   Forces and Energy Requirements

The ease of machining a material is also reflected in the *specific cutting energy,* the energy required to remove a unit volume of it. This energy is affected by the flow stress of the material at the strain rates and temperatures prevailing in the shear zone. Strain rates can be very high, because large shear strains (in units of natural strain, of the order of 1 to 2) are produced in a relatively narrow (0.005 in or 0.1 mm) shear zone; even the mean values of strain rates, taken across the zone of intense deformation, reach several thousand per second. Such high strain rates affect (usually increase) the flow stress of most materials even at cold-working temperatures, while the high temperature rise lowers the flow stress. Therefore, only flow-stress values determined at the appropriate (and often unknown) temperatures and strain rates are of any value for calculations of energy requirements. Even then, predictions of the shear angle and the shear zone width are needed before a reasonable estimate can be made.

Altogether, it is much easier to measure specific energy requirements under typical cutting conditions and use these as the basis of judging the power required for a given process. Some guide values are given in Table 8.2 for a 1 mm (0.040 in) undeformed chip thickness. The cutting energy has two components. The specific energy of chip formation proper is not affected by chip thickness because the greater cutting force developed in a deeper cut needs to act only over a shorter cutting distance to remove a unit of volume. In contrast, the ploughing and flank-friction forces (Sec. 8.1.2) are constant, independent of depth of cut, and account for an increasing proportion of the specific cutting energy as the chip thickness diminishes and the total distance traveled by the tool increases. The total specific cutting energy E_c for any undeformed chip thickness h_c may be calculated on the basis of various published data; it appears that most of the values fit a power law

$$E_c = E_1 h_c{}^a \qquad (8.6)$$

TABLE 8.2 APPROXIMATE SPECIFIC ENERGY REQUIREMENTS FOR CUTTING*

Undeformed Chip Thickness (feed): 1 mm (0.040 in)

Material	Hardness		Specific energy E_1	
	BHN	R_c	hp·min/in³	W·s/mm³
Steels (all)	85–200		0.5	1.4
		35–40	0.6	1.6
		40–50	0.7	1.9
		50–55	0.9	2.4
		55–58	1.5	4.0
Stainless steels	135–275		0.5	1.4
		30–45	0.6	1.6
Cast irons (all)	110–190		0.3	0.8
	190–320		0.6	1.6
Titanium	250–375		0.5	1.4
Superalloys (Ni and Co)	200–360		1.1	3.0
Al alloys	30–150 (500kg)		0.12	0.35
Mg alloys	40–90 (500kg)		0.08	0.22
Copper		$80R_B$	0.45	1.2
Copper alloys		$10–80R_B$	0.3	0.8
		$80–100R_B$	0.45	1.2

* Extrapolated from data in *Machining Data Handbook,* 2d ed., Machinability Data Center, Metcut Research Associates, Cincinnati, Ohio, 1972.

where E_1 is taken from Table 8.2 for a chip thickness $h_c = 1$ mm, h_c is the chip thickness in mm, and the exponent has a value of $a = -0.4$.

The power to be developed by the machine tool can then be estimated if the *rate of material removal* V_t and the efficiency of the machine tool η (usually around 0.7 to 0.8) are known:

$$hp = \frac{E_c V_t}{\eta} \quad (hp \cdot min/in^3)(in^3/min) \tag{8.7a}$$

or, in units of kilowatts,

$$kW = \frac{E_c V_t}{\eta} \quad (W \cdot s/mm^3) \, (mm^3/s) \, (10^{-3}) \tag{8.7b}$$

Since power divided by speed is force, the cutting force F_c can be easily calculated in units of pounds if the cutting speed v is in units of fpm:

$$P_c = \frac{33,000 \; hp}{v} \quad (lb) \tag{8.8a}$$

or, in newtons if the cutting speed v is in m/s:

$$P_c = \frac{kW}{v} \quad (kN) \tag{8.8b}$$

This force will have to be resisted by the tool holder and the machine.

8.3 CUTTING TOOLS

Specific features of cutting tools are varied to suit the process, but some basic characteristics are common to all.

8.3.1 Tool Materials

In general, one expects the *tool material* to have properties just opposite to those of the workpiece: high strength should now combine with as high toughness (ductility) as possible. High strength (or high hardness) at operating temperatures assures that the tool geometry is maintained under the extreme conditions presented by the chip formation process, and it also aids in resisting wear. Toughness is needed to survive impact loading that may occur even in continuous chip formation processes when the tool encounters a harder localized spot in the workpiece material.

A further important property is low adhesion to the workpiece material to avoid localized welding, diffusion, and subsequent rapid wear. Paradoxically, high adhesion is desirable when a secondary shear zone is to be stabilized; however, a diffusion barrier is then needed. Low hardness and high adhesion allow distortion of the tool profile, rounding of the tool nose, and gradual flank wear. Chipping is a sign of inadequate toughness, and total fracture may occur with very brittle tool materials.

In general, the hardness and heat resistance of materials can be increased only at the expense of toughness; therefore there is no absolute best tool material available. In the following paragraphs, the most important tool materials will be discussed in order of rising temperature resistance.

Tool steels

Steels derive their hardness partially from the martensitic transformation and partly from carbides of alloying elements stable at high temperatures. Martensite itself softens (tempers) above 250°C; therefore carbon steels are suitable only for the machining of wood, and then only at low production rates. *High-carbon steel* hand reamers are sometimes made for metal cutting.

The vast majority of tool steels is in the *high-speed steel* (HSS) category, which comprises steels of the molybdenum (M1, M2, etc., typically with 0.8% C, 4% Cr, 5–8% Mo, 0–6% W, and 1–2% V) and tungsten (most frequently T1, with 0.7% C, 4% Cr, 18% W, and 1% V) types. The carbides produced by the alloying elements now amount to 10–20 percent of the volume and allow repeated heating and cooling to 550°C without any loss in hardness. Even higher temperatures are permissible with the addition of 5–8% Co, sometimes coupled with increased carbon content (M40 and T15 grades).

All these steels can be hot rolled or forged to a dimension from which the cutting tool can be readily manufactured by conventional machining techniques. Final heat treatment imparts their great strength coupled with a reasonable toughness. They can be repeatedly reground and they remain the general workhorse of the metal-cutting industry, especially for drills, reamers, broaches, and also other forms of tooling (Fig. 8.8) at moderate cutting speeds. Improvements in metal cleanliness (Sec. 2.3.2) have resulted in great advances in their quality and some grades are made by consolidation of prealloyed powder (Sec. 7.5), assuring more uniform distribution of finer carbides.

Cast carbides

When the carbides reach very high proportions, the tool material is not hot-workable any more and must be cast to shape. The matrix is usually

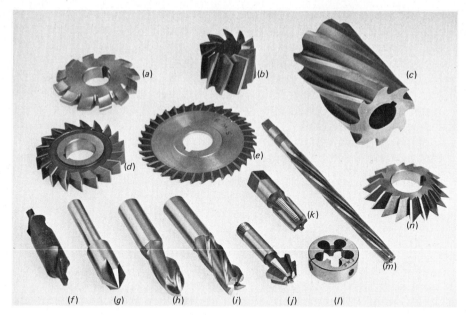

FIG. 8.8 Some high-speed steel (HSS) tools commonly encountered: (a) gear-tooth cutter, (b) shell-end mill, (c) slab mill, (d) side mill, (e) slotting mill, (f) combined drill and countersink, (g) countersink, (h) ball-end mill, (i) square-end mill, (j) single-angle cutter, (k) tap, (l) thread-cutting die, (m) reamer, and (n) angular cutter.

Co (around 45%) in which carbides of Cr (around 30%) and W (around 18%), formed with 2 to 3% C, are embedded. Softening is very gradual; therefore high cutting temperatures are permissible but ductility is now much reduced.

Sintered carbides

Much more widespread is the use of *sintered carbides* produced by powder-metallurgy techniques (Sec. 7.6). The matrix is always cobalt, 3–6% for greater hardness, from 6–30% for greater toughness. The carbide phase may be all WC for nonferrous metals and gray cast iron (C1 and C2 grades), but diffusion would destroy these grades rapidly in the cutting of steel. Therefore, 10–40% TiC or TaC (or both) is added to grades (C4 to C8) destined for the machining of steel and, because of the same diffusion danger, of malleable and spheroidal cast iron. All sintered carbides soften only gradually and work best at a higher temperature (over 600°C).

Coated carbides

Ideally, the tool should possess a very hard, nonreactive surface that also acts as a diffusion barrier. This aim can be achieved by combining the virtues of sintered tungsten carbide with those of a thin (5 μm) *coating* of a ceramic such as TiC, TiN, or Al_2O_3. Even higher performance is achieved with *cubic boron nitride,* second in hardness only to diamond. Most recently, polycrystalline diamond coatings have become available.

Coated carbides are used in machining superalloys (at 2–3 times the usual speeds) and very hard nonferrous materials such as hypereutectic Al-Si alloys, and in production turning and milling of steels and cast irons.

Ceramic tools

Instead of being used as a coating, *ceramics* such as Al_2O_3 may also be made into solid tool inserts. They are suitable for very high speeds but only at light and continuous loads.

The hardest material, *diamond,* in the form of a natural or manufactured *single crystal* or, more recently, as a sintered *polycrystalline* tool tip outperforms all other materials on hard and highly abrasive workpieces.

8.3.2 Tool Construction

High-speed steels have sufficient toughness to be made into *monolithic tools.* Solid sintered carbide tools can be made and are sometimes used, but the risk of total fracture is great and the cost can become high. Therefore their broadest application is in the form of *tool inserts,* which are either *brazed* (Fig. 8.9a) or *clamped* (Fig. 8.9b) to a tough steel body. Specially constructed cutters (*indexable cutters,* Fig. 8.10) permit moving the

FIG. 8.9 Turning tools with (a) brazed and (b) clamped carbide inserts; (c) carbide inserts with preformed chip-breaker grooves.

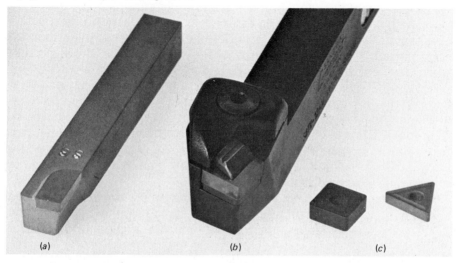

(a)   (b)   (c)

Direction of rotation

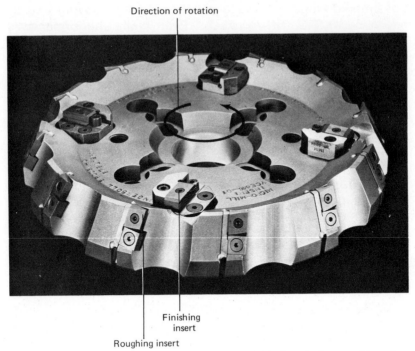

Finishing
insert

Roughing insert

FIG. 8.10 An indexable face mill; peripheral carbide inserts for roughing, face inserts (replaced by dummy inserts during roughing) for finishing. (Ingersoll Cutting Tool Division, Rockford, Ill.)

insert to compensate for wear, and can thus be used for extended periods of time. Ceramic tools are always made as inserts.

High-speed steel and many WC tools are *reground* several hundred times in the machine shop. Some WC and most ceramic tools are of the *throwaway* type, and are made so as to have several usable cutting edges.

As discussed in Sec. 8.1.1, a large positive rake angle shortens the shear zone and reduces the energy consumption. This, however, also weakens the tool; therefore large rake angles are permissible only for cutting lower-strength materials with a tough tool material (see Table 8.3). Other tool materials, particularly the more brittle varieties, must be made with a small positive, zero, or even *negative rake angle*. A three-cornered cutting insert can then have six usable cutting edges (as in Fig. 8.9c). In cutting with a negative rake angle, the force pushing the tool out of the workpiece is large, and vibrations are easily generated; therefore, an extremely stiff machine tool is needed.

The unbroken chip formed by ductile materials, even if of a helical shape, is a nuisance. It is difficult to remove, it may clog up the work zone, and it may present dangers to the tooling, machine, and operator alike. A

TABLE 8.3 TYPICAL SINGLE-POINT CUTTING TOOL ANGLES*

Workpiece material	BHN	High Speed Steel					Brazed WC			Throwaway		All WC		
		Back rake	Side rake	End relief	Side relief	Edge	Back rake	Side rake	Edge	Back rake	Side rake	End relief	Side relief	Edge
Zn alloy	80–100	10	10	12	4	5	5	5	5	0	5	5	5	15
Al, Mg alloy		20	15	12	10	5	3	15	5	0	5	5	5	15
Cu alloy		5	10	8	8	5	0	8	5	0	5	5	5	15
Steels	<225	10	12	5	5	15	0	6	15	−5	−5	5	5	15
	to 325	8	10	5	5	15	0	6	15	−5	−5	5	5	15
	to 425	0	10	5	5	15	0	6	15	−5	−5	5	5	15
	>425	0	10	5	5	15	−5	−5	15	−5	−5	5	5	15
Stainless														
Ferritic		5	8	5	5	15	0	6	15	0	5	5	5	15
Austenitic		0	10	5	5	15	0	6	15	5	5	5	5	15
Martensitic		0	10	5	5	15	0	6	15	−5	−5	5	5	15
Cast iron	<300	5	10	5	5	15	0	6	15	−5	−5	5	5	15
	>300	5	15	5	5	15	−5	−5	15	−5	−5	5	5	15
Superalloy		0	10	5	5	15	0	6	15	0	5	5	5	45
Ti alloy		0	5	5	5	15	0	6	15	−5	−5	5	5	5
Thermoplastic		0	0	20–30	15–20	10	0	0	10	0	0	20–30	15–20	10
Thermosetting		0	0	20–30	15–20	10	0	15	10	0	15	5	5	15

* Extracted from *Machining Data Handbook*, 2d ed., Machinability Data Center, Metcut Research Associates, Cincinnati, Ohio, 1972.

partial remedy is found in *chip breakers,* which are designed to impart additional strain to the chip, causing it to break into shorter lengths or at least curl up into tight coils that break frequently. A chip breaker is formed either by giving the rake face a curvature (*groove type,* Fig. 8.9*c*) or by attaching a separate chip breaker to the rake face (*obstruction type,* Fig. 8.9*b*).

8.4 METHODS OF MACHINING A SHAPE

Irrespective of the machining process employed, the shape of the workpiece may be produced by two basically different techniques: *forming* and *generating.*

8.4.1 Forming

A shape is said to be formed when the cutting tool possesses the finished contour of the workpiece. All that is necessary, in addition to the relative movement required to produce the chip (the *primary motion*), is to feed (*plunge*) the tool in depth. It is immaterial how the primary motion is generated. The workpiece can be rotated against a stationary tool (*turning,* Fig. 8.11*a*), or the workpiece and tool can be moved relative to each other in a linear motion (*shaping* or *planing,* Fig. 8.11*b*), or the tool can be rotated against a stationary workpiece (milling and *drilling,* Fig. 8.11*c*) or against a rotating workpiece (grinding). The accuracy of the surface profile depends mostly on the accuracy of the forming tool.

FIG. 8.11 Forming processes: (*a*) form turning, (*b*) shaping or planing, and (*c*) drilling.

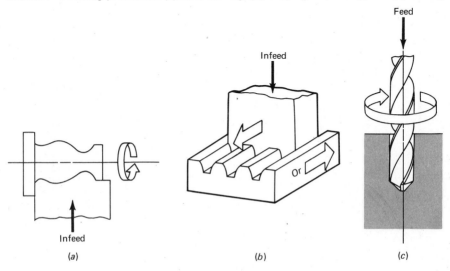

(a)   (b)   (c)

8.4.2 Generating

A surface may be developed by combining several motions that not only accomplish the chip-forming process (primary motion) but also move the point of engagement along the surface (usually described as the *feed motion*). Again, the workpiece may rotate around its axis (as in turning). The tool is set to cut a certain depth while it also receives a longitudinal feed motion. When the workpiece axis and feed direction are parallel, a cylinder is *generated* (Fig. 8.12a), and when they are at an angle, a cone is generated (Fig. 8.12b). If, in addition to the primary and feed motions, the distance of the cutting tool from the workpiece axis is varied in some programmed fashion—e.g., by means of cams, a copying device, or numerical control (NC)—a large variety of shapes can be generated.

When the tool (or the workpiece) is fed perpendicular to the primary linear (shaping or planing) movement, a flat surface is generated (Fig. 8.12c). If the workpiece were rotated on its axis parallel to the tool motion, a cylinder could be machined. More significantly, the workpiece axis could be set at an angle and then a rotational hyperboloid would be generated (Fig. 8.12d). In principle, any surface that can be described by a

FIG. 8.12 Some examples of generating a shape: (a) turning a cylinder and (b) a cone; (c) shaping (planing) a flat and (d) a hyperboloid; (e) milling a pocket; and (f) grinding a flat (principal motions marked with hollow arrows, feed motions with solid arrows).

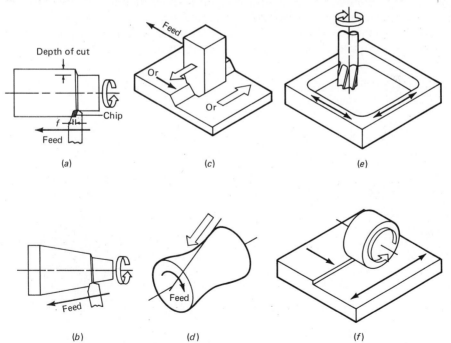

straight generatrix may be produced by this technique. A tool of axial symmetry may rotate while the workpiece is being fed, leading to *milling* (Fig. 8.12*e*) or *grinding* (Fig. 8.12*f*).

Frequently, the principles of forming and generating are *combined* to advantage. Thus, a thread may be cut with a tool of the intended profile fed axially at the appropriate rate (Fig. 8.13*a*). A slot or dovetail may be milled into a workpiece (Fig. 8.13*b*), or a gear may be cut with a *hob* that gradually generates the profile of the gear teeth (Fig. 8.13*c*) while both hob and workpiece rotate.

8.5 SINGLE-POINT MACHINING

It is obvious from the previous discussion that one of the most versatile tools is a *single-point cutting tool* moved in a programmed fashion.

8.5.1 The Tool

The tool must accommodate not only the primary motion (as an orthogonal tool would, Fig. 8.1) but it must also allow for feeding and chip disposal. Therefore the cutting edge is inclined (*oblique cutting,* Fig. 8.14), and the chip is wound into a helix rather than a spiral. The tool must now be relieved both in the direction of feed and on the surface that touches the newly generated surface. Thus, it will have major and minor flank surfaces (Fig. 8.15). Intersections of these with the face of the tool constitute the *major* and *minor cutting edges,* respectively. The

FIG. 8.13 Combined forming and generation of surfaces: (*a*) thread cutting, (*b*) T-slot milling, and (*c*) gear hobbing.

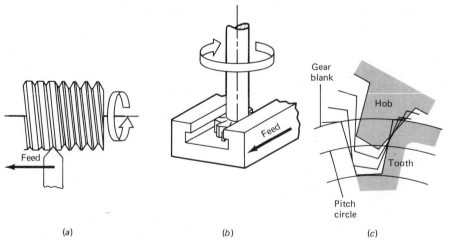

(*a*)   (*b*)   (*c*)

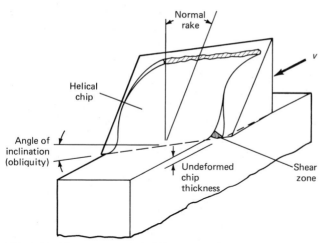

FIG. 8.14 The geometry of oblique cutting.

two cutting edges meet at the *corner* or *nose;* obviously, a completely sharp intersection would generate a sawtooth-type surface with a roughness controlled by the feed rate. A smoother finish may be obtained by rounding the nose with an adequate (typically, $^3/_{64}$ in or 1 mm) radius.

The all-important rake angle should really be measured in a plane perpendicular to the major cutting edge, but, for convenience, all angles are measured in a coordinate system that coincides with the major axes of the tool bit (Fig. 8.15). While this system appears simple, it creates various problems; these are resolved, however, by the new International Standard

FIG. 8.15 Lathe tool designations.

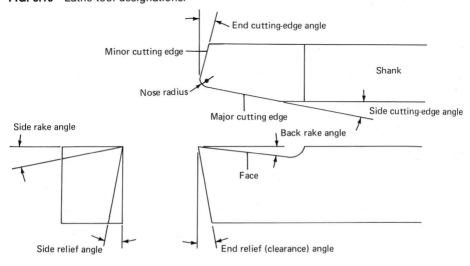

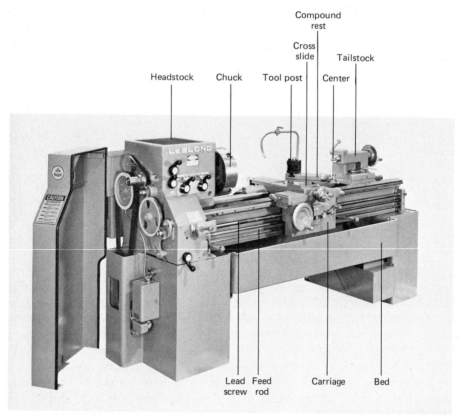

FIG. 8.16 A typical engine lathe. Capacity: 15 in (380 mm) diameter swing; 54 in (1370 mm) length. (LeBlond Inc., Cincinnati, Ohio.)

Organization (ISO) recommendation on cutting tools.[1] In any case, it must be recognized that tool angles have meaning only in relation to the workpiece, after installation in the machine tool.

Some recommendations on cutting-tool angles are contained in Table 8.3. They represent a compromise to give minimum cutting force with maximum tool strength.

8.5.2 Turning

The most widely used machine tool is the *engine lathe* (*center lathe,* Fig. 8.16), which provides a rotary primary motion while the appropriate feed motions are imparted to the tool.

The workpiece must be firmly held, most frequently in a *chuck.* Three-jaw chucks with simultaneous jaw adjustment are self-centering.

[1] Draft International Standard ISO/DIS 3002, 1973.

Other chucks have two, three, or four independently adjustable jaws for holding other than round workpieces. Bars may be held also in *collets,* which consist of a split bushing pushed or pulled against a conical surface. Workpieces of awkward shape are often held by bolts on a *face plate.*

The *headstock* contains the drive mechanism, usually incorporating change gears or a variable-speed drive. Long workpieces are supported at their end with a *center* held in the *tailstock.* The tool itself is held in a *tool post* which allows setting the tool at an angle (horizontally and vertically). The tool post is mounted on a *cross slide* which provides radial tool movement. The cross slide is guided in a *carriage,* which in turn receives support from the *ways* machined in the *bed* that assures rigidity and freedom from vibrations. An overhanging part, the *apron* of the carriage, may be engaged with the *feed rod* to give continuous feed motion, or with a *lead screw* for the cutting of threads. Very long workpieces are secured against excessive deflection by a *center rest.*

Sometimes the tool post sits on a *compound tool rest* which incorporates a slide that can be set at any angle; thus conical surfaces may be formed by hand feeding the tool. A four-way tool post allows quick changing of tools in preset positions, thus speeding up successive operations.

8.5.3 Boring

When the internal surface of a hollow part is turned, the operation is referred to as *boring* (Fig. 8.17a). For short lengths, the tool may be mounted cantilevered in the tool post. Excessive vibration sets in at long lengths and it is then preferable to have the workpiece secured to the lathe bed while the *boring bar,* clamped in jaws at one end and supported in the tailstock at its other end, is driven. A number of patented solutions exist that aim at reducing or damping out vibrations. A special-purpose machine performing a similar operation, but with more firmly guided boring bars, is the *horizontal boring machine.*

Heavy and large-diameter workpieces that need to be machined on both inside and outside surfaces may be better supported on a lathe turned into a vertical position; called a *vertical boring machine,* such a lathe can work on several surfaces of a workpiece fastened on the rotating, horizontal face plate of the machine.

Holes may be produced in solid workpieces by single-point machining techniques resembling boring. In *gun drilling* the cutting forces are balanced by guide pads placed at angles of 90° and 180° to the cutting edge (Fig. 8.17b). To start the hole, a hardened steel guide (*boring bush*) is held against the face of the workpiece. Once the hole has started, the tool guides itself. Larger holes (diameter of ¾ in or 20 mm and over) can be made by *trepanning:* the cutting tool bit is fastened on the end face of a

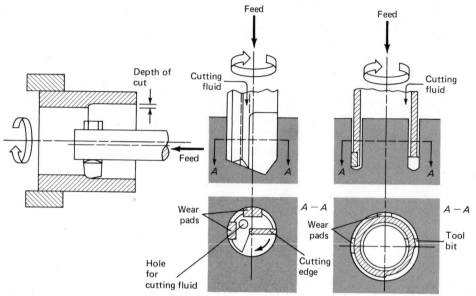

FIG. 8.17 Machining of holes with single-point tools: (*a*) enlarging (improving the surface finish) by boring, (*b*) gun-drilling, and (*c*) trepanning.

tube, and the hole is machined by removing an annulus while leaving a center core (Fig. 8.17*c*). Again, greatly improved patented tool varieties exist. Both techniques are suitable for making relatively deep holes, of a depth-to-diameter ratio of 5 and over. Force-fed cutting fluid assures removal of the chips and is vital to success.

8.5.4 Facing

A plane perpendicular to the lathe axis is produced by moving the single-point tool in the carriage so that the feed motion is toward the center of the lathe (Fig. 8.18*a*). *Parting off* accomplishes the same task but two surfaces are now simultaneously generated (Fig. 8.18*b*).

8.5.5 Forming

This method of producing complex rotational shapes (Fig. 8.11*a*) is fast and efficient, but cutting forces are high and the workpiece could suffer excessive deflection. On cantilevered workpieces the length of the *forming tool* is usually kept to 2.5 times the workpiece diameter. For longer lengths the workpiece is supported by a *backrest* or *roller support,* or, if possible, on a center.

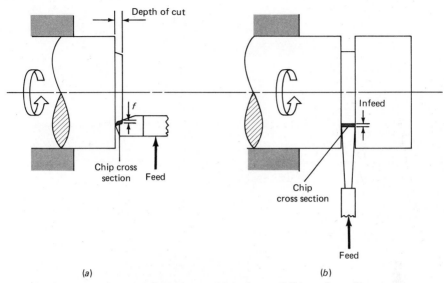

Depth of cut

f

Chip cross
section   Feed

(a)

Infeed

Chip
cross section

Feed

(b)

FIG. 8.18   Generating a flat end face by *(a)* facing and *(b)* parting off.

8.5.6   Automatic Lathe

The hand operation of a lathe requires considerable skill. The talents of a
highly skilled operator are poorly utilized in repetitive production; therefore, various efforts at automation have long been made. Unfortunately,
the terminology has become somewhat confusing. In the context used
here, an *automatic lathe* is similar to an engine lathe, but all movements of
the carriage required to generate the workpiece surface are obtained by
mechanical means.

Radial movement of the tool may be derived from a *cam bar* or a *tracer
template,* or separate drives may be actuated by NC. Alternatively, the motions may be derived from a model of the workpiece using a *copying
arrangement.*

All these machines may be supplied with material by hand, semi-
automatically or fully automatically.

8.5.7   Turret Lathe

When the surface can be generated or formed with relatively simple motions but requires a larger number of tools and operations (such as
turning, facing, boring, and drilling) for completion, the requisite number
of tools can be accommodated by replacing the tailstock of a lathe with a
turret (Fig. 8.19). Equipped with a *quick-clamp device,* a turret brings several (usually six) tools into position very rapidly. All tools are fed in the
axial direction, by moving the turret on a slide (*ram-type lathe*) or, for

FIG. 8.19 A numerically controlled turret lathe, with protective guards removed for photographic purposes. (The Warner & Swasey Co., Cleveland, Ohio.)

heavier work, on a *saddle* which itself moves on the ways (*saddle-type* lathe). Axial feed movement is terminated when a *preset stop* is reached. Four additional tools are mounted in a square turret on the cross slide and two more tools on the rear tool post. The number of possible operations and the variety of combinations is very large, because several tools may be mounted at any one station for *multiple cuts,* or *simultaneous cuts* may be performed at several stations (*combined cut*). Once the machine is set up, it requires relatively little skill to operate.

8.5.8 Automatic Screw Machines

As the name suggests, these machines were originally developed for making screws at high production rates. Cold heading followed by thread rolling has almost eliminated this market, but machines have been developed to mass-produce more complex shapes.

Single-spindle automatics

Single-spindle automatic screw machines fall into two basically different groups. The first group comprises machines based on the principle of the turret lathe, but operator action is replaced by appropriately shaped cams that bring various tools into action at preset times. The *stock* (a bar drawn to close tolerances) is *indexed* forward, with cam-operated *feed fingers,* by the length of one workpiece at the end of each machining cycle.

The other type, the so-called *Swiss automatic,* is radically different in that all tools are operated in the same plane (Fig. 8.20), extremely close to the *guide bushing* through which the rotating bar is continuously fed in a programmed mode. Individual tools are moved radially inward with the

FIG. 8.20 Tooling area of a ⅜-in (10-mm) capacity Swiss-type automatic. (American Bechler Corporation, Norwalk, Conn.)

aid of cams. Since there is no workpiece overhang, parts of any length may be produced to unsurpassed accuracies and tolerances (down to 0.0001 in or 2.5 μm).

Even though several tools may be set to cut at the same time, the total machining time on single-spindle automatics is the sum of individual or simultaneous operations required to finish the part. Productivity may be substantially increased if all operations are simultaneously performed.

Multispindle automatics

In *multispindle automatics* (Fig. 8.21) the head of the lathe is replaced by a *spindle carrier* in which four to eight driven spindles feed and rotate

FIG. 8.21   Six-spindle automatic bar machine, without tooling.   (National Acme, Cleveland, Ohio.)

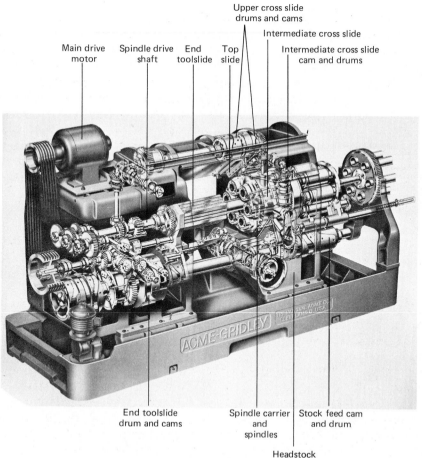

as many bars. The turret is replaced by a *tool slide* on which the appropriate number of *tool holders* (sometimes separately driven) are mounted. Additional tools are engaged radially, by means of *cross slides;* the number of these is sometimes less than the number of spindles (there may be insufficient room for them). The tool slide with the tool holders moves axially forward, and the cross slides move in under cam control, complete their assigned task, withdraw, and the spindle carrier indexes the bars to the next position. Thus, for each engagement of the tools, one part is finished.

Automatic screw machines produce mostly parts of axial symmetry (including threaded parts), but special attachments permit auxiliary operations such as milling or cross-drilling while rotation of one spindle is arrested. Workpieces of irregular shape can be handled on so-called *chucking machines.*

8.5.9 Shaping and Planing

As indicated in Fig. 8.12c, a surface can be generated with a linear primary motion. When the primary motion is imparted to the tool, and the feed to the workpiece, the process is termed shaping (Fig. 8.22a). The tool is moved back and forth by an *overhanging ram,* the deflection of which limits the length of stroke. Longer stroke (of practically unlimited length) can be obtained by having the workpiece attached to a long, horizontal, *reciprocating bed* while the tool is attached to a sturdy column or arch or, rather, a cross rail with a lead screw that generates the feed movement (Fig. 8.22b). This is called planing.

8.6 MULTIPOINT MACHINING

In *multipoint machining* at least two cutting edges of the same tool are simultaneously engaged at any one time.

8.6.1 Drilling

In Sec. 8.5.3 we have already discussed two methods of making deep holes. The vast majority of holes, however, are made by the familiar two-point tool, the *twist drill* (Fig. 8.23a). Two cutting edges are more efficient. Helical *flutes* allow access of cutting fluid, help to move the chip out, and a small *margin* left on the cylindrical surface provides some guidance. Nevertheless, the twist drill has its problems: the two cutting edges must not come together into a point, which, because of its small mass, would quickly overheat and loose its strength. A *chisel edge* is usually left, and no real cutting action then takes place in the center of the hole; the mate-

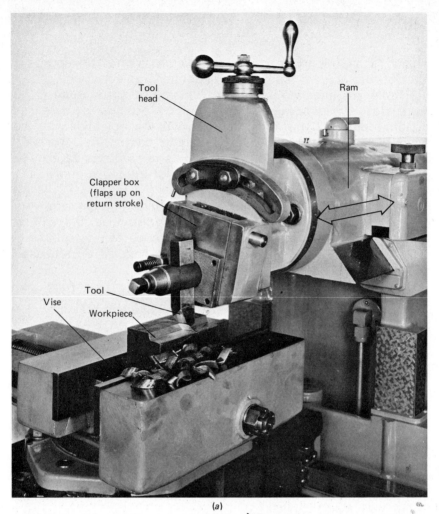

Tool head

Ram

Clapper box (flaps up on return stroke)

Tool

Vise

Workpiece

(a)

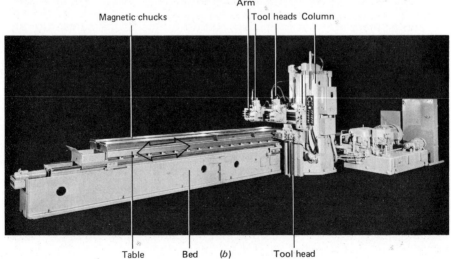

Magnetic chucks

Arm

Tool heads

Column

Table

Bed

(b)

Tool head

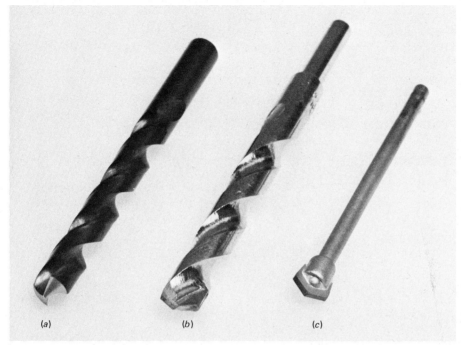

(a)   (b)   (c)

FIG. 8.23 Drills: (a) HSS twist drill, (b) carbide-inserted fluted-shank drill, and (c) carbide-inserted spade drill.

rial must be plastically displaced, to be subsequently cut by the main cutting edges. The surface finish is not as good as that of a bored hole, and the drill begins to wander at greater depths. Nevertheless, the quality is adequate for a great many purposes, in diameters ranging from 0.006 to 3 in (0.15 to 75 mm). Smaller holes, as well as larger ones, and very hard materials are amenable to *spade-drilling* (Fig. 8-23b).

The simplest *drill press* (Fig. 8.24) has a single rotating spindle which is fed axially into a workpiece held rigidly on a table. The swinging arm of a *radial arm drill* provides much greater freedom. When several holes have to be produced in a large number of workpieces, simultaneous drilling in a *multiple-spindle drillhead* or press assures better accuracy of relative hole location. Exceptional accuracy of hole location is achieved on the *jig borer,* which is really a drill press equipped with a high-accuracy table movement in two directions.

Drills can be held in the tailstock of a lathe to machine holes of good concentricity, and drills are important tools for all automatic lathe work.

FIG. 8.22 Machine tools with linear motion: (a) 24-in (600-mm) stroke ram shaper and (b) 36 in × 36 in × 20 ft (0.9 × 0.9 × 6 m) planer with magnetic chucks mounted on table. (Rockford Machine Tool Co., Rockford, Ill.)

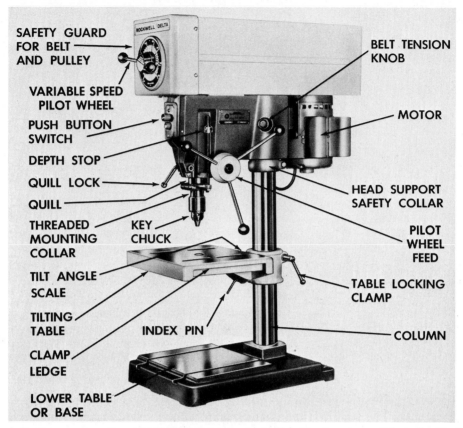

SAFETY GUARD
FOR BELT
AND PULLEY

BELT TENSION
KNOB

VARIABLE SPEED
PILOT WHEEL

MOTOR

PUSH BUTTON
SWITCH

DEPTH STOP

QUILL LOCK

HEAD SUPPORT
SAFETY COLLAR

QUILL

THREADED   KEY
MOUNTING   CHUCK
COLLAR

PILOT
WHEEL
FEED

TILT ANGLE
SCALE

TABLE LOCKING
CLAMP

TILTING
TABLE   INDEX PIN

COLUMN

CLAMP
LEDGE

LOWER TABLE
OR BASE

FIG. 8.24 A drill press of 15-in (380-mm) capacity (drill center to column distance). (Rockwell International Power Tool Division, Pittsburgh, Pa.)

The quality of drilled holes is greatly improved by *reaming,* which could be classified as a milling operation. Seats for countersunk screws are prepared by *spot facing,* essentially an end-milling operation in the plunging mode.

8.6.2 Milling

Milling is one of the most versatile cutting processes, and it is indispensable for the manufacture of parts of nonrotational symmetry. There are innumerable varieties of milling cutter geometries, but basically they can all be classified according to the orientation of their axis of rotation and the cutting edges relative to the workpiece.

Horizontal mills have the axis of the cutter parallel with the workpiece. In *plain* or *slab milling* (Fig. 8.25a) the cutting edges define the surface of a

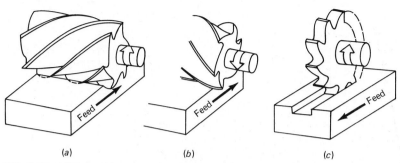

(a)   (b)   (c)

FIG. 8.25 Milling with cutter axes parallel to the workpiece surface: (a) slab (plain) milling, up and (b) down, and (c) slot milling.

cylinder and can be straight (parallel to the cylinder axis) or helical. The cylinder covers the entire width of the workpiece surface. The primary motion is provided by the rotation of the cylinder while feed is imparted to the workpiece. Because both the primary and feed motions are continuous, the chips are thickest on the surface of the workpiece and diminish towards the base of the cut. If the cutter rotates in the direction of the feed motion (*down* or *climb milling,* Fig. 8.25b), the cut begins at the surface with a high initial force. Thus, the machine must be of sturdy construction, but surface quality is better because the workpiece is pressed down on the table. When the cutter rotates against the feed direction (*up* or *conventional milling,* Fig. 8.25a), the tooth engages at a minimum depth and the starting forces are not so high, but the surface will be more wavy and the cutter tries to lift the piece off the table.

When a cylindrical cutter is narrower than the workpiece, the cutting edges must be carried over the end faces of the cylinder (Fig. 8.25c). Because of their action, these are called *slotters* or *slitting cutters.* When only the side teeth are engaged, they become *side-milling cutters.*

Other classes of cutters have their axes of rotation perpendicular to the workpiece surface (Fig. 8.26). When the teeth are at the face perpendicular to the axis, one speaks of *face milling* (8.26a). In many ways, this is similar to machining with many single-point tools moving in circles. Indeed, a variant frequently employed is *fly-cutting,* that is, cutting with a single-point tool fixed to the end of an arm protruding from the perpendicular milling shaft.

Cutting edges carried over onto the cylindrical surface of the cutter lead to the so-called *end mill.* This is one of the most versatile tools, because it can be made to describe any path in the plane of and perpendicular to the workpiece surface. Thus, pockets of any shape, depth, and size can be cut. Even enormous surfaces are sometimes fully machined, as, for example, in making wing skins for aircraft.

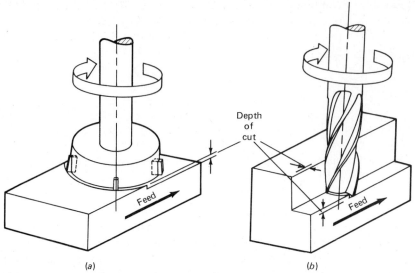

(a)   (b)

FIG. 8.26   Milling with cutter axes perpendicular to the workpiece surface: (a) face milling and (b) end milling.

As might be expected, the first two types of cutter are used primarily on *horizontal milling machines,* with the tool arbor in a horizontal position, supported at both of its ends.   The third and fourth types are cantilevered in *vertical milling machines.*

The various feed motions of a milling machine may be controlled by hand although complex shapes require considerable skill.   For production quantities, the milling machine can be automated to various degrees. *Copy millers* use a model of the finished part to transfer the movement from a *copying head* to the *milling head.*   The fastest development has been in NC machines (Sec. 10.2.1), which move some of the skill into the program-writing stage.   If properly utilized, they speed up production by eliminating much of the setup time and the trial-and-error procedures inevitable with manual control.

8.6.3   Sawing and Filing

A very narrow slitting cutter becomes a *cold saw.*   The teeth need not be deep in the radial direction, and they can be made as *inserts* on a much larger *saw blade.*   For less-demanding applications the very accurate formation of various tool angles may be relaxed, and the *teeth* can be formed by bending them approximately into position.   The basic cutting action of such *circular saws* is still closely related to milling.

When the teeth are laid out into a straight line, one obtains a *hacksaw* or, if the saw blade is flexible and made into an infinite loop, a *band saw.*

A fine-pitch slab mill laid out into a flat becomes a *file. Crosscutting* breaks the individual cutting edges up into a series of teeth.

8.6.4 Broaching

Broaching differs from the processes discussed thus far in that the only motion is the primary, linear motion of the tool; the feed is obtained by placing the teeth progressively deeper, thus each tooth takes off a successive layer of the material (Fig. 8.27). Most of the material is removed by the roughing teeth, which are followed by a number of finishing teeth designed to give the best possible surface finish.

Because there is no feed motion, the shape of the broach determines the shape of the part (pure forming). A separate broach has to be made up for each shape and size; therefore, broaching is primarily a method of mass production. The workpiece must be rigidly held and the broach firmly guided. Rigidity of the machine tool is particularly important when a *flat* surface is broached, since the broach would be lifted out of the workpiece by the cutting forces. *Internal broaching* and *external* (*pot*) *broaching* are, to some extent, self-guiding. The machine tool resembles a hydraulic press of long stroke.

FIG. 8.27 A broach used for finishing the internal profile of hammers for air-powered impact wrenches. (Apex Broach & Machine Co., Detroit, Michigan.)

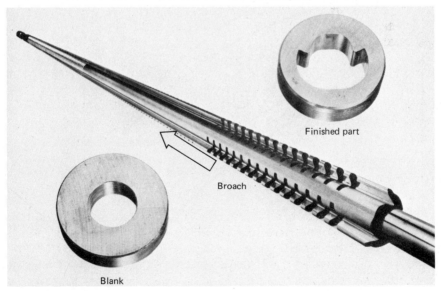

Finished part

Broach

Blank

8.7 CUTTING CONDITIONS AND ECONOMY

Machining, while often indispensable, is potentially very expensive. It is, therefore, of utmost importance that *optimum conditions,* resulting in *minimum cost,* should be chosen.

8.7.1 Cutting Speeds and Feeds

The discussions of the system (Sec. 8.1.3) and of tool materials (Sec. 8.3.1) have shown that temperature is an absolute limiting factor. This translates into rate of work done, which is a function of speed and cutting force. It is found, therefore, that for any one class of tool and workpiece material, the permissible maximum speed drops with increasing workpiece material hardness.

FIG. 8.28 Typical speeds and feeds for roughing ferrous materials with a 0.150 in (3.8 mm) depth of cut. Increase speed by 20 percent for throwaway WC; reduce speed by 20–30 percent for austenitic stainless steels and for tool steels over 1 percent carbon.

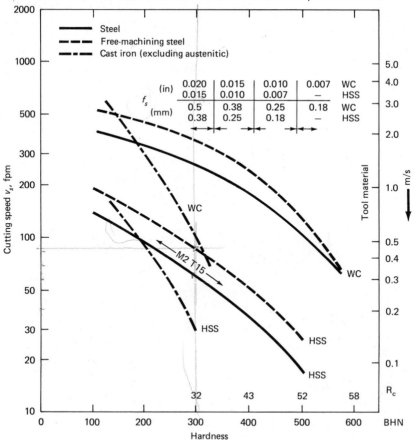

The data summarized in Figs. 8.28 and 8.29 provide a conservative starting point for *rough turning* with a 0.150 in (4 mm) depth of cut (indicated in Figs. 8.12a and 8.18a) and a typical tool life of 1 to 2 hours for high-speed steel (HSS) or brazed carbide (WC) tools. If it is found that the actual tool life is much longer, the operation may be speeded up, first of all by optimizing feed (which determines the undeformed chip thickness), remembering that heavier cuts are more efficient [Eq. (8.6)]. Higher cutting speeds generate more heat and should be used only when tool life is still excessive.

While these figures are valid primarily for rough turning and boring, they can be also used as a guide for *finish turning* and other processes.

FIG. 8.29 Typical speeds and feeds for roughing nonferrous materials with a 0.150 in (3.8 mm) depth of cut (FM means free machining).

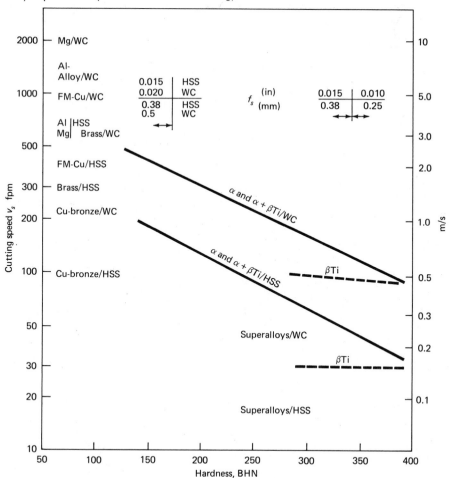

The speed v and feed f for any particular process are found by multiplying v_s and f_s (from Figs. 8.28 and 8.29) by the factors Z_v and Z_f, respectively (Table 8.4).

For drilling (HSS twist drill), take $v \simeq 0.7 v_s$ for ferrous and $v \simeq 0.5 v_s$ for nonferrous materials. Feed is typically a function of drill diameter D and is $0.02D$ per revolution for free-machining materials, $0.01D$ per revolution for tougher or harder materials, and $0.005D$ per revolution for very hard ($R_C > 45$) materials.

For broaching (HSS), speeds range from 40 fpm (0.2 m/s) on free-machining material to 5 fpm (0.025 m/s) on hard material, while the undeformed chip thickness reduces from 0.005 to 0.002 in (0.12 to 0.05 mm) per tooth.

Ceramic tools can be tried first at speeds recommended for WC in finishing cuts, and then raised to higher speeds (up to 3 times higher).

Diamond tools are suitable only for finishing cuts of 0.002 to 0.008 in (0.05 to 0.2 mm) depth at 0.0008 to 0.002 in (0.02 to 0.05 mm) feed and speeds of 800 to 2800 fpm (4 to 14 m/s).

The data given here are sufficiently accurate for initial planning purposes. Detailed recommendations on initial cutting conditions are given in the *Machining Data Handbook.* Optimum conditions depend, however, on many factors, including the rigidity of the tooling, workpiece holder, and machine tool. It is not uncommon for practical metal removal rates to reach twice the recommended values.

8.7.2 Choice of Machine Tool

With this information in hand, the machining process can be planned, and the appropriate equipment chosen. The following steps should be followed:

1 Considering the material depth to be removed and the surface finish required, choose a depth of cut and find the appropriate cutting speed v and feed f, together with the class of tool (HSS or WC) and the cutting fluid (from Table 8.1).

2 Calculate the volume V to be removed in the cut; e.g., for turning,

$$V = \frac{(d_0^2 - d_1^2)\pi l}{4} \tag{8.9}$$

where d_0 is the diameter before turning, d_1 is the diameter after turning, and l is the length of cut.

3 Calculate the chip removal rate V_c ($=$ cutting speed $\times$ chip cross-sectional area); e.g., in turning,

TABLE 8.4 SPEEDS AND FEEDS IN VARIOUS METAL-CUTTING OPERATIONS*

Process	Z_v (speed† $v = v_s Z_v$)	Depth of cut		Z_f (feed† $f = f_s Z_f$)	Other
		in	mm		
Rough turning	1	0.15	4	1	
Finish turning	1.2–1.3	0.025	0.65	0.5	
Form tools, cutoff	0.7				Infeed 0.1f–0.2f
Shaping	0.7	0.15	4		Feed: 0.060–0.020 in HSS, 0.080–0.050 in WC (× 2 on Cu, Al and Mg)
Planing	0.7	0.15	4		
Face milling	1	0.15	4	0.8–1‡	
Slab milling	1	0.15	4	0.5‡	
Side and slot milling	0.5–0.7	0.15	4	0.5‡	
End mill, peripheral	1	0.05	1.2	0.5–0.25‡	for 1-in-dia. cutter
End mill, slotting	1	0.05	1.2	0.2‡	
Threading, tapping	0.5–0.25				slower for coarser thread

* Approximate values, compiled from *Machining Data Handbook*, 2d ed., Machinability Data Center, Metcut Research Associates, Cincinnati, Ohio, 1972.

† Take v_s and f_s from Fig. 8.28 or 8.29.

‡ Feed per tooth

$$V_c = \frac{vf(d_0 - d_1)}{2} \tag{8.10}$$

4 From these, the net machining (cutting) time is

$$t_c = \frac{V}{V_c} \tag{8.11}$$

5 After finding the specific energy consumption from Eq. (8.6), obtain the power of the machine tool from Eq. (8.7).

The variety of commercially available equipment is immense, in terms of the types of operations and sizes. Table 8.5 gives but a general review of machine tools commonly manufactured; special-purpose machines, some enormous, others of minute size, are in operation. Thus, a lathe (made by Farrell Co.) for turning steam turbine rotors has a swing of 75 in (1.9 m) and a bed 46 ft (14 m) long, and a six-gantry NC milling machine (made by Cincinnati Milacron Inc.) for sculpturing airplane parts has a table 160 in (4 m) wide and 360 ft (110 m) long.

It may turn out, of course, that no machine tool of sufficient power is available, and then the speed or feed, or both, must be cut.

8.7.3 Cutting Economy

The broader issues of selecting the most economic process for producing a given part will be discussed in Chap. 11. Machining is unique, however, in that even if the process types and sequence are determined, it is still necessary to select the most economical process conditions. This aspect of economy will be discussed here.

The choice of the tool material is dictated by a number of considerations (Sec. 8.3), not the least of which is total cost (not only of the tool itself but also of the entire process). Therefore it may be necessary to perform an economic analysis for all technically attractive tool materials and then choose the one that assures lowest overall cost; the lubricant is then defined by the tool material and cutting speed (Table 8.1).

The total time for dealing with one part (the *floor-to-floor time*), t_f, is composed of the *loading and unloading time* t_1 and the actual *cutting time* t_c. The first is constant, while the second is a function of cutting speed, Eq. (8.11). Assuming that the operator's pay, including overhead, is R_0 per unit time, and the machine is charged (on the basis of considerations discussed in Chap. 11) at the rate of R_m per unit time, the cost associated with time t_f is

$$C_f = (R_0 + R_m)(t_1 + t_c) \tag{8.12}$$

TABLE 8.5 CHARACTERISTICS OF CUTTING EQUIPMENT*

Machine tool	Workpiece, max. dimensions[†]			Main motion, max.[‡]			Other
	Width, m	Diameter, m	Length, m	Speed, m/s	RPM	Drive kW	
Lathe, center		0.1–2	0.3–5		3000–100	1–70	
Turret (bar)		0.02–0.3	0.1–1.5		3000–300	1–60	
Automatic (single-spindle)		0.01–0.15	0.05–0.3		9000–500	1–40	
Automatic (multispindle)		0.01–0.15	0.1–0.3		4500–300	5–50	
Automatic (screw)		0.01–0.1	0.03–0.3		10,000–1500	2–20	
Automatic (Swiss)		0.005–0.03	0.05–0.3				
Boring machine, horizontal		0.5–1.5	0.4–2		1000–150	2–70	
Vertical		1–6	0.7–2.5		300–30	20–200	
Shaper	0.2–0.8		0.15–1	0.4–1		1–7	
Planer	0.6–2.7		1–10	0.5–1.7		10–100	
Drills					12,000–400	<1–10	Drill dia.: 0.3–100 mm
Milling machine	0.1–0.4		0.5–2.5		4000–1000	1–20	
Broaching machine			8–24>	0.2–0.02		1–40	Stroke: 0.5–2 m and up
Grinder, surface	0.1–0.9		0.2–6			<1–30	
Cylindrical		0.02–0.8	0.2–6			<1–20	
Centerless		0.01–0.3				<1–30	

* Commercially available equipment, selected sizes from *Machine Tool Specification Manual*, Maclean-Hunter, London, 1963.

† The range indicates the maximum dimensions taken by equipment of various sizes; smaller workpieces are usually accommodated.

‡ Speeds indicated are maximum for equipment of different sizes; variable speed is usually provided down to 1/50 or 1/200 of maximum.

The production cost must include an allowance for the tool cost.   If one tool costs C_t, takes time t_{ch} to change, and cuts N_t pieces before it has to be removed, the total cost per piece C_{tp} is a fraction $1/N_t = n_t$ of the total cost

$$C_{tp} = n_t [t_{ch} (R_0 + R_m) + C_t] \qquad (8.13)$$

The total *production cost* is the sum of Eqs. (8.12) and (8.13)

$$C_{pr} = C_f + C_{tp} = (R_0 + R_m)(t_1 + t_c) + n_t[t_{ch}(R_0 + R_m) + C_t] \qquad (8.14)$$

From Eqs. (8.10) and (8.11), the cutting time is

$$t_c = \frac{V}{V_c} = \frac{2V}{vf(d_0 - d_1)} = \frac{B}{v} \qquad (8.15)$$

where $B = 2V/f(d_0 - d_1)$ is constant (and is the length of cut).

The number of pieces N_t cut with one tool is a function of tool life t from Taylor's equation

$$vt^n = C \qquad (8.5)$$

Since by convention C is taken in units of feet per minute for 1 min tool life (the reference life t_r), the equation should really be written

$$vt^n = Ct_r{}^n \qquad (8.16)$$

The number of pieces cut

$$N_t = \frac{t}{t_c} = \frac{t_r}{t_c}\left(\frac{C}{v}\right)^{1/n} \qquad (8.17)$$

Equation (8.17) combined with Eq. (8.15) becomes

$$\frac{1}{N_t} = n_t = \frac{B}{vt_r\left(\dfrac{C}{v}\right)^{1/n}} = \frac{B}{t_r C^{1/n}} v^{(1 - n)/n} \qquad (8.18)$$

With this and Eq. (8.15), Eq. (8.14) can be rewritten

$$C_{pr} = (R_0 + R_m)\left(t_1 + \frac{B}{v}\right) + \frac{B}{t_r C^{1/n}} [t_{ch}(R_0 + R_m) + C_t] v^{(1 - n)/n} \qquad (8.19)$$

The production cost C_{pr} is minimum where the first derivative with respect to v is zero:

$$\frac{dC_{pr}}{dv} = 0 = -\frac{(R_0 + R_m)B}{v^2} + \frac{1-n}{n}\frac{B}{t_r C^{1/n}} t_{ch}(R_0 + R_m) + C_t v^{(1 - 2n)/n}$$

Rearranging, we get the *cutting speed for minimum cost*

$$v_c = C \left[\frac{n}{1-n} \left(\frac{t_r(R_0 + R_m)}{t_{ch}(R_0 + R_m) + C_t} \right) \right]^n \tag{8.20}$$

Thus, the most economical speed increases with increasing C and n (as one might expect) and also with R_0 and R_m, i.e., increasing labor and equipment costs. Conversely, there is not much point in increasing speeds if much time t_{ch} is wasted on tool changing or, as evident from Eq. (8.19) and Fig. 8.30, if the nonproductive time t_1 of loading and unloading is a large fraction of the total time. All too often, this point is overlooked and the effort spent in increasing metal removal rates could be better spent on improving material movement and providing workpiece-locating devices (jigs and fixtures).

8.8   RANDOM-POINT MACHINING

The term *abrasive machining* usually describes processes in which the individual cutting edges are not only *randomly distributed* but also more or

FIG. 8.30   Optimum cutting speed for lowest-cost production.   (See Example 8.1.)

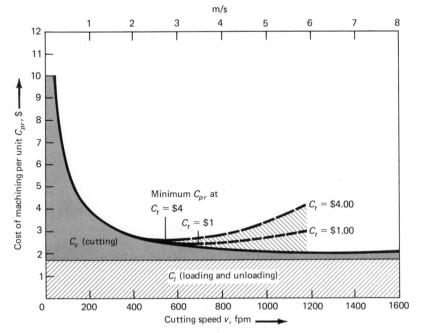

less *randomly oriented.* This changes the character of the cutting process substantially.

8.8.1 Abrasive Machining

The process is most readily understood with reference to *grinding,* in which *abrasive grit* (hard particles with sharp edges) is bonded into a *wheel.* The orientation of individual grains is mostly a matter of chance; therefore a grain may encounter the workpiece surface with a positive, zero, or, most likely, negative rake angle (Fig. 8.31*a*), with consequences depending on the depth of engagement. A slight engagement is accommodated by purely elastic deformation of the workpiece, grit, and binder, and only frictional heat is generated. Deeper penetration causes the grit to plough through the workpiece surface, pushing material to the side, but also lifting out some material, depending on the effective rake angle and the properties of the workpiece material. Much work is expended in a small space; therefore some material may even melt. The proportion of encounters resulting in chip formation decreases as the grit looses its sharp edges. This natural wear must be compensated either by allowing the worn grit to be dislodged, or by inducing fracture of the grit, thus exposing fresh edges. Either way, there is weight loss, and the efficiency of grinding can be expressed by the *grinding ratio* (the volume of material removed divided by the volume of wheel lost); it ranges from 5–1000.

Poorly controlled grinding gives rise to harmful *residual stress distribution* (Fig. 8.31*b*) and, in heat-treatable steels, to martensite formation.

8.8.2 Abrasives

To perform its function, the abrasive grit obviously must possess not only a high hardness but also a favorable, many-faceted shape with sharp edges.

FIG. 8.31 The process of grinding: (*a*) schematically and (*b*) the residual stress distribution observed under unfavorable conditions.

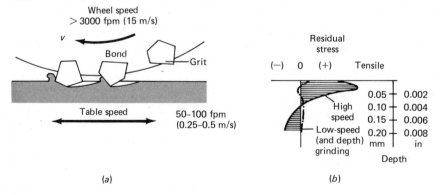

The grain size is important in relation to the depth of engagement and the surface finish to be produced.

Naturally occurring abrasives (various forms of SiO_2 and Al_2O_3) find only limited use, mostly on softer materials or in finishing processes. The vast majority of industrial abrasives are manufactured. *Silicon carbide* (SiC) and *alumina* (Al_2O_3) are supplemented with natural or manufactured diamond for machining the hardest materials, including other abrasives. Manufacturing is controlled to give abrasives the most desirable crystalline structure, grain shape, and size distribution.

8.8.3   Bonded Abrasives

In the majority of abrasive operations, the abrasive is made into a tool of some definite geometry.

Grinding

Grinding wheels are always of axial symmetry, and are carefully balanced for rotation at high speeds. The *bonding agent* is chosen to give greatest efficiency yet permit the worn grit to fall out.

The majority of grinding wheels are bonded with a glass. Wheels with such *vitrified bonds* are the strongest and hardest, and the composition of the glass can be adjusted over a wide range of strengths. The so-called *silicate wheel,* bonded with water glass, is the softest. In general, the harder the material to be ground, the softer the bond chosen.

Organic bonding agents are of lower strength but possess a wide range of properties. *Resinoid wheels* are bonded with thermosetting resins and can be readily reinforced with steel rings, or fiber glass or other fibers, to increase their flexural strength. More flexible polymers such as *shellac* or *rubber* can be made into very thin *cutoff wheels.*

The size and size distribution of the grit and the openness of the structure all contribute to determining grinding performance; therefore standard grinding wheel designations refer to all these factors (Fig. 8.32). *Operating (surface) speeds* are usually between 4000 and 6000 fpm (20–30 m/s), although a dramatic drop in wheel wear is found at 2 to 4 times these speeds. Such speeds require specially constructed wheels.

An indispensable part of the grinding system is the *grinding fluid* (Table 8.1), which keeps the surface cool and prevents burning and cracking of the surfaces of hard materials. It also affects the cutting process itself as well as the wear of the grinding wheel (and thus the grinding ratio). The vital importance of grinding fluids is shown by the fact that metal removal rates can be increased two- or three-fold when the liquid is pumped under high pressure into the grinding zone.

The geometry of grinding can be as varied as that of other machining

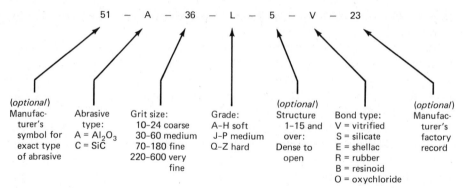

51 — A — 36 — L — 5 — V — 23

| (*optional*) Manufacturer's symbol for exact type of abrasive | Abrasive type: A = Al_2O_3 C = SiC | Grit size: 10-24 coarse 30-60 medium 70-180 fine 220-600 very fine | Grade: A-H soft J-P medium Q-Z hard | (*optional*) Structure 1-15 and over: Dense to open | Bond type: V = vitrified S = silicate E = shellac R = rubber B = resinoid O = oxychloride | (*optional*) Manufacturer's factory record |

FIG. 8.32 Grinding wheel nomenclature. (From ANSI B74.13–1970, American Society of Mechanical Engineers, New York.)

processes (Fig. 8.33). *Surface grinding* may be practiced with the cylindrical surface (Fig. 8.33*a*) of a wheel (as in slab milling), but the wheel is usually narrower than the workpiece and is *cross-fed* (it *traverses* the width). Alternatively, the entire width may be ground with the annular end face of a *cup-shaped wheel* (Fig. 8.33*b*); this resembles face milling. *Cylindrical grinding* (Fig. 8.33*c*) is similar in its results to turning, except that the fast-rotating grinding wheel now works on the surface of the more slowly rotating part, and individual cuts are short. *Internal grinding* (Fig. 33*d*) is the same process, but with a small wheel working on the cavity of the workpiece. Cylindrical parts of great accuracy are obtained at high

FIG. 8.33 Various grinding processes: (*a*) surface grinding, horizontal spindle; (*b*) surface grinding, vertical spindle; (*c*) cylindrical grinding; (*d*) internal grinding; (*e*) centerless grinding; and (*f*) form (plunge) grinding.

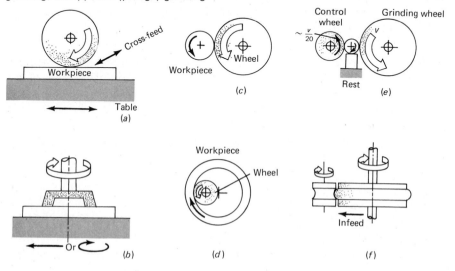

rates in *centerless grinding* (Fig. 8.33*e*): the workpiece is lightly supported on a *work rest* while the grinding pressure is taken up by the *regulating wheel* that rotates at about $1/20$ the grinding wheel speed. Apart from generating basic geometric surfaces, grinding can be used for finishing many parts of complex shape, including threads and gears. As in other machining, both forming (Fig. 8.33*f*) and generation of surfaces is practiced (Figs. 8.11 to 8.13).

Even the best-chosen grinding wheel becomes gradually dull, or clogged (*loaded*) with the workpiece material, and *dressing* of the wheel becomes necessary. This is usually accomplished in the grinder itself, so that alignment and wheel balance is not lost. The wheel is dressed by cutting with a diamond tip. Vitrified wheels may also be dressed by pressing a high-strength steel roller against the surface; such *crush-dressing* is very fast and particularly economical for forming wheels.

Coated abrasives

The abrasive grit may be bonded to a paper, fabric, or metal surface with a water-soluble or water-resistant adhesive. The grit is spaced out in a controlled manner and is often deposited electrostatically to align its sharp edges perpendicular to the backing surface.

While grinding and abrasive processes in general have traditionally been used for finishing, with only 0.0005–0.005 in (0.01–0.1 mm) material removed per pass, they are becoming important machining processes in their own right. Grinding wheels and, particularly, coated abrasives with proper support can be used to remove up to 0.25 in (6 mm) of material at high rates, replacing turning, planing, or milling in mass production—e.g., in machining the gasket surfaces of engine blocks and cylinder heads. Rough grinding (*snagging*) of castings and forgings, to remove ingates or flash, has long been practiced.

An approximate estimate of the required *grinding power* may be obtained from Table 8.6 for a 1 mm depth of grind. Smaller depths require higher specific energies, approximately according to Eq. (8.6).

Honing

Just as the keenest edge on a knife is obtained by finishing it with a stone, so *honing* is the finishing process that gives the smoothest finish and often the greatest accuracy. The abrasive is made into a slab (*stone, stick*), and is moved in a rapid oscillating motion over the surface (usually a hole) to be finished (Fig. 8.34*a*). A *honing fluid* is applied to wash out abraded particles. *Superfinishing* is a variant in which oscillating motion is imparted to the fairly large stone, and the surface pressure is kept very low. Thus, as the surface becomes flatter, it builds up its own hydro-

TABLE 8.6 APPROXIMATE SPECIFIC ENERGY REQUIREMENTS FOR SURFACE GRINDING*

Depth of Grinding: 1 mm (0.040 in)

Material	Hardness, BHN	Specific energy, E_1	
		hp·min/in³	W·s/mm³
1020 steel	110	4.6	13
Cast iron	215	3.9	11
Titanium alloy	300	4.7	13
Superalloy	340	5.0	14
Tool steel (T15)	67 R_c	5.5	15
Aluminum	150 (500 kg)	2.2	6

* Extracted from *Machining Data Handbook,* 2d ed., Machinability Data Center, Metcut Research Associates, Cincinnati, Ohio, 1972.

dynamic lubricant film which terminates the action of the abrasive (Fig. 8.34*b*).

8.8.4 Machining with Loose Abrasives

Many finishing processes apply a loose powder or slurry of the abrasive to the interface between the workpiece and a tool. Thus the process becomes a case of three-body wear.

Lapping

In its best-known form, a relatively soft (e.g., cast iron) table is rotated in a horizontal plane; the workpieces are loaded on the surface (some-

FIG. 8.34 Schematic illustration of random-motion abrasive machining with bonded abrasives: (*a*) honing and (*b*) superfinishing.

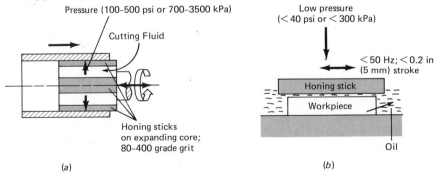

Pressure (100–500 psi or 700–3500 kPa)

Cutting Fluid

Honing sticks on expanding core; 80–400 grade grit

(*a*)

Low pressure (< 40 psi or < 300 kPa)

< 50 Hz; < 0.2 in (5 mm) stroke

Honing stick

Workpiece

Oil

(*b*)

times in cages driven by a gear from the center of the rotating table), and a *slurry* of the abrasive in oil is applied to the surface (Fig. 8.35*a*). The workpiece describes a planetary movement and acquires a very uniformly machined, random finish of excellent flatness. When the soft surface (the *lap*) is made into a three-dimensional shape, curved surfaces (e.g., glass lenses) may be lapped. It is also a very fast process for breaking in mating gears or worms, thus eliminating the need for a running-in period (e.g., for the hypoid gears of an automotive rear axle).

Less frequently, the lap is made of a bonded abrasive and the process is then really similar to finish grinding at a slow speed.

Ultrasonic machining

A rather specialized process, *ultrasonic machining* utilizes the small-amplitude (about 0.04 mm), high-frequency (about 20,000 Hz) vibrations produced by an ultrasonic wave generator to drive the *form tool* made of a not-too-hard metal. Abrasive grit is supplied in a slurry to the interface, and the workpiece surface is gradually eroded (Fig. 8.35*b*).

Buffing, polishing, and burnishing

In most such processes, the abrasive is applied to a *soft surface,* for example, the cylindrical surface of a wheel composed of felt or other fabric (Fig. 8.35*c*). For *buffing,* the abrasive is in a semisoft binder, while the abrasive may be used dry or in oil, water, or other carrier/lubricant for *polishing* purposes. Both buffing and polishing are capable of producing surfaces of high reflectivity which, however, should be attributed not to greater smoothness but to a smearing of surface layers. Burnishing (Sec. 4.8.6) accomplishes the same result faster, especially in holes, and when

FIG. 8.35 Machining with loose (unbonded) abrasives: (*a*) lapping, (*b*) ultrasonic machining, and (*c*) buffing.

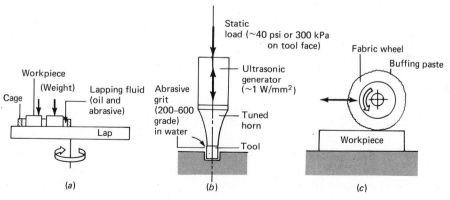

the burnishing roll penetrates to some depth, also produces a more favorable residual stress pattern. Therefore it is preferred for applications where fatigue resistance is important (e.g., the journal radii on crankshafts or the finish rolling of gears).

Barrel finishing

A completely random process, *barrel finishing* or *tumbling* is of great value in removing burrs and fins from workpieces and, generally, in improving their surface appearance. In principle, the workpieces are placed into a barrel (often with a many-sided cross section to prevent bunching up of workpieces). When the drum is rotated, mutual impact removes surface protuberances. Much-improved finish is obtained when a *tumbling medium* is added, either in a liquid carrier (*wet*) or by itself (*dry*). The medium is chosen according to the intended purpose and may range from metallic or nonmetallic balls to chips, stones, and conventional abrasives.

Grit blasting

All contact with the tool or other workpiece is eliminated in *grit-* or *shot-blasting*. The abrasives or particles (*shot*) of some other material are hurled at the workpiece surface at high enough velocities to remove surface films, such as oxides, and impart to the surface a uniformly mat, indented appearance. Excessive impact velocities could cause damage, but properly controlled shot-blasting promotes slight plastic surface deformation and residual compressive stresses which, as noted in Sec. 4.3.3, improve fatigue resistance. The required velocities may be produced by compressed air or a wheel rotating at very high speed, with or without paddles. Contact with the wheel accelerates the grit particles.

A tightly controlled *jet* of dry abrasives is sometimes used to cut slots or holes into very hard materials.

A special form of grit blasting is *hydrohoning* in which the abrasive medium is suspended in a liquid, which is then directed onto the surface in the form of a high-pressure jet.

8.9 ELECTRICAL AND CHEMICAL MACHINING

All machining processes discussed up to this point have in common the formation of chips, even if this is sometimes only imperfectly accomplished. There are a number of processes, some of them not very new, that remove material purely by electrical or chemical action, or both. Most of these processes have gained wider application primarily because of the impetus provided by the aerospace industry; therefore they have often been denoted as *nonconventional, nontraditional,* or alternative processes. Chipless machining is sometimes used, but is an ambiguous term which, rather as in German, can also refer to plastic deformation processes.

8.9.1 Chemical Machining (CM or CHM)

It has been known for many years that most metals (and also some ceramics) are attacked by specific chemicals, typically, acids or alkalies. The metal is dissolved atom by atom and converted into a soluble compound over the entire exposed surface. If parts of the surface are protected by a tape or by a paint or wax layer resistant to the chemical, various shapes can be *etched.* The principle has been exploited for hundreds of years by the artist and the printer for *engraving;* more recently, industry has made use of it to thin down workpieces, remove pockets of material (*chemical milling* of wing skins and other aircraft components), or cut through thin sheet (*chemical blanking* of printed circuit boards and other parts). The etchant dissolves material in all directions; therefore, it undercuts to approximately the same width as the depth of cut (Fig. 8.36a).

 Metal removal rates are given in Table 8.7.

8.9.2 Electrochemical Machining (ECM)

The rate of material dissolution is greatly increased by the application of direct electric current. The process is the reverse application of electroforming (Sec. 7.9.1). The workpiece, which must be conductive, is made the anode, while the cathode is either a distantly mounted flat plate (*electrochemical milling*) or a negative of the shape to be produced (ECM). In the latter process, which is much more widespread, the cathode is fed into the workpiece at a controlled rate (Fig. 8.36b). The electrolyte is circu-

FIG. 8.36 Chemical and electrical machining processes: (a) chemical, (b) electrochemical, and (c) electrodischarge machining.

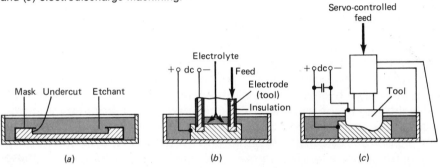

TABLE 8.7   CHEMICAL AND ELECTRICAL MACHINING PROCESSES*

	Chemical machining	Electrochemical machining	Electrodischarge machining
Metal removal rate	0.0005–0.003 in/min (0.012–0.07 mm/min)	0.1 in³/1000 A·min (1.5 cm³/1000 A·min)	0.01–25 in³/h (0.15–400 cm³/h)
Surface finish			
μin AA	90 on Al 60 on steel 25 on Ti	4–50	30 at 0.015 in³/h 200 at 0.5 in³/h 400 at 3.0 in³/h
μm AA	2 on Al 1.5 on steel 0.6 on Ti	0.1–1.4	0.75 at 0.25 cm³/h 5 at 8 cm³/h 10 at 50 cm³/h
Electric current:			
Volts		4–24	< 300
Amperes		50–40,000	0.1–500
Frequency		dc	500,000–200 Hz

* Extracted from *Machining Data Handbook,* 2d ed., Machinability Data Center, Metcut Research Associates, Cincinnati, Ohio, 1972.

lated, often through the cathode, and the machine tool is rigidly constructed to prevent vibration and consequent inaccuracies. Rough guidelines on process variables are given in Table 8.7.

A variant of the process, *electrochemical grinding,* uses a conductive grinding wheel (copper-bonded Al_2O_3 or metal-bonded diamond) as the cathode.

Since metal removal occurs in the ionic state, the hardness of the material is of no consequence in either CHM or ECM.

On some metals an insulating oxide film may build up during electrochemical machining. This can be broken down by intermittent spark discharges produced by an ac or pulsed dc circuit.

8.9.3   Electrodischarge Machining (EDM)

In this process, chemical action is abandoned, and metal is removed (Fig. 8.36c) by the intense heat of *electric sparks.* The workpiece and the cathode (tool), made of metal or graphite, are submerged in a *dielectric fluid,* commonly a hydrocarbon oil (mineral oil). A direct current at up to 300 V is applied to the system, which includes a capacitor in parallel with the spark gap. At low voltages the fluid acts as an insulator; as the voltage in the capacitor builds up, the fluid suffers dielectric breakdown (large numbers of electrons appear in the conduction band) and a spark passes through the gap, vaporizing some of both the workpiece and tool materials. Deionization of the dielectric reestablishes the insulating film, the current flow drops to zero and the capacitor can begin charging again. The cycle is repeated at a rate of 200–500,000 Hz. Metal removal rate is a function of *current density* (Table 8.7); higher current densities create rougher surfaces and also extend deeper the *heat-affected* (damage) *zone* of typically 0.0001–0.005 in (2–120 μm) depth. Therefore, it is customary to end the cut at a low current density. Material removal is not limited to the workpiece; the tool is eroded too. Under optimum conditions, the *wear ratio* (workpiece volume to tool volume removed) is 3:1 with metallic tools and from 3:1 to 100:1 with graphite electrodes (roughing is sometimes done with a *no-wear* EDM process in which the polarity is reversed and the graphite anode suffers no weight loss).

The process is insensitive to the hardness of the material; therefore it has found wide application for *sinking cavities* into hardened steel die blocks, thus avoiding distortion that would be inevitable if the die were milled and then heat treated. The copying milling machines take over the task of machining the electrodes.

EDM (together with ECM) is capable of dealing with seemingly impossible tasks. For example, cooling holes of small diameter (under 1 mm) and very great length can be cut into turbine blades made of superalloys, using tungsten wire as the electrode.

8.9.4   High-Energy Beam Machining

Various materials can be machined—mostly cut or drilled—by melting and vaporizing the substance in a controlled manner. This can be achieved with an *electron beam* (in vacuum) or with a *laser beam* (in air). The latter is particularly useful for materials that are otherwise difficult to machine, for example, plastics and ceramics.

8.10   MACHINING OF POLYMERS AND CERAMICS

While our discussion has centered on metals, the principles discussed can be applied, with appropriate modifications, to plastics and ceramics too.

8.10.1   Machining of Plastics

Even though plastics do not possess a crystalline structure of the kind that one finds in metals, they can be separated by shear stresses just as metals are. Therefore chip-forming processes can be applied, if allowance is made for the property differences.

Compared with metals, plastics have a low elastic modulus and deflect easily under the cutting forces; therefore they must be carefully supported.

Because of viscoelastic behavior, some of the local elastic deformation induced by the cutting edge is regained when the load is removed. This *elastic recovery* calls for large relief angles, and for a tool setting closer than the finished size of the part.

In general, plastics have low thermal conductivity; therefore the heat buildup in the cutting zone is not distributed over the body and the cut surface may *overheat*. In a thermoplastic resin, the glass-transition temperature T_g may be reached and the surface smeared or damaged, while thermal breakdown and cracking may occur in thermosetting resins. Therefore, friction must be reduced by polishing and honing the active tool faces and by applying a blast of air or a liquid coolant (preferably water-based, unless the plastic is attacked by it). Since the shear zone is shortened and the cutting energy is reduced with a large rake angle (Sec. 8.1.1), cutting tools are made with as large a rake angle (Table 8.3) as possible without changing the cutting mechanism into cleaving, in which coarse, disjointed fragments are lifted up and a very poor surface is produced.

Plastics can be surprisingly difficult to machine when reinforced by fillers. Glass fibers are particularly hard on the tool and it is not uncommon that only WC or diamond tools can stand up. Noncontacting techniques such as laser-beam drilling or cutting offer promise, but have not found extensive application yet.

Twist drills should have wide, polished flutes, a low ($< 30°$) helix angle and a 60–90° point angle, particularly for the softer plastics.

In general, molding and forming methods (Sec. 6.1) produce an acceptable surface and tolerance, and design usually aims at avoiding subsequent machining. Occasionally, however, machining is a viable alternative to molding (e.g., for PTFE, which is a sintered product and not moldable by the usual techniques).

8.10.2 Machining of Ceramics

Most ceramics are hard and act as abrasives themselves. Therefore their machining is very often limited to abrasion by a yet harder ceramic. Thus diamond can be used to dress grinding wheels or to finish tool bits or ceramic (e.g., Al_2O_3) components. All abrasive processes, including grinding, lapping, polishing, ultrasonic machining, gritblasting, and hydrohoning, are employed for both overall finishing and localized shaping of ceramic (including glass) parts. Ceramics that are susceptible to chemical attack (as glass is to HF acid) can be chemically machined (etched).

Metal-matrix composites, such as WC tool bits, can be conventionally ground with diamond, or the electrical conductivity of the matrix can be exploited by electrodischarge or electrochemical machining and grinding.

8.11 PROCESS LIMITATIONS AND DESIGN ASPECTS

Apart from shape limitations (to be discussed in Sec. 11.2.1), there are also dimensional limitations arising from the elastic deflection of thin shells and slender shafts under the forces imposed by the cutting tool and the work-holding devices. For the same force, the deflection increases for a material of lower elastic modulus, and better means of support become necessary. Further limitations are mostly economic in nature: even though it may be possible to mill or turn a very hard material, the cost becomes extremely high (see Table 11.8). Similarly, machining of a very complex shape may be possible, but only by a sequence of expensive operations.

Design must, as always, take into account the manufacturing process. The sequence by which a part may be machined must be envisaged, and design features must be provided that assure easy machining, preferably without the need of transferring the part to a different machine tool. In general, the following points need consideration:

1 The workpiece must have a *reference surface* (an external or internal cylindrical surface, flat base, or other surface suitable for holding it on the machine tool or in a fixture).

2 If at all possible, the part shape should allow finishing with a single setup; if the part needs gripping in a second, different position, one of the already-machined surfaces should become the reference surface.

3 The shape of slender parts should permit adequate additional support against deflection.

4 Deflections of the tool in drilling, boring, or milling of internal holes and recesses limit the depth-to-diameter (or width) ratio. Deep recesses or holes call for special (and more expensive) techniques, or a sacrifice in tolerances.

5 Internal undercuts (Fig. 8.37a) can be machined if not too deep, but they increase costs.

6 Radii (unless set by stress-concentration considerations) should accommodate the most natural cutting tool radius: the nose radius of the tool in turning and shaping, the radius of the cutter in milling a pocket, the sharp edge of the cutter in slot milling, or the rounded edge of a slightly worn tool in EDM.

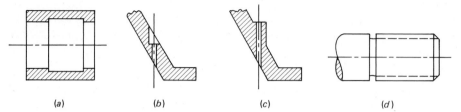

FIG. 8.37   Some features of design for machining: (a) material undercut, (b) hole at an angle to surface, spotfaced and (c) redesigned, and (d) runout for thread-cutting tool.

7   Features at an angle to the main machine-movement direction call for a more complex machine, transfer to a different machine, interruption of the main machining action, and special tooling or special attachments, and should be avoided.

8   Holes and pockets placed at an angle to the workpiece surface deflect the tool and necessitate a separate operation such as spot-facing (Fig. 8.37b) or redesign of the part (Fig. 8.37c).

9   One must not forget that a tool cannot be retracted immediately and appropriate runout provisions must be made (Fig. 8.37d).

8.12   SUMMARY

Material removal is, in a sense, admission of defeat: one resorts to it when other processes fail to provide the requisite shape, tolerances, and surface finish, or when the number of parts is too small to justify a more economical process.   How often we fail is shown by the fact that more man-hours are spent on machining than on other unit processing methods. Machining is still indispensable for finishing the bores of engine cylinders and the surfaces of some shafts; journal bearings; bearing balls, rollers, and races; accurately fitting machine parts; and parts of complex configuration in general.   It can be also economical, but some basic points must be observed for both economy and quality:

1   Machinability of materials is highly variable, and a condition assuring low UTS, n, and q should be chosen whenever possible.

2   Tool wear and temperature are limiting factors; for fast material removal, the greatest feed tolerated by the machine tool, tool, and workpiece should be chosen, along with a speed suited to the tool material.

3   Finishing must be directed toward producing a surface that is within tolerances, has the proper roughness and texture, and is free of damage (tears, cracks, smeared zones, excessive work hardening, and changes in heat-treatment conditions) and of harmful residual stresses.

4 All manufacturing processes must be treated as a system in which the workpiece and tool materials, the lubricant (coolant), process geometry, and machine tool characteristics interact. This approach is essential in machining because some of the interactions may easily go undetected, to the detriment of quality, production rates, and economy.

Further Reading

DETAILED PROCESS DESCRIPTIONS:
   Metals Handbook, 8th ed., American Society for Metals, Metals Park, Ohio, vol. 3, *Machining,* 1967.

RECOMMENDATIONS FOR MACHINING CONDITIONS:
   Machining Data Handbook, 2d ed., Machinability Data Center, Metcut Research Associates, Inc., Cincinnati, Ohio, 1972.

TEXTS:
   ARMAREGO, E. J. A., and R. H. BROWN: *The Machining of Metals,* Prentice-Hall, Englewood Cliffs, N.J., 1969.
   BLACK, P. H.: *Theory of Metal Cutting,* McGraw-Hill, New York, 1961.
   BOOTHROYD, G.: *Fundamentals of Metal Machining and Machine Tools,* Scripta/McGraw-Hill, Washington, D.C., 1975.
   KRONENBERG, M.: *Machining Science and Application,* Pergamon, Oxford, 1966.
   ZOREV, N. N.: *Metal Cutting Mechanics,* Pergamon, New York, 1966.
   See also Chap. 4 references: Cook, Kalpakjian, Lissaman and Martin.

SPECIALIZED BOOKS:
   BHATTACHARYA, A., and I. HAM: *Design of Cutting Tools: Use of Metal Cutting Theory,* American Society of Tool and Manufacturing Engineers, Dearborn, Mich., 1969.
   HOWE, R. E. (ed.): *Producibility/Machinability of Space-Age and Conventional Materials,* American Society of Tool and Manufacturing Engineers, Dearborn, Mich., 1968.
   KOBALYASHI, A.: *Machining of Plastics,* McGraw-Hill, New York, 1967.
   MCGEOUGH, J. A.: *Principles of Electrochemical Machining,* Chapman and Hall, London, 1974.
   NORDHOFF, W. A.: *Machine Shop Estimating,* McGraw-Hill, New York, 1960.
   SHAW, M. C. (ed.): *New Developments in Grinding,* Carnegie Press, Pittsburgh, 1972.
   SPRINGBORN, R. K. (ed.): *Cutting and Grinding Fluids: Selection and Application,* American Society of Tool and Manufacturing Engineers, Dearborn, Mich., 1967.
   SPRINGBORN, R. K. (ed.): *Non-Traditional Machining Processes,* American Society of Tool and Manufacturing Engineers, Dearborn, Mich., 1967.
   SWINEHART, H. J. (ed.): *Cutting Tool Material Selection,* American Society of Tool and Manufacturing Engineers, Dearborn, Mich., 1968.

SWINEHART, H. J. (ed.): *Gundrilling, Trepanning and Deep Hole Machining,* American Society of Tool and Manufacturing Engineers, Dearborn, Mich., 1967.

WILSON, F. W. (ed.): *Machining the Space-Age Metals,* American Society of Tool and Manufacturing Engineers, Dearborn, Mich., 1965.

ZLATIN, N., M. FIELD, and J. F. KAHLES: *Machining of Malleable Iron,* Malleable Founders' Society, Cleveland, Ohio, 1971.

Machinability, Special Report 94, The Iron and Steel Institute, London, 1967.

Machine Tool Specification Manual, Maclean-Hunter, London, 1963.

Examples

8.1 Bars of free-machining steel (BHN 180) of diameter $d_0 = 4$ in and length $l_0 = 40$ in are rough turned with throwaway WC tool inserts. Calculate the variation of production cost C_{pr} with cutting speed for $R_0 = \$7.20/h$, $R_m = \$5.40/h$, $C_t = \$4.00/edge$, $t_1 = 8$ min, and $t_{ch} = 10$ min.

Solution. First, find cutting speed v. From Fig. 8.28, $v_s = 500$ fpm with 0.150 in depth of cut and $f = 0.020$ in/rev. From Sec. 8.1.3, Tool Wear, $n = 0.25$ and $C = 3.5\, v_s = 1750$; $t_r = 1$ min.

Next calculate the cutting time t_c. For this, the volume removed is needed:

$$d_1 = 4.00 - (2 \times 0.15) = 3.7 \text{ in}$$

$$V = \left(\frac{4^2 \pi}{4} - \frac{3.7^2 \pi}{4} \right) 40 = 72 \text{ in}^3$$

From Eq. (8.15),

$$B = \frac{2 \times 72}{0.02(4.0 - 3.7)} = 24{,}000 \text{ in} = 2000 \text{ ft}$$

$$t_c = \frac{B}{v} = \frac{2000}{v} \text{ min}$$

The net cost of cutting

$$C_c = \frac{B(R_0 + R_m)}{v} = \frac{2000(7.20 + 5.40)}{60v} = \frac{420}{v}$$

and the cost of loading and unloading

$$C_1 = (R_0 + R_m)t_1 = \$1.68$$

are plotted in Fig. 8.30.

The total production cost per piece, C_{pr}, is next calculated from Eq. (8.18) and is plotted in a solid line. Recalculation with $C_t = \$1.00/\text{edge}$ shows (broken line in Fig. 8.30) the magnitude of speed increase one could afford if the cost of tooling were lower. Obviously, the loading time t_1 is the most likely target of production organization efforts.

8.2 Holes of 10 mm diameter are to be drilled, with a twist drill, into the material of Example 8.1. Determine the recommended cutting speed and feed.

From Fig. 8.28, for HSS, $v_s = 0.75$ m/s.

From Sec. 8.7.1, $v = 0.7 v_s \approx 0.5$ m/s.

For $D = 10$ mm, the rotational frequency

$$N = \frac{v}{D\pi} = \frac{0.5 \times 1000 \times 60}{10\pi} = 955 \text{ rpm}$$

The feed is 0.02 D/rev or 0.2 mm/rev (and, because there are two cutting edges, $f = 0.2/2 = 0.1$ mm/edge).

8.3 Section 8.1.3 suggests a way of finding C for Taylor's equation, Eq. (8.5). Since v_s in Figs. 8.28 and 8.29 is based on a tool life of 1–2 h or, on the average, $t = 90$ min, what is the implied value of n?

Solution: From Eq. (8.5) or Eq. (8.16)

$$v_s\, 90^n = C$$

for HSS: $v_s\, 90^n = 1.75 v_s;\ n = 0.125$

for WC: $v_s\, 90^n = 3.5 v_s;\ n = 0.28$

Problems

8.1 Take $n = 0.15, 0.3,$ and 0.6 for HSS, WC, and ceramic tooling, respectively. (*a*) Find the change in tool life when increasing the cutting speed v_c to twice its initial value v_i. (*b*) From the answer to (*a*) would you recommend experimentation with higher cutting speeds with HSS in preference to WC?

8.2 The full form of Taylor's equation, Eq. (8.16), is valid in any measurement system. In SI units, v and v_r will be in m/s, and t and t_r in s. However, a reference tool life of $t_r = 1$ s would be quite unreasonable; therefore $t_r = 60$ s is used. (*a*) Develop a general formula for converting C into v_r (in SI units). (*b*) If $C = 500$ fpm, what is the value of v_r?

8.3 The cutting speed for minimum-cost v_c is given by Eq. (8.20). (*a*) Substitute v_c into Eq. (8.5) and express the cutting time t for minimum-cost operation; denote it t_c. (*b*) Using $n = 0.12, 0.25,$ and 0.5 for HSS, WC, and ceramic, respectively, express the relative magnitudes of t_c (assuming identical tool and tool-changing costs). (c) Find t_c for a WC tool with $R_0 = \$10.00/h$, $R_m = \$4.00/h$, $C_t = \$0.50$, and $t_{ch} = 1$ min.

8.4 Holes of 0.25 in diameter and 1 in depth are to be drilled into 304 stainless steel. Determine the recommended (a) cutting speed (fpm), (b) drill rpm, (c) feed (in/rev), (d) feed rate (in/min), and (e) power requirement (hp). Check whether (f) the recommended values are feasible on commercially available equipment.

8.5 Holes of 12 mm diameter and 36 mm depth are to be drilled into 7075-T6 aluminum alloy. Determine the recommended (a) cutting speed (m/s), (b) drill rpm, (c) feed (mm/rev), (d) feed rate (m/min or m/s), and (e) power requirement (kW). Check whether (f) the recommended values are feasible on commercially available equipment.

8.6 A rectangular slab of 4 × 10 × 1 in dimensions, of 2024-T6 aluminum alloy, is to be face milled on the two larger surfaces. Determine, for a finishing cut, the (a) speed, (b) feed, and (c) net cutting time t_c (remember that the face-milling cutter has a diameter D_m and that it is usual, for uniformity of surface finish, to run the cutter clear off the ends of the workpiece).

8.7 The threaded bushing shown in the illustration is to be made in large quantities (over 10,000 pieces/month). Determine the optimum screw-machine operation sequence by going through the following steps: (a) clarify missing dimensions, tolerances, and surface finish specifications (in the absence of consultation with the designer, make common-sense assumptions), (b) choose the starting material dimensions and metallurgical condition, (c) determine which way the part should be turned to allow finishing from the bar, (d) select the basic operations required, listing the appropriate speeds and feeds and machining times, (e) determine the operation sequence, keeping in mind the possibility of simultaneous machining and of spreading a lengthy operation over several positions.

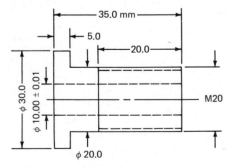

Material: AISI 1117 steel

8.8 Repeat the above procedure for the instrument screw shown.

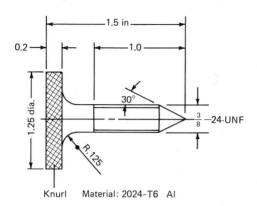

Knurl   Material: 2024–T6 Al

8.9 Rank the following materials in order of their anticipated machina-bility: 1008 steel, 1045 steel, 302 stainless steel, Muntz metal, 1100 Al, and nickel, all in the annealed condition. In formulating your judgments, refer to the appropriate equilibrium diagrams, and to numerical data con-tained in this volume.

8.10 Establish the lowest-cost cutting speed for rough turning a cast iron of 200 BHN with WC tooling. Cost data are the same as in Exam-ple 8.1.

8.11 A hardened tool steel (R_c 55) surface of 2 × 10 in dimensions is to be ground to a surface finish of 16–20 μin AA. Past experience shows that a suitable wheel (A-46-H-V) will produce the requisite finish with a cross-feed of 0.020 in/pass, and a grinding depth of 0.003 in. If the table speed is 60 fpm, and the wheel is 0.5 in wide, calculate (a) the number of passes taken (count table movement back and forth as separate passes, and take into account that the wheel starts and finishes outside the workpiece width); (b) the total time taken, remembering that the wheel must run out at each end of the workpiece (make sketch to calculate wheel position at end); (c) the material removed, total and per minute; (d) the horsepower requirement.

8.12 The shape of the part in Example 11.1 (as defined by the solid lines) is to be sunk into a forging-die block by electrodischarge machining. Nine percent of the total material is removed at a high rate but the final one percent is removed at a low rate to produce a finish of 30 μin AA and freedom from surface damage. Calculate the total machining time.

9

JOINING PROCESSES

Joining differs from previously discussed processes in that it takes parts produced by other unit processes and unites them into a more complex part; therefore it could also be regarded as a method of assembly. Some joints are purely mechanical (Fig. 9.1), and within this category the nonpermanent ones are properly classified as assembly. Other joints establish permanent interatomic bonds, and only these will be discussed here in any detail.

9.1 MECHANICAL JOINING

Disregarding the nonpermanent screw joint, the most common mechanical fastener is the *rivet*. Whether it be solid or hollow (Fig. 9.2a and b), it makes a joint by clamping the two parts between heads. One head is usually formed in a prior operation; the resulting rivet is fed through predrilled or punched holes, and the second head is produced by upsetting, either cold or hot (for upsetting see Sec. 4.4.1). On a hollow rivet, the head is formed by flaring, an operation related to flanging of a tube (Sec. 5.3).

Thin sheets can be joined without preliminary drilling by *stitching* or *stapling* (Fig. 9.2c); stapling is extensively used to fasten sheet to wooden backing. *Seams* (Fig. 9.2d) produced by a sequence of bends (Sec. 5.3), may be made permanent, with or without fillers such as adhesives or solder.

Shrinking a sleeve onto a core is applicable mostly to round parts, and the compressive stress necessary to maintain a permanent joint is attained either by heating the sleeve (and/or cooling the core) or by swaging (Sec. 4.4.4).

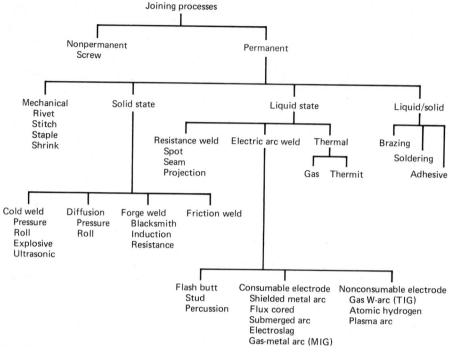

FIG. 9.1   Classification of joining processes.

9.2   SOLID-STATE BONDING

Interatomic bonds may be established by bringing atoms of two surfaces in close enough proximity to assure adhesion (Sec. 2.4).   As indicated previously, it is essential that the surfaces should be free of contaminants (oxides, adsorbed gas films, or lubricant residues) that might prevent the formation of interatomic bonds.   Relative movement of the surfaces under

FIG. 9.2   Permanent mechanical joints: (a) rivet, (b) tubular rivet, (c) staple, and (d) seam.

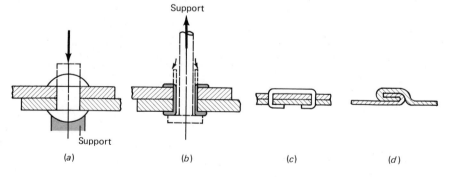

pressure and controlled roughness are helpful in breaking through surface films. While theoretically no *pressure* would be required for *bonding,*
in practice a certain normal pressure is necessary to assure conformance
of the contacting surfaces. Heat is not an essential part of the basic
bonding process, but softening of the materials and diffusion of atoms
help to achieve the atomic registry necessary to create the bond. Diffusion is objectionable, however, when two dissimilar materials form intermetallic compounds that embrittle the joint.

As indicated in Sec. 2.4, greatest adhesion is attained between similar
materials and those that form solid solutions. In principle, however, any
material can be bonded, and solid-state bonding is often applied when
other techniques fail.

9.2.1 Cold-Welding

The term *cold-welding* is used loosely to describe processing at room temperature.

The usual cold-welding process applies pressure to the pieces to be
joined. The harmful effect of interface films is minimized when substantial relative movement and expansion of surfaces is assured, as in *roll
bonding* (Fig. 9.3*a*).

Deformation of the interface is assured in *explosive welding* by
placing the sheets or plates at an angle to each other (Fig. 9.3*b*). When
the explosive mat placed on top of the inclined plate (fly plate) is deto-
nated, the sheet joins to the other by forming tight whirls or *vortexes* at the
interface. This technique is useful also for *in situ* expansion of tubes into
the head plates of boilers.

Relative movement of the interface can also be induced by *tangential
ultrasonic vibration* (Fig. 9.3*c*). There is no massive deformation, and the
process is suitable for delicate instrument components.

FIG. 9.3 Solid-state joining, cold: (*a*) roll bonding, (*b*) explosive joining, and (*c*) ultrasonic
welding.

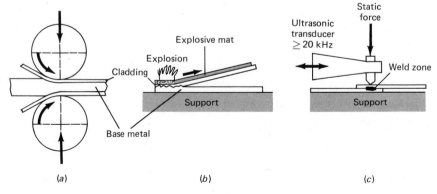

9.2.2 Diffusion Bonding

Generally, better bonding is obtained when the temperature is high enough to assure diffusion—typically, above $0.5T_m$ (Sec. 2.1.8). The goldsmith has for centuries made *filled gold* by placing a weight on top of a *sandwich* composed of a silver or copper *core* with gold *face sheets* (Fig. 9.4a). When this is held in a *furnace* for a prolonged time, a permanent *bond* is obtained. Large airframe parts are now produced by this technique.

Simultaneous deformation greatly accelerates bonding, and *hot roll-bonding* is used extensively to create composites of low cost and high performance. Thus, pure aluminum is *clad* to precipitation-hardened aluminum, or stainless steel to mild steel for corrosion protection, or nickel to a copper core to provide the required magnetic properties of United States quarters and dimes.

9.2.3 Forge Welding

The only industrial welding process until the turn of the century, *forge welding* assures a bond by combining high temperatures with substantial local deformation in the joint. The hot, preshaped workpieces, usually of steel, are forged together to squeeze out oxides, slag, and contaminants, and assure interatomic bonding (Fig. 9.4b).

Forge welding in which the ends of workpieces are pressed together axially (*butt welding*) is possible, but the joint quality tends to be poor. In a more recent variant of the process, heat is provided by induction to minimize oxidation (Fig. 9.4c). Much less deformation is sufficient; therefore butt welding of workpieces becomes practicable. The heat may also be generated by passing a current through the compressed faces (Fig. 9.4d). *Electric butt welding,* however, has now largely been replaced by flash butt welding (Sec. 9.3.5).

FIG. 9.4 Solid-state joining, hot: (a) diffusion bonding, (b) forge welding, (c) induction welding, and (d) resistance welding.

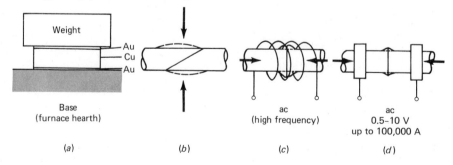

9.2.4 Friction Welding

The frictional energy generated when two bodies slide on each other is transformed into heat; when the rate of movement is high and the heat is contained in a narrow zone, welding occurs. In the practical form of *friction welding* (Fig. 9.5), one part is firmly held while the other (usually of axial symmetry) is rotated under simultaneous application of axial pressure. The temperature rises, partially formed welded spots are sheared, surface films are disrupted, and the rotation is suddenly arrested when the entire surface has welded. Some of the softened metal is squeezed out into a *flash,* but it is not fully clear whether melting also takes place. The heated zone is very thin; therefore dissimilar metals are easily joined, for example, mild steel shanks can be fastened to high-speed-steel tool ends.

9.3 LIQUID-STATE WELDING

In the great majority of applications, the interatomic bond is established by melting. When the workpiece materials and the *filler* (if used at all) have similar compositions and melting points, the process is referred to as *welding.* The principles of processes will be discussed here; process characteristics are summarized in Table 11.7.

9.3.1 The Welded Joint

A welded joint is far from homogeneous. The large amount of heat supplied to melt the joint also causes metallurgical changes in the adjacent nonmolten, *heat-affected zone.* Since metals have a high thermal conductivity, fast cooling rates (typically of the order 100°C/s) are induced and proper control of the process calls for a thorough familiarity with nonequilibrium metallurgy. Here we will have to compromise and be satisfied with

FIG. 9.5 Solid-state (and possibly liquid-state) joining by friction welding (numbers are typical of steels).

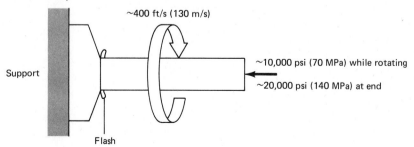

a cursory examination, based to a great extent on concepts discussed in Chaps. 2 and 3, which could be usefully reviewed at this time. Some acquaintance with material effects is nevertheless indispensable if limitations of various processes are to be appreciated.

Single-phase materials

A section (Fig. 9.6) through the joint in a pure metal, such as Al or Cu, welded with a rod of identical composition, shows the coarse, columnar grains typical of as-cast structures (as in Fig. 2.2). The applied heat has melted some of the workpiece material, and this, together with the material of the *filler rod,* has solidified rapidly in the mold formed by the solid workpiece.

If the workpiece material was originally cold-worked and therefore of highly elongated grains, the heat-affected zone will show recrystallization. For a given cold work, the grain size increases with increasing annealing temperature (Fig. 2.13); therefore the very coarse grains found at the melt face gradually change to finer ones until, at the edge of the heat-affected zone, only partial recrystallization is evident. If the workpiece has been annealed, the further heat input during welding just coarsens the grains. In either case, the strength of the coarse-grain structure will be low (Sec. 2.1.4).

FIG. 9.6 Structure and properties of a welded joint in a pure metal (or single-phase material). (HAZ, heat-affected zone.)

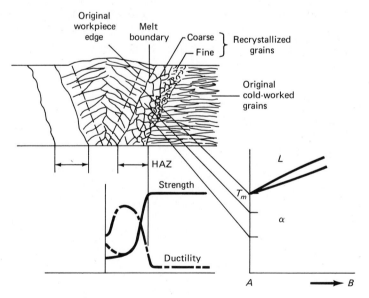

Two-phase materials

Many technically important materials derive their strength from pre-cipitation hardening (Sec. 2.2.7). They can be readily welded in the an-nealed state, and the effects of welding are mostly washed out on subse-quent solution treatment and aging of the hole weldment. If its size is too big for this, fusion welding is possible in either the solution-treated or aged conditions, but the strength advantages of heat treatment are lost in the *weld zone* (Fig. 9.7). The weld itself contains the intermetallic constit-uent in a coarse form, resulting in low strength and ductility. The immedi-ately adjacent heat-affected zone has been rapidly heated and quenched, but also aged by *back-heating* during cooling; therefore it is fully heat-treated but of a coarse grain, with high strength and moderate ductility. Farther away, the original structure becomes overaged and soft. Apart from the poor strength of the joint, the varying composition can also lead to corrosion problems, e.g., in Al and Mg alloys. A compromise solution is to use a different filler rod.

Materials of eutectic composition present little problem because of their favorable solidification mode (Sec. 2.2.4).

Phase transformations lead to complex changes. To take a medium-carbon steel in the annealed state as an example (Fig. 9.8), the weld itself has the usual coarse, cast structure. Next to it the material has been heated high into the austenitic range and cooled from there, resulting in a coarse grain. With decreasing temperatures, the austenite grains also be-come finer; thus finer-grain ferrite and pearlite are found in the trans-

FIG. 9.7 Structure and properties of a welded joint in a precipitation-hardened alloy.

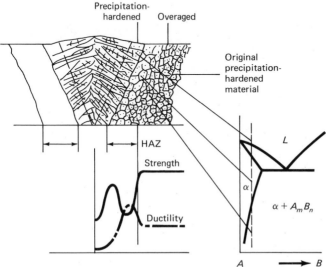

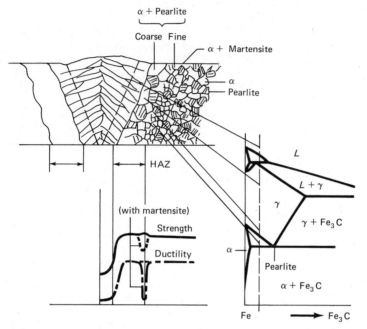

FIG. 9.8   Structure and properties of a weld in a medium-carbon steel.

formed structure. The edge of the heat-affected zone has been heated just above the eutectoid temperature and only the pearlite transformed. Fast cooling rates will convert any austenite into martensite, and a large drop in strength and ductility may result. With higher (above 0.35%) carbon content, martensite will inevitably form, as it will if the steel is also alloyed. In general, increasing hardenability means increasing danger of martensite formation and therefore decreasing weldability. Martensite is not only hard and brittle, but its formation proceeds with a *volume increase* which imposes further stresses on the structure. Preheating below the transformation temperature slows cooling to miss the nose of the TTT curve (Fig. 2.24); if this is not possible, *postwelding heat treatment* is essential.

Dissimilar materials

The situation becomes even more complicated when two dissimilar materials are used (either in the workpiece or the filler). The heating and cooling history is compounded by alloying, and the weld zone traverses the equilibrium diagram in both the temperature and concentration directions. Nonequilibrium effects complicate the matter further.

As might be expected, materials that form solid solutions present no problems. Eutectics tend to be more brittle, although they may still be acceptable when both phases in the eutectic are ductile (Sec. 2.2.3). Inter-metallic compounds invariably embrittle the structure to make it useless; thus, for example, copper cannot be joined to iron. It is possible, how-ever, to use a *mutually compatible interface,* in this case, nickel.

In many instances the melting-point difference between dissimilar materials is very large, and then it is customary to classify the process as brazing or soldering (Sec. 9.4).

Weld quality

In addition to the above metallurgical changes, there are other compli-cating factors to consider. Fusion welding is a melting process and must be controlled accordingly (Sec. 3.1.1). Unwanted reactions with the atmo-sphere are prevented by sealing off the melt zone with a vacuum, a protec-tive (inert) atmosphere, or a slag (which is chosen to double as a solvent for the oxides present on the workpiece and is, therefore, called a flux). Surface films, especially organic ones that could prevent bonding or would lead to gas porosity, are kept out by *chemical* or *mechanical prepa-ration of the surfaces.* As in casting, care is taken to prevent entrapment of gas, slag or other inclusions (Sec. 3.1.2). Solidification shrinkage (Sec. 2.1.1) coupled with solid shrinkage imposes *internal tensile stresses* on the structure (Fig. 9.9), and may lead to cracking. *Preheating* of the structure reduces differential shrinkage, and also helps to attain fusion tempera-tures in materials of high heat conductivity.

The properties of a pure metal weld can be improved only by strain hardening, induced by hammering (*peening*) or rolling of the *weld bead.* If the metal undergoes allotropic transformation on heating (as Fe and Ti do), complete re-heat-treating is possible. Even without it, the weld material will be transformed if deposited in more than one pass (*multipass welding*).

Welds in a solid-solution alloy show the effects of nonequilibrium solidification (Sec. 2.2.5). Coring in itself is harmless but a low-melting grain-boundary eutectic could result in cracking of the joint on cooling

FIG. 9.9 Distortion and residual tensile stresses generated in a structure by liquid-phase welding.

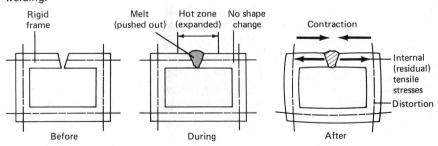

(hot shortness) while intermetallic compounds would make the joint brittle. Preheating minimizes internal stresses but even then it may be necessary to use a less-alloyed, more ductile filler that reduces hot shortness. Residual stresses and their harmful effects can be lessened, and, at the same time, a more homogeneous structure can be obtained by *postwelding stress-relief* heat treatment or even homogenization of the entire welded structure.

9.3.2 Weldability

It is obvious from the foregoing discussion that the term *weldability* denotes an extremely complex collection of technological properties. Some of the factors of importance are:

1 Melting temperature determines—together with specific heat and latent heat of fusion—the requisite *heat input.*

2 High thermal conductivity allows the heat to dissipate and therefore requires a higher *rate of heat input* and leads to more rapid cooling.

3 High thermal expansion results in greater distortion, residual stresses, and greater danger of cracking in a hot-short material.

4 Surface contaminants including oxides, oils, dirt, paint, metal platings and coatings incompatible with the workpiece material, atmospheric contaminants, etc., all mitigate against the success of welding, as do gas-forming reactions.

5 Metallurgical transformations, discussed earlier, are of great importance, especially if they lead to brittle phases such as martensite.

6 The absolute and relative thicknesses of parts to be joined and the design of the joint have a powerful influence on heating and cooling, and thus on weldability.

7 The welding process itself imposes varying conditions, and thus the weldability of the same material is different in different processes.

9.3.3 Weldable Materials

Generalizations are more dangerous for welding than for other processes, but some guidelines are still necessary.

Low-melting materials

Tin and lead are easily welded, provided that the heat input is kept low enough to prevent overheating. Zinc, on the other hand, is one of the most difficult materials to weld, because it oxidizes readily and also *vaporizes* at a low temperature (at 1663°F or 906°C). It can be resistance welded and stud welded, although soft soldering is more usual.

Light metals

Aluminum and magnesium share a number of characteristics, including an oxide film that is usually removed with a powerful flux, which in turn must be washed off after welding to prevent corrosion. Moisture (H_2O) must be kept out, because it reacts to give an oxide and H_2 (which embrittles the joint by causing porosity).

The high thermal conductivity and specific heat, yet low melting point, require high rates of heat input and adequate precautions against overheating. The high thermal expansion coefficient necessitates preheating of hot-short materials. Because of the difficulties encountered with precipitation-hardening materials, such alloys are often heat treated after welding or, if this is not possible, a different filler is used (very often, Al-Si for aluminum alloys).

Copper-base alloys

Deoxidized copper is readily welded, especially if the filler contains phosphorus to provide instant deoxidation. Tough-pitch copper cannot be welded because its oxygen content (typically 0.15 percent) reacts with hydrogen and CO to form water and CO_2 respectively, both of which embrittle the joint by generating porosity.

Brasses can be welded but zinc losses are inevitable; therefore either the filler is enriched in zinc, or Al or Si is added to form an oxide that reduces evaporation.

Tin bronze has a very wide solidification range and is thus exceedingly hot-short. Phosphorus in the welding rod prevents oxidation, while postheating is necessary to dissolve the brittle nonequilibrium intermetallic phase. Aluminum bronzes present no problem but the oxide formed must be fluxed, just as with pure aluminum.

Steels

Steels are readily welded, but martensite formation is a danger. Preheating and, if possible, postheating are necessary when martensite formation is unavoidable.

Alternatively, the structure may be heated into the austenitic range, cooled to a temperature above M_s (Fig. 2.24), and welded before transformation begins. The completed structure is then cooled. Such *step welding* makes even tool steels amenable to welding. A further danger in steels is *hydrogen pickup* which, particularly in the presence of marten-site, accumulates in microcracks and leads to embrittlement. Sulfur creates porosity and brittleness, and, while welding of resulfurized free-cutting steels is possible, they are often brazed in preference.

Stainless steels always contain chromium, which forms an extremely dense Cr_2O_3 film. Welding conditions must be chosen to prevent its formation. Apart from this, austenitic steels (containing both Cr and Ni) are weldable, but chromium carbides formed at high temperatures reduce the chrome content below the level required for corrosion protection, and subsequent corrosion (weld decay failure) is a danger. To avoid this, the carbon content should be very low, or the steel must be stabilized (Ti, Mo, or Nb added to form stable carbides), or the structure must be heated above 1000°C after welding, and then quenched to retain the redissolved carbides in solution. Stainless steels containing only chromium are either ferritic or martensitic in structure. Ferritic steels (over 16% Cr) can be welded but the coarse grain will weaken the joint. Martensitic steels form a martensite with a hardness depending on C content; careful preheating is followed by postheating to above 700°C to change the martensite into ductile ferrite with embedded chromium carbide precipitates.

Cast iron

Of the various classes of cast irons (Sec. 3.1.4), gray iron can be welded, but loss of Si must be prevented; otherwise, the iron turns into the brittle white form on cooling. Therefore preheating and slow cooling are essential, and the welding rod is enriched in silicon. Nodular cast iron can be welded if Mg is incorporated in the rod to assure spheroidal graphite formation. Malleable iron cannot be welded because it will revert to white iron; if necessary, it is brazed.

High-temperature materials

Nickel and its solid-solution alloys are readily welded. The precipitation-hardening superalloys contain chromium, and the oxide must be fluxed or its formation prevented.

Refractory metal alloys (W, Mo, Nb) can be welded but the volatility of the oxides makes special techniques (e.g., electron-beam welding) mandatory.

9.3.4 Resistance Welding

Electric resistance welding relies on the heat generated at the interface between the two parts to be joined. Surface cleanliness is important but not quite as vital as in solid-state joining, because actual melting takes place at the interface and some of the contaminants are expelled from the melt. According to Joule's law, the *heat generated* is

$$J = I^2Rt \tag{9.1}$$

where I is the current in amperes, R is the resistance in ohms, and t is the duration of current application in seconds. Since heat must be concentrated in the weld zone, resistance away from this zone should be low, especially at the point where the current is supplied to the workpiece by the *electrodes*. High heat conductivity and specific heat call for very high currents to prevent dissipation of heat. The voltage can be low, typically 0.5–10V.

Spot welding

Because of the widespread application of sheet metal parts, *spot welding* has acquired a prominent position, from attaching handles of cookware to assembling whole automobile bodies. Two, usually water-cooled, electrodes of high conductivity (made of copper with some Cd, Cr, or Be) press the two sheets together (Fig. 9.10a). The current is then applied for one or several cycles, the interface heats up and a molten pool (*weld nugget*) is formed. The pressure is released only after the current has been turned off. The sheet surface shows a light depression and discoloration. *Multiple electrodes* (sometimes several hundred) are used for simultaneous welding of large assemblies.

FIG. 9.10 Main features of resistance welding: (a) spot, (b) seam, and (c) projection welding (data are for 0.040-in or 1-mm thick low-carbon steel sheet).

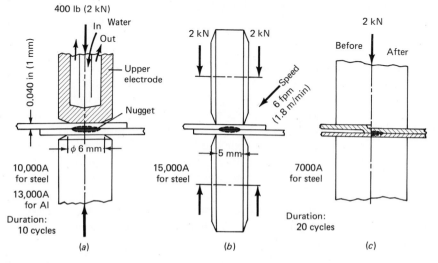

A series of spot welds may be made along a line much more rapidly if the electrode is in the form of *rollers* (Fig. 9.10*b*). The current is switched on and off in a planned succession, giving uniform spacing of spot welds. When the alternating current is left on, a spot weld is made every time the current reaches its peak value, and the welds are spaced close enough to give a gas- and liquid-tight joint. Such *seam welding* is one of the methods of producing the body of a can.

The extent of the weld zone is better controlled, and several welds can be made simultaneously with a single electrode, when small dimples or *projections* are embossed (Sec. 5.5.1) or coined (Sec. 4.4.3) on one of the sheets. When the current is applied, the projection softens and is pushed back in place by the electrode pressure as the weld nugget is formed (Fig. 9.10*c*). Projections forged or machined on solid bodies allow their welding to a sheet or other solid body.

9.3.5   Electric Arc Welding

This differs from electric resistance welding in that a *sustained arc* generates the heat for melting the workpiece (and, if used, the filler rod) material. High temperatures are maintained for a longer time; therefore complete protection from the atmosphere is essential. A further important requirement, at least with some processes and materials, is the provision of a flux that dissolves oxides and removes them from the melt zone. Very broadly, *arc-welding processes* comprise both *consumable and nonconsumable electrode* methods.

Flash butt welding

Butt welding in the general sense means joining the end faces of two bodies. In the narrower sense, it is applied to resistance welding in which, just as in spot welding, the pressure is applied prior to switching on the current (Sec. 9.2.3).

Much more widespread today is *flash butt welding,* in which the current is applied during approach, thus *arcing* takes place between the closest surface irregularities. Preheating may be obtained by resistance heating, but the end faces are then again separated and arcing provides the heat for melting (therefore the process links resistance welding with liquid-phase processes). The end faces to be joined are often chamfered, so that melting moves from the center outward to squeeze out contaminants into the flash (Fig. 9.11*a*). A substantial length may be *burnt off* to assure a good weld. For uniform heating the two parts have equal cross-sectional areas, but their composition can be different, for example, in joining a low-carbon steel shank to a HSS tool bit. Rings, tubes, and sheet structures are often welded edge-to-edge.

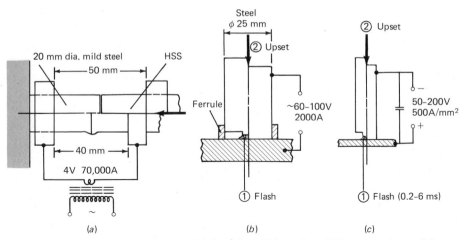

FIG. 9.11   Flash-welding processes: (*a*) flash-butt, (*b*) stud, and (*c*) percussion welding.

A variant of the process, *stud welding* (Fig. 9.11*b*), allows dissimilar areas to be joined by maintaining a controlled arc (with the aid of a ceramic *shielding ferrule*), whereupon the pressure is applied. Very rapid heating with a sudden temperature rise is obtained in *percussion welding* which utilizes the energy stored in a condenser (Fig. 9.11*c*) for heating. Discharge takes place just before or during impact. The intense, localized heat allows the joining of widely differing cross sections and also dissimi- lar materials. Studs can be welded to thin sheets even with a PTFE coating on the other side. These two processes are properly classified as pure arc welding.

Consumable electrode welding

Shielded Metal-Arc Welding.   The arc is struck between the filler wire or rod (consumable electrode) and the workpieces to be joined. A coating applied to the outside of the wire (*coated electrode*) provides shielding and fluxing and also generates readily *ionized gases* that *stabilize* the arc. Since the coating is brittle, only straight *sticks* (of typically 18 in or 450 mm length) can be used, making this process a relatively slow but versatile and inexpensive hand operation. During welding, the electrode melts at a rate of approximately 10 in/min (250 mm/min) while the coating melts into a slag (Fig. 9.12*a*), which must be removed if more than one pass is required to build up the full weld thickness. The current may be ac or dc. In the latter case, the electrode may be negative (*straight polarity*) or positive (*reverse polarity,* giving deeper penetration because the positive gas ions do not hinder the flow of metal). The depth of the heat zone depends on polarity, current intensity, and voltage.

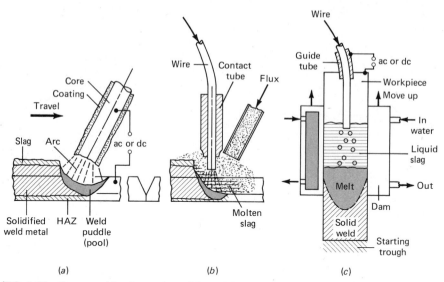

FIG. 9.12 Consumable-electrode welding: (*a*) shielded metal-arc, (*b*) submerged arc, and (*c*) electroslag welding.

Flux-Cored Arc Welding.   Basically the same result is obtained with the flux inside a tube.   However, the wire can now be *coiled,* and automatic, *continuous welding* becomes possible.   Sometimes additional shielding is provided with a gas, and then the process resembles gas metal-arc welding, to be discussed later.

Submerged-Arc Welding.   The consumable electrode is now the bare filler wire and the flux is supplied from a hopper quite independently (Fig. 9.12*b*) and in a thick layer that covers the arc.   It is primarily an automatic welding process with higher travel speeds.   The flux shields the arc, and allows high currents and great penetration depths.

Electroslag Welding.   In this process, used extensively for welding thick (1 in or 25 mm and over) plates and structures, electrode wire is fed into a molten slag pool (Fig. 9.12*c*).   An arc is drawn initially but is then snuffed out by the slag, and the heat of fusion is provided by resistance heating in the slag.   Watercooled copper *shoes* (*dams*) close off the space between the parts to prevent the melt and slag from running off.

Gas Metal-Arc Welding.   The consumable metal electrode is now shielded by an *inert gas,* thus the acronym MIG welding (Fig. 9.13*a*).   No slag is formed and several layers can be built up without intermediate cleaning.   Argon is suitable for all materials; helium is sometimes preferred (because of its higher ionization potential and therefore higher rate of heat generation) for Al and Cu welding.   Metal is transferred from the

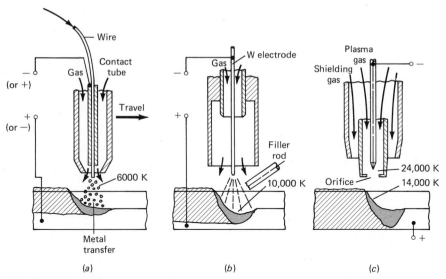

FIG. 9.13 Gas-shielded arc-welding processes: (*a*) gas metal-arc, (*b*) gas tungsten-arc, and (*c*) plasma-arc welding.

electrode to the weld zone in a fine spray at high current densities, and in blobs at low currents and voltages. Particularly with CO_2 shielding gas, a *short-circuiting* mode of operation occurs. This is preferable for thin sections or sheets because of the lower heat. The advantage of CO_2 (with small amounts of oxygen) for steel is that it allows higher transfer rates. Coiling of the wire electrode allows automation.

Electrogas welding is an outgrowth of electroslag welding. Protection is provided by gas (typically 80% Ar, 20% CO_2) and the molten pool is again retained with shoes.

Nonconsumable electrode welding

Gas Tungsten-Arc Welding. The arc is maintained between the workpiece and a tungsten electrode protected by an inert gas (TIG welding). A filler may or may not be used (Fig. 9.13*b*). The process demands considerable skill, but produces very high quality welds on almost any material, and it is particularly suitable for thinner gages (below 0.25 in or 6 mm). The protective atmosphere is argon, which maintains a stable arc, or helium for deeper penetration and a hotter arc, or a mixture of the two. To strike an arc, electron emission and ionization of the gas is initiated by withdrawing the electrode from the work surface in a controlled manner, or with the aid of an *initiating arc. High-frequency current* superimposed on the alternating or direct welding current helps to start the arc and also stabilizes it. Current pulses of *controlled rise rate,* produced with advanced electronic techniques, improve the quality of the weld. Both hand and automatic operations are possible. The filler rod, when used, is drawn away from the liquid weld pool, resulting in leftward welding for a right-handed welder.

It is also possible to draw an arc between two tungsten or carbon electrodes. Such *twin-arc welding* was the forerunner of present-day nonconsumable electrode processes, and is now used with hydrogen, which dissociates into the *atomic* form and assures high-quality welds in many materials.

Plasma-Arc Welding. In the space between the tungsten electrode tip and the workpiece, some of the gas becomes ionized (its atoms lose electrons). The ionized gas plasma (the fourth state of matter) gets hotter by resistance heating from the current passing through it. If the arc is constrained by an *orifice,* the proportion of ionized gas increases and plasma-arc welding is created (Fig. 9.13*c*), which provides an intense source of heat and greater arc stability. It is particularly useful for the welding of thin sheets. Again, filler metal may be used if an extra material supply is needed.

9.3.6   Thermal Welding

The heat required for fusion may be provided by a *chemical heat source.*

Gas welding

In the most widespread form, *oxyacetylene welding,* heat is produced by combustion of acetylene (C_2H_2) with oxygen. Both are stored at high pressure in *gas tanks* and are united in the *welding torch.* After ignition, a temperature of approximately 3700°K is generated in the *flame.* Three zones may be distinguished (Fig. 9.14*a*). Primary combustion takes place in the *inner zone* and generates two-thirds of the heat by the reaction

$$2C_2H_2 + 2O_2 \longrightarrow 4CO + 2H_2$$

These reaction products predominate in the *second zone,* and thus provide a reducing atmosphere favorable for the welding of steel. Complete combustion takes place in the *outer envelope:*

$$4CO + 2O_2 \longrightarrow 4CO_2$$

$$2H_2 + O_2 \longrightarrow 2H_2O$$

The relatively low flame temperature, the ability to change the flame from oxidizing to neutral and even reducing, and the flexibility of manual control make the process suitable for all but the refractory materials. The

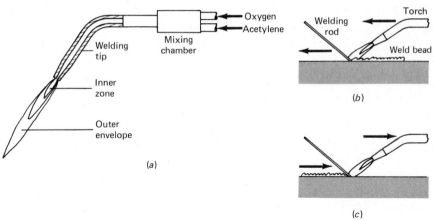

FIG. 9.14   Gas welding: (*a*) main features of oxyacetylene welding, and definitions of (*b*) forehand and (*c*) backhand welding.

welder (assumed to be right-handed) holds the torch at an angle to the surface, and may pull the filler rod (with its flux coating) away from the *weld puddle* in *leftward* or *forehand welding* (Fig. 9.14*b*), thus preheating the joint area and obtaining a relatively wide joint.   When the filler is moved above the weld bead (*rightward* or *backward welding,* Fig. 9.14*c*), the weld puddle is kept hot for a longer time and a narrower weld, often of better quality, results.

Other gases, including natural gas, are sometimes also used.

Thermit welding

This process finds application for joining heavy (minimum 10 in^2 or 6000 mm^2 area) sections such as rails in the field.   A sand mold complete with sprue, gate, and riser is built around the joint area.   *Thermit powder* is ignited in a crucible placed on top of the mold and the reaction

$$3Fe_3O_4 + 8Al \longrightarrow 4Al_2O_3 \text{ (slag)} + 9Fe$$

results in pure iron which is tapped, through a bottom hole of the crucible, directly into the mold.   After solidification the mold is destroyed and the still-hot excess steel is chiseled off.

9.3.7   High-Energy Beam Welding

The heat of fusion may be provided by converting the energy of impinging *electron* or *light beams* into heat.

Electron beam welding

The *electron gun* (Fig. 9.15a) is similar to a vacuum tube. The cathode emits a mass of electrons that are accelerated and focused to a 0.01–0.03 in (0.25–1 mm) diameter beam of high energy density (typically 1 MW/in^2 or 1.5 kW/mm^2). This is sufficient to melt and vaporize the workpiece material and thus fill a narrow weld gap even without a filler (although filler rods may be used).

Deepest penetration and best weld quality are obtained when the workpiece too is enclosed in *high* (10^{-2}–10^{-3} Pa) *vacuum,* but pumping down the *welding chamber* takes several minutes. *Medium* (1–10 Pa) *vacuum* still permits welding many metals with a pumping time of less than 1 min. With specially constructed vacuum traps, the electron beam can emerge from the gun into the air, and such *out-of-chamber welding* still permits high-quality welds in many materials.

The process is extremely adaptable and excels in welding thin gages, dissimilar thicknesses of metals, and hardened or high-temperature materials, and lends itself to automatic control.

Laser beam welding

An even more concentrated beam is produced, but at a lower overall efficiency, with the *laser beam* (Fig. 9.15b). A CO$_2$ laser pumped with 500

FIG. 9.15 Principles of high-energy-beam welding: (a) with an electron beam and (b) with a laser.

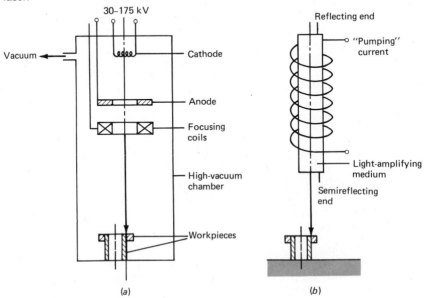

W emits far-infrared light (10.6 μm wavelength) and develops a peak energy density of 50 MW/in² (80 kW/mm²), yet the heat-affected zone is only 0.002 to 0.004 in (0.05–0.1 mm) wide. Oxygen blown on the surface of metals reduces heat reflection and increases material removal rates by oxidation; inert gas increases heat transfer for nonmetals.

The laser has the advantage that vacuum is not necessary and it is finding limited but growing application, particularly for thin-gage metals. Welding speeds of about 100 ipm (2500 mm/min) are achieved on steel sheet 0.060 in (1.5 mm) thick.

9.4 LIQUID–SOLID-STATE BONDING

When the workpiece and filler metal have substantially different melting points, the joint is established without the workpiece material being melted, and adhesion developed in the absence of contaminant films is the main source of strength.

9.4.1 Brazing

A common feature of all *brazing processes* is the very small gap between the parts to be joined, which is then filled out by *capillary action* (Fig. 9.16).

Surface cleanliness and adhesion can be assured simply by a suitable (typically reducing or neutral) *furnace* atmosphere in the *brazing* of steel with copper at around 2000°F (1200°C). This is a mass-production process for assemblies that can be passed through a furnace without jigging. The clearance is essentially zero or even negative, thus assuring good fit.

In all other processes a flux is necessary to provide cleansing action and thus assure wetting and adhesion (Sec. 2.4) of the surfaces. The most common *filler material* for steel is 60/40 Cu-Zn brass or a variant of it (brazing temperature around 1800°F or 1000°C) or silver-base alloys (*silver brazing,* around 1400°F or 800°C). The necessary heat may be imparted by

FIG. 9.16 Joining of three components by brazing. (By permission, from *Metals Handbook,* vol. 6, Copyright, American Society for Metals, 1971.)

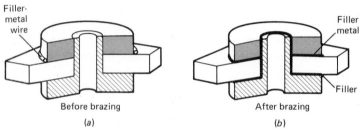

Before brazing
(a)

After brazing
(b)

dipping the assembly into molten salt (which may also perform the fluxing function); such *dip brazing* can be automated for mass production. Heating with a torch (*torch brazing*) may be either manual or automatic, and localizes the heat effects. *Resistance brazing* uses the same equipment as resistance welding (Sec. 9.3.4), but a filler is placed between the joints. The heat-affected zone is also minimized by *induction brazing* which, however, is not suitable for Al or Mg alloys because the melting point of the filler is very close to that of the workpiece material. Close temperature control and active fluxes that remove the oxide are essential for success with these metals.

It should be noted that *braze-welding* differs from brazing in that a much wider gap is filled up with the brazing metal, and thus capillary action plays no part.

9.4.2 Soldering

Filler metals of even lower melting point (usually tin-lead alloys) are used in *soldering;* thus the joint is of much lesser strength, but parts can be joined without exposing them to excessive heat. Fluxing is again essential, usually with a zinc chloride solution for general work, or with organic rosins for electrical connections where corrosion could create a severe conductivity drop.

9.4.3 Adhesive Joints

Natural or manufactured polymeric materials, both of the thermosetting and thermoplastic varieties, as well as aqueous or organic solutions of various long-chain molecules, can be used to establish *adhesive joints* of great permanency. The joint strength relies on surface bonding, usually of the van der Waals variety; therefore *removal of surface films* is imperative. Increased strength is often obtained by *roughening* the surfaces to be joined, thus enhancing bond strength by mechanical interlocking of the adhesive to the surface.

9.5 JOINING OF PLASTICS

Of the organic polymers, some thermoplastics may be joined by techniques similar to cold-welding, that is, applying pressure and assuring some deformation to break through surface contaminant films. Diffusion and thus bonding are much accelerated when the material is heated into the plastic range by a hot iron, roller, or even hot air (PVC, PE, polypropylene), or by dielectric or radio-frequency heating (PVC), ultrasonics, or friction welding. Liquid-phase welding is also possible, using filler rods of the same material, if necessary. Because of the poor heat conductivity of plastics, the process is slow and a low-temperature heat source (hot air or gas flame) must be applied for longer times.

Nylon and acetal may be cemented with adhesives, while PMMA and PVC are also joined with solvents.

9.6   HARD-FACING

An important nonjoining application of various welding processes is *hard-facing,* in which a lower strength but tough body is coated with a very hard (and sometimes brittle) surface layer. It is used not only for repair but also for the initial manufacture of cutting tools, rock drills, forging dies, and, in general, in applications requiring wear resistance. The alloys used for hard facing resemble cutting tool materials (Sec. 8.3.1). They often have a high alloying-element concentration and cannot be manufactured into welding rods; the ingredients are therefore incorporated in the flux coating or packed inside tubular rods, and the alloy is formed in the welding process itself. Arc as well as oxyacetylene welding may be used.

Ceramic coatings are deposited from a *plasma arc* or with a special *detonation gun.* In the latter, the powder is shot onto the surface in a thin (5 μm) film by detonating an acetylene-oxygen mixture.

9.7   CUTTING

A very important application of fusion welding processes accomplishes an exactly opposite function: workpieces of varying shape are *cut* out from sheet, plate, and even very heavy sections. The heat required for melting may be provided by an electric arc, a high energy beam, or a chemical heat source.

Ferrous materials, preheated to 1600°F (900°C) burn if oxygen is blown on them. *Oxygen cutting* is widespread in steel mills for cleaning up (*scarfing*) surfaces and cutting off billets, and, in heavy machine and ship-building, for the cutting of plates.

Cutting processes are readily automated, and, with suitable tracer mechanisms or NC, they can become competitive with various sheet blanking operations.

9.8   DESIGN ASPECTS

The design of weldments is very much a function of the particular process applied. Thus, the cross sections to be joined must be equal when heat generation is a function of cross section (as in butt welding and flash butt welding), while they may be quite dissimilar when a high-energy external source is used (as in plasma-arc and electron-beam welding). In processes where a weld bead is deposited (as in most arc- and gas-welding processes) the bead should be positioned so as to be subjected to minimum bending stresses, and the joint should have properly prepared grooves (Fig.9.17):

1 Butt joints may be formed with square grooves (Fig. 9.17*a*) in thinner sheet and plate, but grooves shaped for greater weld penetration and controlled bead formation are essential for thicker gages (Fig. 9.17*b* and *c*).

2 Corner and T joints are similarly prepared (Fig. 9.17*d*).

3 The proper groove geometry is obtained by bending the edges of sheets for edge joints (Fig. 9.17*e*) made with a filler.

4 A *backing strip* is often used to seal off a wide groove temporarily.

9.9 SUMMARY

Joining expands the scope of all manufacturing processes; castings, forgings, extrusions, plates, sheet metal, and machined parts can all be joined to make more complex shapes or larger structures. Welded constructional girders, machine frames, automobile bodies, tubing and piping

FIG. 9.17 Some examples of grooves for arc and gas welding: (*a*) square, (*b*) single-V, (*c*) double-V, (*d*) corner, and (*e*) edge grooves.

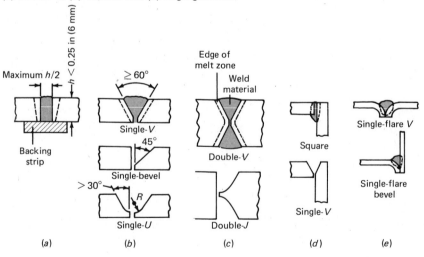

of all sizes, containers, and cans are all around us. Brazed bicycle frames, soldered plumbing joints, and adhesive-bonded structures abound. Welding is often the most economical and practical repair for broken machinery.

Joining processes, by their very definition, provide a transition between two not necessarily similar materials. Quality control is even more important than in other processes because oxidation, surface films, slag inclusions, porosity, gaps, undercuts, hot cracks, cold cracks (embrittlement), and residual stresses could cause dangerous delayed failures. Nevertheless, good-quality joints, sometimes equal to the parent (workpiece) material in strength, can be obtained through a variety of means:

1 Solid-phase welding relies entirely on adhesion; while it is extremely sensitive to surface contaminants, it does allow joining of a very wide range of similar and dissimilar materials.

2 Highly localized melting in resistance welding represents a transition from solid- to liquid-phase processes; it is still sensitive to contaminants but the heat-affected zone is small.

3 Deep or through-the-thickness melting in liquid-phase welding processes broadens the heat-affected zone in the parent metal. Surface preparation, weld geometry, protection atmospheres and/or slag and fluxes, and the rate of heating and cooling must be simultaneously controlled.

4 Joining of solid workpieces (without melting) with a lower-melting or adhesive substance again relies on adhesion, making surface cleanliness the critical factor.

Particular attention must be paid to the following: (1) localized heating and cooling makes welded structures susceptible to distortion and cracking under the influence of internal stresses; (2) welding processes generate very high temperatures, and eye protection is absolutely essential.

Further Reading

DETAILED PROCESS DESCRIPTIONS:
CAGLE, C. V. (ed.): *Handbook of Adhesive Bonding,* McGraw-Hill, New York, 1973.
Metals Handbook, 8th ed., American Society for Metals, Metals Park, Ohio, vol. 6, *Welding and Brazing,* 1971.
Welding Handbook, 6th ed. (in five sections), American Welding Society, New York, from 1968 on.

PROCESS DETAILS, MOSTLY FOR THE LABORATORY SCALE:

CARMICHAEL, D. C.: Gas-Pressure Bonding Techniques, in *Techniques of Metals Research,* R. F. Bunshah (ed.), Interscience, New York, 1968, vol. 1, pt. 3, pp. 1739–1778.

MONROE, R. E.: Joining Methods, ibid., pp. 1611–1660.

SPECIALIZED BOOKS:

CAGLE, C. V.: *Adhesive Bonding Techniques and Applications,* McGraw-Hill, New York, 1968.

GIACHINO, J. W., W. WEEKS, and G. S. JOHNSON: *Welding Technology,* American Technical Society, Chicago, 1968.

KOENIGSBERGER, F., and J. R. ADAIR: *Welding Technology,* Hart, New York, 1968.

LANCASTER, J. F.: *The Metallurgy of Welding, Brazing and Soldering,* Allen and Unwin, 1965.

LINNERT, G. E.: *Welding Metallurgy, Carbon and Alloy Steels,* 3d ed., American Welding Society, New York, 1965.

MOSS, J. B.: *Science and Metallurgy for Students of Welding,* Butterworths, London, 1967.

PATTON, W. J.: *The Science and Practice of Welding,* Prentice-Hall, Englewood Cliffs, N.J., 1967.

PHILLIPS, A. L. (ed.): *Current Welding Processes,* American Welding Society, New York, 1964.

ROMANS, D., and E. N. SIMONS: *Welding Processes and Technology,* Pitman, London, 1974.

Fastener Standards, 5th ed., Industrial Fasteners Institute, Cleveland, Ohio, 1970.

Problems

9.1 A small, complex aircraft part is to be manufactured of 6061 aluminum alloy. The highest strength obtainable with this alloy must be maintained in the entire part. (a) Check in the literature (e.g., *Metals Handbook,* vol. 1) what metallurgical condition will give the highest strength. (b) Determine whether this material would permit manufacturing the part by welding two less complex parts together. (c) If a way can be found, specify the best welding process (in justifying your choice, work by the process of elimination described in Prob. 3.3). (d) Describe what postwelding treatment, if any, is needed.

9.2 It is proposed to join aluminum to low-carbon steel. After reviewing the equilibrium diagram, consider the feasibility of establishing a reasonably ductile joint by: (a) cold-welding, (b) diffusion bonding, (c) spot welding, and (d) flash butt welding. Justify your judgements.

9.3 A single-phase resistance-welding machine is used to join two steel sheet parts with 20 projection welds made simultaneously. One assembly is joined every 10 s. If the steel sheets are 1 mm thick and individual projection welds are similar to the one shown in Fig. 9.10c, calculate: (a) the welding current required; (b) the duty cycle, expressed as the fraction (or percentage) of time the current is on; (c) the required kVA for each welding cycle, if the desired current is attained at a voltage of 4 V; (d) the kVA rating of the transformer, taking into account the duty cycle; (e) the kW rating, allowing for a power factor of 0.5 (for phase-shift due to inductive loading in the transformer); (f) the electrical energy consumption for each weld in kWh; and (g) the force requirement.

9.4   Two 5052 Al-alloy sheets of 1 mm thickness are to be joined by spot welding.   (a) Review the appropriate phase diagram and deduce if the operation is feasible.   (b) If the answer to (a) is yes, calculate the transformer rating and the power consumption as in Problem 9.3, but with one spot weld made every 2 s.

9.5   Two 0.25-in-thick low-carbon steel plates are to be joined by oxyacetylene welding in a single-V groove (Fig. 9.17).   Using data from Table 11.7, determine (a) whether this is feasible, and, if the answer is yes, (b) what welding rate (in/min) one might expect at the top metal deposition rate.

9.6   Stainless steel sheets of 0.4 mm thickness are to be joined (butted) on their edges.   (a) What processes could be considered, and (b) which of these is likely to be most frequently available.

10

MANUFACTURING SYSTEMS

The emphasis in this book is on unit processes needed to manufacture hardware. Without them there is no manufacturing, yet on their own they cannot provide the finished goods. As indicated in Fig. 1.2, there are further, indispensable elements of making a manufacturing enterprise into a viable undertaking. Each of these elements has developed into separate and sometimes highly specialized activities, and we cannot do more than discuss the general relation of these activities to unit processes; in doing so, concepts will be defined and their influence on the choice of unit processes will be discussed.

10.1 PRODUCTION CHARACTERISTICS

Two important factors in the choice of the process are the total *number of parts* produced and the *rate of production* (i.e., the number of units produced in a time period such as an hour, month, or year). The *batch size* or *lot size* is the number of units produced in an uninterrupted run. It is customary to speak of *small-batch* (1 to 100 units), *batch* (over 100), and *mass* (over 100,000 or even 1 million units) *production*. In general, a larger batch size justifies the choice of processes with inherently higher production rates. The total production quantity and the production rate together define the justifiable expenditure on special machinery and tooling. When the total production quantity is insufficient to keep a production unit continuously occupied, the lot size will be determined by weighing the cost of *setting up* (changing over) against the cost of *stocking* (warehousing) parts between production runs. All these decisions affect costs, profitability, and productivity, as can best be demonstrated in a simple (and oversimplified) example.

A ubiquitous manufactured product is the automobile. Half the total weight is sheet metal. To take one part, the outer door panel is a fairly sophisticated product with a shallow curvature, complex outer profile, and several cutouts. Several car models of a manufacturer share the same door, and the production quantity over a typical 7-year model period may account to, say, 6 million units for a mass-produced car. This justifies a special-purpose production line (a transfer line) in which several presses perform the required sequence of operations, with mechanical means of passing the part from one press to another. The line is tended by three operators in each of three shifts, and their productivity can be expressed as the number of units produced per year (Table 10.1). A more generally applicable measure of *labor productivity* is value produced, and if we assume (without further justification) that the door panel is worth $3.00 for the manufacturer, productivity can be expressed in value produced per operator (Table 10.1).

At the other end of the scale, some luxury cars are made by specialized companies in small numbers, say 10,000 units over 10 years. Equipment and tooling costs have to be reduced by stretch-forming and low-cost blanking (such as rubber forming) techniques, for example. The plant operates perhaps on only one shift and we can assume that five operators produce the requisite 1000 door panels per year. This may, of course, be split up into several lots spread over the year, and we assume that equivalent parts are produced for the rest of the time for a total of 1.2×10^4 units per year. The door panel, in the luxury market, is now worth $26.00 to the manufacturer, and productivities can again be calculated (Table 10.1).

When productivities are compared on either a unit or dollar basis, there would seem to be no room for the batch-produced automobile even in the luxury market. However, our calculation ignored several important factors. First of all, the nine operators of the transfer line are backed up by a team of, say, 18 people in maintenance, supervision, quality control, and various levels of management. The five operators of the batch production plant perform some of these functions themselves and are supported by only two people. Productivities now change (Table 10.1).

These numbers reveal nothing about *costs* or *profits.* For these, the means of production must also be considered. *Capital* is invested in the purchase of presses and auxiliary equipment, their installation, and the requisite buildings and services, and must be regained through some time period that formed the basis of investment decision. The actual annual repayment (amortization) depends on interest rates, tax treatment, and accounting procedures. The same argument applies to the cost of tooling, except that it has to be repaid over a shorter time period.

The *production cost* per unit is still vastly different (Table 10.1) between mass and batch production, and the difference remains when the cost of material (somewhat higher in batch production because of the

TABLE 10.1 COMPARISON OF PRODUCTIVITIES AND COSTS IN A SHEET-METALWORKING OPERATION

	Mass production	Batch production
Production, units	6×10^6	1.2×10^4
Value generated, $	1.8×10^7	3.1×10^5
Number of operators	9	5
Number of employees	27	7
Productivity, units/operator	6.7×10^5	2.4×10^3
$/operator	2×10^6	6.25×10^4
Productivity, units/employee	2.2×10^5	1.7×10^3
$/employee	6.4×10^5	4.4×10^4
Capital outlay, total $	10×10^6	4×10^5
Amortization, $/annum	4×10^6	8×10^4
Tooling cost, total $	5×10^5	1×10^5
Amortization, $/annum	3×10^5	4×10^4
Cost of tool changes, $/annum	1×10^4	1×10^4
Labor cost, $/employee	2×10^4	2×10^4
Total $/annum	5.4×10^5	1.4×10^5
Production cost, $/annum	4×10^6	8×10^4
	$+ \quad 3 \times 10^5$	$+ \quad 4 \times 10^4$
	$+ \quad 1 \times 10^4$	$+ \quad 1 \times 10^4$
	$+ \; 5.4 \times 10^5$	$+ \; 1.4 \times 10^5$
	4.85×10^6	2.7×10^5
Production cost, $/unit	0.86	22.50
Material cost, $/unit	1.50	2.00
Total cost, $/unit	2.36	24.50

Note: All values are fictitious and bear no relation to any existing manufacturer's operation.

greater losses in rubber blanking and stretching) is added. After deducting other costs not accounted for in Table 10.1 (e.g., carrying an inventory of parts), the *profit* for the two operations is obtained.

10.2 MECHANIZATION AND AUTOMATION

The above example shows the benefits obtained through mechanization of mass-production processes. Before going further, it is necessary to clarify the terms mechanization and automation.

   Mechanization means that something is done or operated by machinery and not by hand: industrial development has been mostly a story of mechanization. Feedback is not provided and thus one deals with open-loop systems.

   Automation means a system in which many or all of the processes in the production, movement, and inspection of parts and material are automatically performed or controlled by *self-operating devices.* This implies that the essential elements of automation include—in addition to mechanization—*sensing* and *feedback devices* and some degree of *decision making;* thus, a closed-loop system is created.

Frequently, the meaning of automation is loosely expanded to encompass mechanization, and then the term describes all actions that make the life of the worker easier while also increasing productivity. A careful distinction must be made, however, between productivity of labor and productivity of capital. Mechanization and automation, implemented sensibly, increase the productivity of labor; in the example of the transfer line, 3 or 4 times as many operators would be needed to move the part from press to press by hand. The *productivity of capital* (holding the quality and quantity of other inputs fixed) increases only if the efficiency of new capital goods (in this instance, of the mechanical handling devices) grows faster than the price of these goods. Capital productivity is much more difficult to measure and, therefore, more controversial than labor productivity. What is more easily seen is that there must be an *economic limit* on automation where further investment yields *diminishing returns* in terms of cost reduction.

There are a great many levels of mechanization and automation, and each of them has its rightful place, dictated often by factors other than economy. Such factors are more consistent quality, reduced operator fatigue, and a better working environment (less exposure to noise, noxious fumes, gases, and hazards). Great impetus was given by legislation [such as the Occupational Health and Safety Act (OSHA) in the United States and similar measures elsewhere], the concern for the environment [Environmental Protection Agency (EPA) regulations], and the need for better material and energy utilization.

Automation reduces the number of operators and also demands a different mix of skills, shifting some of the production functions to specialists while removing certain special skills from the operator level, or replacing the operator with a machine supervisor. This creates its own problems, not the least of which are operator boredom and mismatch between the output of the educational system and the needs of industry.

10.2.1 Mechanization and Automation of Unit Processes

We have already encountered numerous examples of mechanization.

Mechanization for mass production

The engine lathe is the classic example of a manually operated machine, but even there the feed becomes mechanized when the feed-rod is engaged. A further step is the turret lathe, on which the travel of the tool is mechanically limited. The automatic screw machine with its cam-actuated tools and sequencing represents a high degree of mechanization. Similarly, the process of upsetting nail heads was once a manual job; in the nail-making machine no human interference is needed or indeed possible, apart from setting up the tooling, feeding the end of a wire coil, and turning on and shutting off the machine.

These so-called *automatics* are a testimony to the ingenuity of the mechanical engineer. They utilize purely mechanical means of control, and offer unparalleled reliability and very high production rates. They are, however, quite inflexible, and while reprogramming and retooling is routinely done, they are time-consuming and limit application to mass or large-batch production.

Numerical control

Numerical control (NC) of machinery is a development of the 1950s that continues to cause changes of vast technological and social consequence. It brings the benefits of mass-production technology to small-batch production. Spreading from the aerospace industry, it has already revolutionized some job-shop operations.

Before NC can be applied, the machine tool itself must be changed. The fixed gears (Fig. 10.1a), clutches, cams, and tracer templates of the mechanized machine and the hand-wheels of others are all dispensed with. For example, *backlash-free lead screws* may now be fitted to the table of a machine tool and actuated with stepping *motors* that move the table in small increments of, say, 0.0001 in (0.0025 mm), the number of increments depending on the number of pulses received. By presetting the number of pulses, the machine-tool operator controls a mechanized, *open-loop control machine tool* (Fig. 10.1b) for, say, drilling holes or making spot welds in locations specified on a part drawing. Most machinery suitable for NC goes a step further: the drive of the feed screw (or the valve of a hydraulic cylinder, etc.) is controlled by electric, hydraulic, or pneumatic (fluidic) means while the position of the table is continuously monitored by a high-resolution *rotary* or *linear transducer* (Fig. 10.1b). The signal from this transducer is processed by a *comparator* (if necessary, after conversion into a digital form) which compares it with the *control signal,* and then issues an *error signal* to correct the position and/or speed of the table. Thus, an essential element of automation, namely feedback, is present in *closed-loop control.* If the position of the table is simply displayed on a digital readout, a *digital-control machine tool* is obtained which offers more accurate positioning and rapid actuation than its mechanically actuated predecessor; this is actually a recent development.

The original NC concept goes one step further and replaces operator actions with *digitized commands* from a *punched tape* (or *card,* or *mag-*

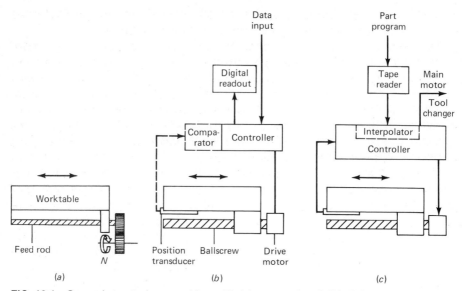

FIG. 10.1 One-axis control on machine with (*a*) conventional, (*b*) digital open-loop control (broken lines for closed-loop control), and (*c*) numerical point-to-point control (broken lines for continuous control).

netic tape). The command may specify the position of the table on *x-y* coordinates, for example, for drilling or punching holes, setting wires, or bending tubes, by a sequence of discrete movements (*point-to-point system*). Greater flexibility is offered by the *continuous-path* (*contouring*) system. The position of the tool relative to the workpiece is again specified by a series of coordinates but the control system is now designed to follow the required curved path between adjacent coordinate points either in very small linear increments, or along continuous circular or parabolic curve segments (*linear, circular,* or *parabolic interpolation,* respectively, Fig. 10.1*c*). It is then possible to mill a pocket of complex contour, or to weld a seam or cut a plate along a tortuous path.

A machine with controls for the *x* and *y* directions would be described as a *two-axis NC machine*. Vertical movement of the table (or of the tool) would add the third axis, while tilting around one or two axes (either of the table or of the tool head) would make it into a four- or five-axis machine. A five-axis milling machine is able to contour a spatial surface.

It is an easy matter to punch on the tape commands for actuating the machine tool or to bring into position various tools mounted on a turret or stored in a *tool magazine*. Multiple operations may be performed and parts of complex geometries produced in a short time, particularly on multispindle machines with automatic tool-changing facilities (*machining centers*) that have access to the workpiece on all sides but the base face.

One should note, however, that the dimensions of the cutting tool must still be separately determined and entered into the program.

Simple point-to-point programs can be developed by hand, but more complex ones, as well as almost all contouring programs, are developed with the aid of computers (Fig. 10.2a). Specialized *programming languages* (e.g., APT, for Automatically Programmed Tool) have been developed that facilitate generation of tapes with relatively little knowledge of computers. The part is described in a simplified Englishlike language; the computer translates this input into numbers and computing instructions and performs the calculations. The output must then be translated, with the aid of *postprocessor programs,* into a form acceptable to the particular machine tool control. As opposed to *part programming,* programming of the computer and the postprocessor calls for substantial computing background. All the process information, and decisions on speed, feed, and operational sequences, still must come from an intimate knowledge of the process. If so desired, Taylor's equation, Eq. (8.5), can be incorporated to compensate for tool wear.

FIG. 10.2 Main features of (*a*) computer numerical control and (*b*) computer-aided manufacturing.

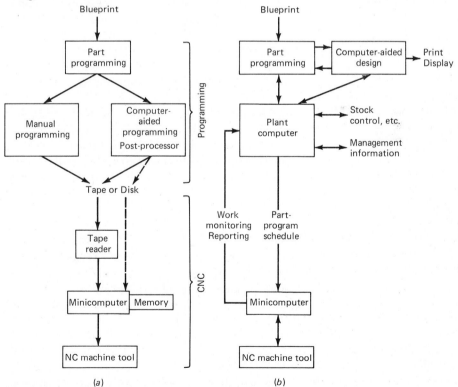

(*a*)

(*b*)

CNC, DNC, and CAM

The flexibility of an NC machine tool is greatly increased if the hard-wired controls are replaced with a programmable mini- or micro-computer (CNC for *Computer Numerical Control,* Fig. 10.2a). For repetitive production, their memory can be utilized for temporary storage of the part program being run, but tape (or disk) is still the medium of transmitting the program from the programming computer to the dedicated computer of the machine tool. Direct link is established between the two computers in DNC (*Direct Numerical Control*), but the expenses of this can usually be justified only if the system is expanded (Fig. 10.2b) to encompass process management and control (CAM, *Computer-Aided Manufacturing,* sometimes combined with CAD, *Computer-Aided Design*).

Adaptive control

A further step toward true automation is taken by *adaptive control* (AC). Long employed in some primary industries (e.g., in the rolling of flat products), the vital element of decision-making is added. In the simplest form, for example, the torque on a drill is measured and speed and feed, or both, are adjusted within preprogrammed limits. A more complex case is milling, where feed (and/or speed) and depth of cut are controlled in response to signals received from torque, force, and vibration transducers, and—at present to only a limited extent—dimensional transducers. Limits set by the process (surface finish, cutting force and torque, maximum speed, feeds, and depth of cut) and by the machine tool (vibration and elastic deflections) are preprogrammed.

Since there is so much that is still unknown about the effect of process variables on the finished part, adaptive control cannot grow until a more quantitative understanding of processes is gained and process reliability is improved. Among others, a major problem is on-line measurement of dimensions without which the wear of tools cannot be compensated for with any great accuracy.

10.2.2 Mechanization and Automation of Material Movement

Closer observation of many seemingly well-run processes reveals that the part is being worked upon perhaps only 5 percent of the time, the rest being spent in moving the workpiece and tooling. Recognition of this fact had led to attempts at *mechanization* and *automation of material movement.*

Automation in mass production

As with the production process itself, handling and moving the part has first been tackled in mass production. Some prime examples are progressive and transfer dies (Sec. 5.5.4) in sheet metalworking; upsetters and cold headers (Sec. 4.6.3) in bulk deformation; automatic screw machines (Sec. 8.5.8) in machining; and bottle-making machines for glass and plastic containers. All these are characterized by a large-volume production of identical parts which simplifies the problem of holding the part firmly, in a fixed position relative to the tool. The part is moved from station to station by mechanical means and the only manual movement remaining is that of loading a new length of bar or strip.

The problem is greater but not insurmountable when irregularly shaped parts are to be made in large quantities, typically by the process of machining. After establishing a reference surface (*qualifying the part*), for example, by high-speed machining of a flat, the part can be clamped to a base (or fixture) and moved past a number of tools that may be accommodated in a single machine (usually, in a carrousel-type arrangement, as in chucking automatics) or past a line of special machine heads designed to perform one or several operations at every station (*transfer lines,* as used for machining automotive engine blocks). The flexibility of such lines can be increased by the use of standardized *modular components,* quick-change tool heads, and provision for performing the more time-consuming operations on a *branch line.*

Mechanical loading and transfer devices are used to advantage for moving components of highly varied shape from one machine to another. Special jaws grip the part, lift, move, and turn it on arms, and place it into the new work position. Travel distance, direction, sequence, and speed are controlled mechanically, electromechanically, or with fluidic controls; dead-stops and limit switches are widely applied. Such *mechanical arms* or *iron hands* can be reprogrammed, but not very rapidly.

Automation in batch production

Many examples of small-batch production are to be found in job shops, die-making shops and in the aerospace and machine-tool industries. The accuracy of the final product depends not only on the machine but also on the skill of the operator who has to set up the workpiece for, typically, machining or welding. The task is greatly facilitated by *jigs* (which hold the workpiece or are held on it, and incorporate a guide for the tool) and *fixtures* (which firmly hold one or more workpieces in the correct position in relation to the machine tool bed and to each other). The more complex fixtures can become too expensive (or too cumbersome to handle) for short production runs, and it is here that the greatest benefits of NC are reaped.

Once the workpiece is clamped in place on the NC machine tool table and a reference point is established, machining, bending, welding, cutting, etc., proceed with great accuracy and repeatability. Nonproductive setup time is practically nil; therefore, NC can become economical even for very small lots (Fig. 10.3) widely separated in time. If part programming is cheap and tape proving is performed off the machine tool (e.g., by cathode-ray tube or automatic plotter display), NC is justifiable even for one-off production.

Loading and unloading, and movement of material from one machine to the next, are still very time-consuming, and recent efforts aim at automating these functions. Mechanical arms with more flexible programming capability are required. Very often, electromechanical or fluidic devices are perfectly adequate, but for greatest freedom of movement and ease of preprogramming, NC *industrial robots* (Fig. 10.4) are needed. Programming is often done by *teaching* the robot, that is, taking it through the sequence of movements by manual control so that it may then replay the program any number of times, without human fatigue or error. Such robots can perform not only the relatively simple tasks of loading and unloading die-casting and plastic-molding machines, forging presses, welders, and other machine tools, but they can also guide a welding gun over a spatial track required to weld, say, pipes into a T joint. Most of these robots are still without feedback sensors and all parts and tools must be located in exact positions. There is, however, work in progress to endow them with *artificial intelligence,* so that they may grope for misplaced (or randomly placed) parts and *recognize* parts of different shapes.

The ultimate development is the fully automated factory where unit processes, material movement, operational sequence, tool changing, and even assembly are accomplished without operator assistance.

FIG. 10.3 Typical trends in the economy of manufacturing as a function of batch size (without carrying costs for inventory).

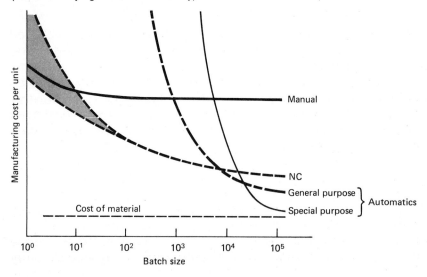

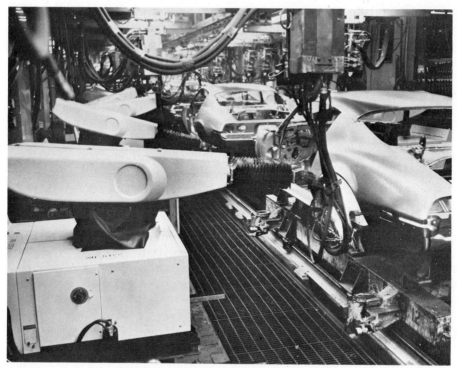

FIG. 10.4 Industrial robots performing automatic welding of automotive underbodies along an automatic assembly line. (Unimation Inc., Danbury, Conn.)

10.3 ASSEMBLY

In the final phase of manufacturing, individual components are assembled into the end product. This presents a wide range of problems, depending on production quantities.

Manual assembly is still the only method practicable for small-batch production. In larger production quantities, the repetitive nature of work, the danger of making errors when hundreds of parts are involved, and the low overall efficiency has led to early attempts at organizing and mechanizing assembly operations.

Assembly lines

In assembling a complex machine, great progress can be made by breaking down the operation into smaller units; this also facilitates material handling by assuring that all parts can be supplied in their proper place and sequence. This concept led to the *assembly line,* pioneered by Henry

Ford in 1913. The units to be assembled move on a conveyer at a preset rate while operators stationed along the conveyer perform their assigned tasks. This assembly method, more than any individual advance, has made possible the mass production of consumer goods that previously were regarded as a luxury. However, the monotony of work has led to some dissatisfaction with the system, resulting in attempts at replacing it with alternative yet similarly productive methods. One potentially attractive solution entrusts an entire assembly (e.g., an automobile engine) to a group of operators who are given considerable freedom in organizing themselves and who also perform the quality-control function. The alternative, of course, is mechanization and automation.

Mechanization of assembly

Some types of assembly operations lend themselves to fairly simple mechanical methods of assembly. Thus, screws or bolts can be driven and parts placed, crimped, or riveted with mechanical devices. The cost is reduced while productivity and consistency of product increase, but only if the reliability of mechanization is very high. The cost of off-line repairs can quickly cancel all savings. A crucial factor of success is *in-line inspection* to pinpoint and remove imperfect assemblies, either during or at the end of the assembly operation. Many of the elements used in automatic assembly are the same as in mechanized production and workpiece handling, and may be purely mechanical, electromechanical, or fluidic, or numerically or computer controlled.

Small parts are often handled effectively with simple mechanical devices. Vibratory belts and bowls, reciprocating forks, rotary disks, or magnetic devices are combined with simple but ingenious work-orientation devices from which the parts progress through feed tracks to the assembly machine, where a metering device (such as a mechanically actuated escapement) assures release of the correct number of parts at the proper time. During assembly the unit may move continuously; *indexing workheads* move with it and retract after completing their task, backtrack, and repeat the operation on the next unit to come along. Alternatively, the line itself indexes and *stationary workheads* perform the operation while the line is at rest. Indexing may take place at a preset rate, or it may be triggered by a signal received on completing the operation. In the latter case, buffer storage between stations equalizes production rates. If one of the operations is more time-consuming, a branch feed line can be set up for it.

In all instances, assembly may be performed *in line,* that is, along a conveyor on which parts move (if necessary, on pallets that assure accurate positioning). Alternatively, the assembly machine may be of the *rotary* kind with a carrousel carrying the unit from station to station.

10.4   PRODUCTION ORGANIZATION

A plant equipped with efficient machine tools and served by the best means of material movement and assembly can still lose money. For profitable production, the physical means of production must be backed up by an effective *production planning and control organization.*

10.4.1   Company Organization

Neither company organizations nor the terminology used to describe organizational elements are standardized. Nevertheless, the *organization chart* shown in Fig. 10.5 is fairly typical.

All manufacturing concerns have, at the company level, finance, personnel, purchasing, and sales departments. *Research* and *development,* and *marketing* and *market development,* found in larger concerns, are essential for their growth. Depending on the nature of operations, there may also be *product engineering* which takes care of the development, design, testing, and evaluation of new products, ideally with the full involvement of the manufacturing group.

At the plant level, all facets of production are under the direction of the plant manager. Actual production is headed by the *manufacturing* superintendent, with the assistance of superintendents, general foremen, and foremen. Their job is to keep production going at peak efficiency, a task so complex as to need support from several other departments free from everyday pressures:

FIG. 10.5   Typical organizational structure.

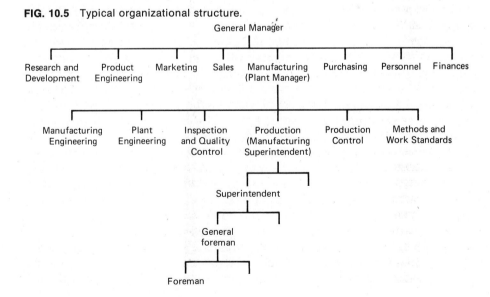

1 *Production control* determines economical lot sizes; establishes schedules for manufacturing and assembly; controls the inventory of raw materials and in-process parts; dispatches materials, tooling, and equipment to the plant locations specified by the schedule; keeps track of progress; and often provides program evaluation review.

2 The *methods and work standards* group provides not only time and motion studies and sets norms, but very often analyzes the whole process (sometimes attached to or working with the process engineering department).

3 *Plant engineering* is responsible for preventive maintenance of equipment, replacement of machinery, and provision of services (power, heating, lighting, etc.).

Inspection and quality control are a vital activity and will be discussed in a separate section.

10.4.2 Manufacturing Engineering

Central to our considerations is the *manufacturing engineering* (also called *process engineering*) group, usually headed by the chief engineer. It is in this group that technological awareness resides, and the competence of its people determines whether the company will be competitive and profitable. Typical tasks include evaluation of manufacturing feasibility and cost, selection of optimum processes and process sequences, equipment, tooling, jigs and fixtures (their design and manufacturing, and control of the tool room), material movement methods and equipment, and optimum plant layout. In addition to specifications for unit processes discussed in this book, the group also issues specifications for auxiliary and finishing processes such as heat treatment, cleaning, painting, plating, coating, and, in general, treating of individual parts and finished assemblies. Cost analysis (Sec. 11.3) is often performed here, and the group also has a central function in value analysis (Sec. 11.4). It cooperates very closely with research and development specialists.

With increasing mechanization and automation, and particularly with the special demands set by numerical and computer control, the activities of the process engineering department account for a substantial share of the total production cost, and there are convincing reasons why at least some of these activities should be regarded as direct rather than indirect labor (Sec. 11.3.2). The true *cost effectiveness* of new manufacturing technologies can then be judged, and the need for a thorough reorganization of production facilities revealed. This is particularly true of numerical control, which often results in no savings unless the entire manufacturing concept is changed. This may mean reorganization of the methods of production preparation, and very often involves physical reorganization of the entire plant as well.

10.4.3 Plant Layout

Plant layout is a relatively simple matter for mass production: even the casual observer will note that material must flow in a logical sequence through various stages of production until it emerges as the finished product. Thus, an in-line arrangement is most frequently chosen. In so choosing, one trades flexibility for increased efficiency, and the approach is not feasible for small-batch production.

Manufacturing plants catering to varied, variable, and often unforeseen needs tend to grow in a higgledy-piggledy fashion. New production equipment is placed where room is found for them; finally, material movement becomes so disorganized and time-consuming that the benefits of more productive equipment are totally lost. Separating machine tools of a kind (e.g., lathes, drill presses, etc.) into groups offers improvements of often doubtful value.

In small-batch production the production-line concept is next to useless, and a more rational approach is found in the application of *group technology.* This concept states that many problems have similar features, and if these problems are solved together, great efficiency results. In applying the concept to manufacturing, individual parts are analyzed in terms of the operations required to make them. To the first order of approximation, this means that their *shape* must be *classified,* because similar shapes tend to be produced by similar techniques. This need has led to efforts at typifying parts according to their shapes and geometric relationships and, while no universally accepted system exists yet, some progress has been made. Once parts belonging to the same *group* (or *family*) are identified, reorganization of production equipment can be justified if total production is sufficiently large. All the equipment necessary to produce the finished parts is grouped into a *cell,* comprising, for example, a complex (and expensive) machine (such as a NC machining center) supported by several special-purpose and therefore lower-cost machines. The part then can be transferred with minimum movement and wasted time from one unit to the other (Fig. 10.6). Alternatively, the machines may be laid out along a conveyor line in the sequence of operations, and thus a transition is created between cell and the modular-construction transfer line (Sec. 10.2.2). The flexibility of cell arrangement is, to some extent, still maintained.

Group technology offers its greatest benefits when it is extended to all phases of production and production preparation, including drafting (and computer-aided design) and part programming for NC.

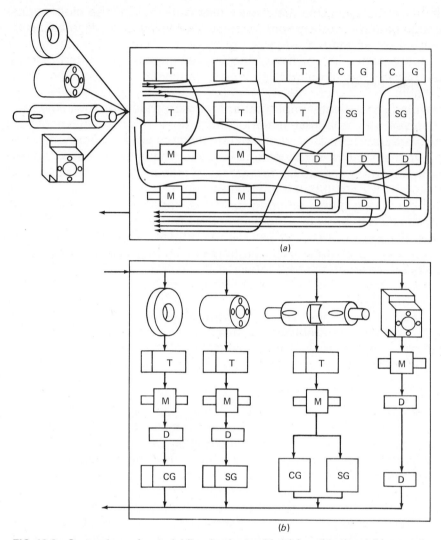

FIG. 10.6 Comparison of material flow in plants with (*a*) functional and (*b*) group layout. T, turning; M, milling; D, drilling; SG, surface grinding; CG, cylindrical grinding. (C. C. Gallagher and W. A. Knight, *Group Technology,* Butterworth, London, 1973, p. 2.)

10.4.4 Choice of Optimum Manufacturing Methods

It is clear from the preceding discussion that, even if the optimum manufacturing process is selected (on the basis of considerations given in Chap. 11), the choice of optimum production rates and associated machinery requires still further considerations.

In general, it can be said that manual control results in smallest capital and largest labor outlay; therefore, it best suits one-off production, although its position is challenged by properly chosen NC systems. For small-batch production, group technology (often employing individual machines with numerical control) is likely to be most economical, while programmable automatics are essentially limited to large batches, and special-purpose automatics are limited to mass-production facilities.

Automobiles, appliances, and consumer goods generally fall into the last two categories and are relatively efficiently produced. Machine tools, off-the-road and railroad equipment, heavy machinery, and aircraft usually fall into the batch-production categories. For this reason, the products of these industries are characterized by a large expenditure for the main parts of complex shape and a relatively small expenditure on the much more numerous, mass-produced, purchased components (Fig. 10.7). Obviously, greatest economy will be assured by organizing effective production of the complex main parts. It is here that numerical and computer control and group technology offer the greatest advantages.

FIG. 10.7 The application of various manufacturing approaches to the components of machinery produced in small lots, by value and number of parts. (H. Opitz and H. P. Wiendahl, *International J. Prod. Res. 9*:181–203, 1971.)

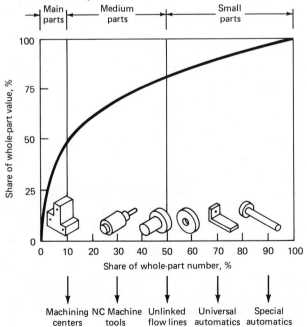

10.5   QUALITY CONTROL

A plant of high output rate is not necessarily productive, because only products that satisfy the application criteria are of any value. Some *reject rate* is often unavoidable but high reject rates can nullify all technological and organizational advances.

Quality control is a task that must be shared by everyone involved in manufacturing. Certain functions can and usually are centralized in a quality control department. These functions include examination of incoming materials and parts, and of finished products, for conformance to specifications, and may require extensive instrumentation and laboratory facilities. In-process quality control is a much more diffused function.

In one-off and small-batch production, the primary duty rests with the operator who has to check all visible or measurable manifestations of quality, if possible, with the aid of *gages* or specialized *instruments.* A 100-percent inspection scheme can, however, slow production excessively, and it is then necessary to arrive at a *sampling strategy.* This may involve checking of every fifth, tenth, etc., part if the anticipated changes are monotonic (e.g., dimensional changes due to tool wear), or checking of randomly picked parts if changes are due to interactions of several random variables (e.g., in casting processes).

The strategy of sampling becomes most important in large-batch and mass production, and *statistical sampling* techniques are usually employed. Ideally, of course, all parts and assemblies should be inspected at every important stage and this can be economically accomplished with *automatic inspection* techniques. Numerical and computer control techniques can be applied; the limiting factor is, just as in adaptive control, the availability of in-process transducers and sensors. These must include not only gaging devices but also *nondestructive methods* of measuring other quality indicators (freedom from cracks, folds, holes, cavities, internal stresses, etc.).

10.6   SUMMARY

Manufacturing industries continually strive to increase productivity and reduce cost, while also improving working conditions. This has led to extensive mechanization and automation of established mass-production techniques.

New technologies can, of course, completely change entire industries. A good example is solid-state technology, which revolutionized the manufacture of electronic and computer hardware. This, in turn, is gradually changing the approach to the automation of small-batch production, and is likely to bring about yet dimly perceived social changes that, in a matter of a few decades, will perhaps be viewed as the second Industrial Revolution.

Further Reading

NUMERICAL AND COMPUTER CONTROL:

CHILDS, J. J.: *Numerical Control Part Programming,* Industrial Press, New York, 1973.

CHILDS, J. J.: *Principles of Numerical Control,* 2d ed., Industrial Press, New York, 1969.

LESLIE, W. H. P. (ed.): *Numerical Control Users' Handbook,* McGraw-Hill, New York, 1970.

OLESTEN, N. O.: *Numerical Control,* Wiley, New York, 1970.

ROBERTS, D., and C. PRENTICE: *Programming for Numerical Control Machines,* McGraw-Hill, New York, 1968.

SIMON, W.: *The Numerical Control of Machine Tools,* Arnold, London, 1973.

WILSON, F. W. (ed.): *Numerical Control in Manufacturing,* Society of Manufacturing Engineers, McGraw-Hill, New York, 1963.

FIXTURES AND ASSEMBLY:

BOOTHROYD, G., and A. H. REDFORD: *Mechanical Assembly,* McGraw-Hill, London, 1968.

GRANT, H. E.: *Jigs and Fixtures,* McGraw-Hill, New York, 1967.

HENRIKSEN, E. K.: *Jig and Fixture Design Manual,* Industrial Press Inc., New York, 1973.

JONES, E. J. H., and R. C. TOWN: *Production Engineering Jig and Tool Design,* Newnes-Butterworths, London, 1972.

TIPPING, W. V.: *Mechanized Assembly,* Business Books, London, 1969.

WILSON, F. W. (ed.): *Handbook of Fixture Design,* American Society of Tool and Manufacturing Engineers, McGraw-Hill, New York, 1962.

GROUP TECHNOLOGY:

GALLAGHER, C. C., and W. A. KNIGHT: *Group Technology,* Butterworths, London, 1973.

EDWARDS, G. A. B.: *Readings in Group Technology: Cellular Systems,* The Machinery Publishing Co., Brighton, Sussex, 1971.

RANSON, G. M.: *Group Technology,* McGraw-Hill, London, 1972.

QUALITY CONTROL:

KIRKPATRICK, E. G.: *Quality Control for Managers and Engineers,* Wiley, New York, 1970.

Automated Inspection for Defects and Dimensions, British Scientific Instrument Research Association, Adam Hilger, London, 1969.

PRODUCTION PLANNING AND MANAGEMENT:

AMRINE, H. T., A. RITCHEY and O. S. HULLEY: *Manufacturing Organization and Management,* 3d ed., Prentice-Hall, Englewood Cliffs, N.J., 1975.

BLANCHARD, B. S.: *Engineering Organization and Management,* Prentice-Hall, Englewood Cliffs, N.J., 1976.

BUFFA, E. S.: *Modern Production Management,* 4th ed., Wiley, New York, 1973.
BURBRIDGE, J. L.: *Production Planning,* Heinemann, New York, 1971.
EILON, S.: *Production Planning and Control,* Macmillan, New York, 1962.
HULL, J. F.: *The Control of Manufacturing,* Gower Press, Epping, Essex, 1973.
LOCKYER, K. G.: *Production Control in Practice,* 2d ed., Pitman, London, 1975.
RIGGS, J. L.: *Production Systems: Planning, Analysis and Control,* Wiley, New York, 1970.

11

COMPETITIVE ASPECTS OF MANUFACTURING PROCESSES

In the introduction (Sec. 1.2) we noted that design criteria, material properties, and manufacturing technology are inextricably interwoven. These points were emphasized, whenever possible, in Chaps. 3–9, which dealt with individual process categories; nevertheless, there is a danger that the principle is obscured by the inevitable mass of details. It is appropriate, therefore, to take a critical bird's eye view of processes.

The crucial point is the one expressed in Sec. 1.1: it is not enough that a process should be capable of producing parts that satisfy service requirements and other constraints; it must do so at *minimum cost*. What this minimum is depends, of course, on the state of technology. It is possible that a less-than-optimum process remains profitable for some time for an individual company because of local economic, trade, labor, or transportation conditions or because of protective trade barriers. In the long run, such conditions may prove harmful; when exposed to the full force of competition, entire industries may wither and disappear.

The necessity of keeping up with (or, preferably, providing leadership in) new developments in a particular field is recognized by many. What is often overlooked is the necessity of considering alternative, *competitive processes* that may encroach on or take over traditional markets. The discussion that follows is based on the premise that there is no absolute and eternal optimum solution to any manufacturing problem, and that all materials and processes have an equal right for at least initial consideration. These initial considerations will be tempered by restraints imposed by part geometry.

11.1   DIMENSIONAL ACCURACY AND SURFACE FINISH

A design decision of great consequence for manufacturing process choice and economy is specification of the required accuracy and surface finish. The subject is, surprisingly enough, seldom covered in design courses, and, because of its manufacturing implications, it will be appropriate to give a brief introduction here.

11.1.1   Tolerances

No process is capable of finishing a part to an absolute dimension, and deviations from the intended size must be allowed even if the part is to mate with another one.

Specification of *dimensions* is a designer function. The *basic size* of one of the mating parts is first determined. In principle, this could be either a hole or a shaft. The manufacture of holes often requires special tooling (drill or punch), and hole diameters can be more difficult to measure during processing; therefore, designers often prefer the *basic hole system* in which a hole of basic size is chosen; then the appropriate shaft dimension is specified to provide the required functional *fit* between the mating parts (Fig. 11.1). Clearance fits allow sliding or rotation, such as is required in a full-fluid bearing (Fig. 11.1*a*). Transition fits provide accurate location with slight clearance or interference, depending on the class of fit (Fig. 11.1*b*). Interference fits assure a negative clearance (interference) and are designed for rigidity and alignment, or even develop a specified pressure (shrink pressure) on the inner part (Fig. 11.1*c*).

The next step is determination of *tolerances,* that is, the permissible difference between the maximum and minimum limits of size. Experience has taught that, in any manufacturing process, dimensional inaccuracies are proportional to the cube root of the absolute size (denoted D, for diameter, in units of in or mm). Therefore, in the ISO (International Standards Organization) system of fits and limits, a *tolerance unit i* is calculated (in units of 10^{-3} in or μm) from

$$i = 0.45D^{1/3} + 0.001D \tag{11.1}$$

and the *grade of tolerance* is then expressed as a standardized multiple of *i* (e.g., the tightest tolerance, designated IT5, implies a standard tolerance of $7i$, while the loosest, IT 16, implies $1000i$).

With reference to the basic hole (with zero lower deviation), the shaft tolerance zones—covering the entire range of fits—occupy one of 28 different positions, each designated by a lowercase letter (or letter combination). Figure 11.2 shows an example for a clearance fit. If the *basic shaft system* is used, the hole size is determined by similar considerations, but the position of the hole tolerance zone is now shown by capital letters.

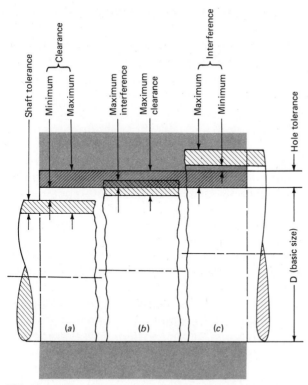

FIG. 11.1 Basic hole system: (a) clearance, (b) transition, and (c) interference fits.

FIG. 11.2 Example of a running fit (ϕ 25 H7/f6, dimensions in mm unless otherwise shown).

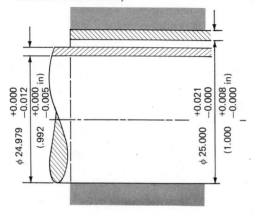

It is obvious from the above description that a tolerance has an absolute value. With reference to the basic size, it can be expressed as deviation in both upper and lower directions (*bilateral tolerancing*), or in only one direction (*unilateral tolerancing*) if the consequences of deviation in that direction are less dangerous.

11.1.2 Surface Roughness

No surface can ever be absolutely smooth and flat (or of cylindrical or other pure geometrical shape).

On the finer scale, a surface exhibits waviness and *roughness.* Both can be quantitatively measured by various techniques, among which the stylus instrument is the most widespread: an arm with a reference rest is drawn across the surface, while a stylus (similar to that of a record player) follows the finer surface details (Fig. 11.3a). The *surface profile* can then be recorded after suitable amplification, usually with a larger gain on the vertical axis (Fig. 11.3b), or the signal may be electronically processed to give numbers that characterize the surface roughness (at least for production-control purposes).

R_t is the *height from maximum peak to deepest trough;* it is important when the roughness is to be removed, for example, by polishing.

R_a (also called the *center-line average* CLA, or *arithmetical average* AA) is the *average deviation* from a mean surface

$$R_a = \frac{1}{l} \int_0^l |y| dl \quad \text{or} \quad R_a = \frac{y_1 + y_2 + y_3 + \cdots + y_n}{n} \tag{11.2}$$

The *root mean square* (RMS) value is frequently preferred in practice

$$RMS = \left(\frac{1}{l} \int_0^l |y^2| dl \right)^{1/2}$$

or

$$RMS = \left[\frac{y_1^2 + y_2^2 + \cdots + y_n^2}{n} \right]^{1/2} \tag{11.3}$$

FIG. 11.3 Surface roughness (a) measured with a stylus instrument, and (b) the recorded trace (note larger vertical magnification).

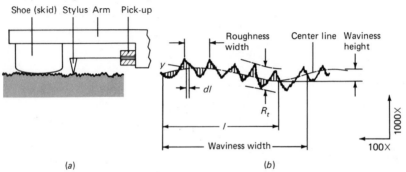

The RMS value (now frowned upon) is rather closely related to the arithmetical or center-line average (CLA $\simeq$ 1.11 $\times$ RMS for a sine wave), and, for a technical surface, the relationship between the various values is fairly well defined (Table 11.1). Convenient units of measurement are the microinch (μin) and micrometer (μm) or nanometer (nm):

1 μin = 0.025 μm = 25 nm = 250 Å

1 μm = 40 μin

Preferred series of roughnesses that form a geometric progression have gained wide acceptance and are used in Fig. 11.7.

Surfaces may exhibit a directionality (*lay*) characteristic of the finishing process (Fig. 11.4). It is *random* and uniform after lapping or shotblasting, and random but not uniform when it was produced by casting or a deformation process such as forging. *Directional finishes* may be regular (periodic, after turning or grinding), or fairly irregular (after rolling or extrusion). The finer details of surface roughness are superimposed on larger-scale periodic or nonperiodic varations (*waviness,* Fig. 11.3). When the surface roughness is measured, the waviness is usually filtered out by electronic processing of the stylus signal, although the allowable waviness is specified (and measured) when it is functionally important.

On drawings, the *roughness limits* are given by a check mark written over the line to which the roughness designation applies (Fig. 11.5). A single roughness number indicates an upper limit, below which any roughness is acceptable; if a minimum roughness is also required, two limits are shown. The waviness, when important, is limited by a number over the horizontal line of the check mark, and the lay by a symbol placed under it.

There is a close relation between roughness and tolerances. In general, tolerances must be in excess of R_t (and waviness, if any), unless the fit is a force-fit and the surface roughness can be at least partially smoothed out in the fitting process. Remembering that $R_t \simeq 10 R_a$, a roughness value

TABLE 11.1 APPROXIMATE RELATIONSHIP OF SURFACE ROUGHNESS VALUES

Type of surface	RMS/CLA	R_t/R_a
Turned	1.1	4–5
Ground	1.2	7–14
Lapped	1.4	7–14
Random	1.25	8

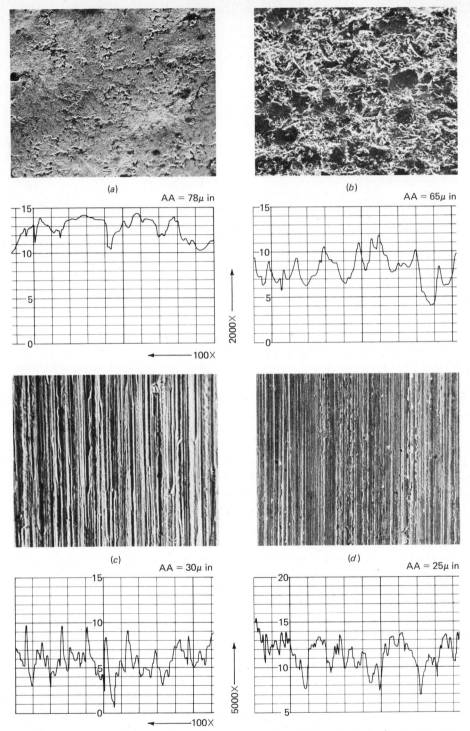

FIG. 11.4 Scanning electron microscope images (all 100X) of typical manufactured surfaces and the corresponding surface profile (Talysurf) recording. Random: (*a*) permanent-mold cast and (*b*) shot-blasted. Directional: (*c*) cold-rolled and (*d*) ground.

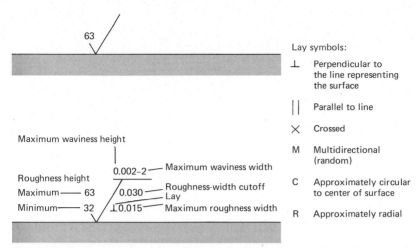

FIG. 11.5 Symbols used to describe surface finish characteristics (the example is given in the inch system, with roughness in units of μin R_a).

R_a of 125 μin (3.2 μm) would exceed a tolerance of 0.001 in (0.025 mm). Therefore, a surface roughness R_a of maximum 23 μ in (0.8 μm) should be used for such a tight tolerance.

It must be remembered that the same numerical RMS or R_a value may be obtained on surfaces of greatly different profile, and highly localized troughs add very little to the average values. Therefore, averages are very often inadequate to describe the surfaces for specific applications, and the problem of surface characterization remains a challenge. Nevertheless, the manufacturing process must be capable of providing a surface suitable for the intended function, and further descriptive terms can be and often are used to elaborate on the required finish.

11.1.3 Shape and Location Deviations

For a part to function properly with respect to other components, it is often necessary to place further restrictions on shape and location.

A simple example would be a rotating hollow shaft (Fig. 11.6). It will run in journal bearings at its ends and support a force-fitted flywheel at its center; appropriate tolerances are given for the A and C portions of the shaft. The nonmating surfaces B could have greatly relaxed tolerances, and dynamic balance could be obtained by drilling holes in the flywheel at appropriate points. It is usually preferable to apply tighter tolerances to B (since it will be machined with the rest, anyway). Of equal importance is specification of hole *concentricity:* maintenance of diametral tolerances alone would not necessarily assure a uniform wall thickness. The same considerations apply to parts not shown in Fig. 11.6; the journal bearings

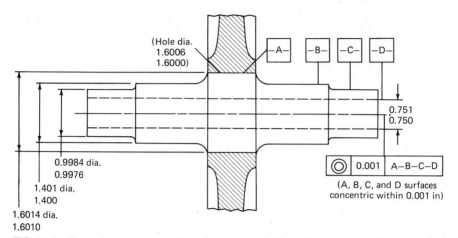

FIG. 11.6 Example to show functionally important dimensions to be held in manufacturing (dimensions in inches).

would need restrictions on OD (for fit into the machine frame), on ID (for assuring the running clearance), and on concentricity or wall-thickness variation (to assure alignment).

On a part of nonrotational symmetry, other qualities—such as flatness or freedom from bow—may be specified.

11.1.4 Economic Aspects

Each manufacturing process is capable of producing a part to a certain finish and tolerance range without extra expense. Some very general guidance is given in Fig. 11.7; tolerances apply to a 1 in (25 mm) dimension and are not necessarily scalable to larger or smaller dimensions for all processes. The typical ranges of individual processes are given by the curved lines; ranges common to several processes are indicated by commas placed between the names of processes. In any given practical instance, the actual (commercial or special) tolerance ranges offered by various industry associations or individual companies should be considered.

If tighter than usual tolerances or smoother surfaces are required, the cost inevitably rises. To take a hot-forged part as an example, the process can be tightened up by more careful preforming in several die cavities, using the finishing die only for a limited amount of work, or by subsequent additional operations such as machining or grinding. Experience shows that cost tends to rise exponentially with tighter tolerances and surface finish (Fig. 11.8), and a cardinal rule of the cost-conscious designer is to specify the loosest possible tolerances and coarsest surfaces that still fulfill the intended function. The specified tolerances should, if possible, be within the range obtainable by the intended manufacturing process (Fig. 11.7) so as to avoid separate finishing operations.

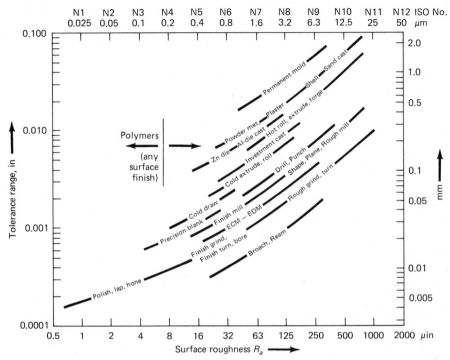

FIG. 11.7 Approximate values of surface roughness and tolerance typically obtained in various processes. Tolerances are for 1-in (25-mm) diameter; for castings and forgings, they apply across the parting line. ECM, electrochemical machining; EDM, electrodischarge machining.

FIG. 11.8 The effect of tolerance and surface finish on cost of (a) turning, (b) milling, and (c) surface grinding. (L. J. Bayer, ASME paper 56-SA.9, 1956.)

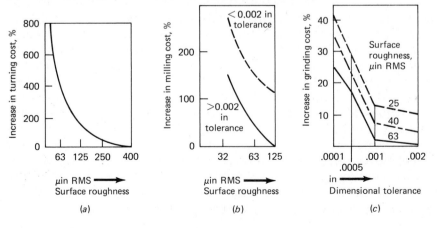

11.2 SHAPE AND DIMENSIONAL LIMITATIONS

From the discussion of various processes, it must be obvious that all of them are subject to limitations on *shape complexity* and *minimum dimensions,* and that these limitations are highly variable for different workpiece materials.

11.2.1 Shape Limitations

The systematic *classification of shapes* (typology) is a developing science, and, as noted in conjunction with group technology (Sec. 10.4.3), no universally applicable and accepted system exists yet. This also means that computerization of process selection is still in its infancy, and creative thinking needs be coupled with an intimate knowledge of various processes if a reasonable choice is to be made. The following material is presented to aid in this initial, approximate process selection.

Before a shape can be described even in general terms, a terminology has to be accepted. Some *basic shapes* are given in Fig. 11.9, together with symbols for defining them. Some additional features that have a meaning only in context with a given process must also be clarified:

1 An *undercut* means a shape element perpendicular (or at some angle) to the main machine or tool motion (Fig. 11.10a).

2 A *reentrant shape* is formed in the main machine-movement direction but opposite the original movement (Fig. 11.10b).

3 A *cross-positioned* element is perpendicular to the main motion (Fig. 11.10c).

With the aid of these definitions, the shape possibilities of various processes are described in Tables 11.2 to 11.7. Much of this is, of course, obvious to anyone familiar with the processes; these tables will hopefully alert the beginner to the alternative processes that could be considered for producing a given shape. To take but one example, a spool shape (Group S3, T2, or F3; Fig. 11.11a) could be machined from the solid or a tube; cast with a horizontal core (Fig. 11.11b) or with a vertical core and a ring-shaped insert core (Fig. 11.11c); forged in the horizontal position with the hole filled out (Fig. 11.11d) or in the vertical position with the hole pre-forged and the outer groove filled out (Fig. 11.11e); forged from a tube by upsetting two flanges in a special tooling; ring-rolled; swaged; centrifugally cast; bent from a U-shaped section and welded; made of a tube with flanges welded to the ends (on the outer surface or the end faces, Fig. 11.11f); friction-welded; made by powder metallurgy (with isostatic compaction in a flexible die, or compaction in a multipiece permanent die); or by a combination of several of these methods. The *optimum method* de-

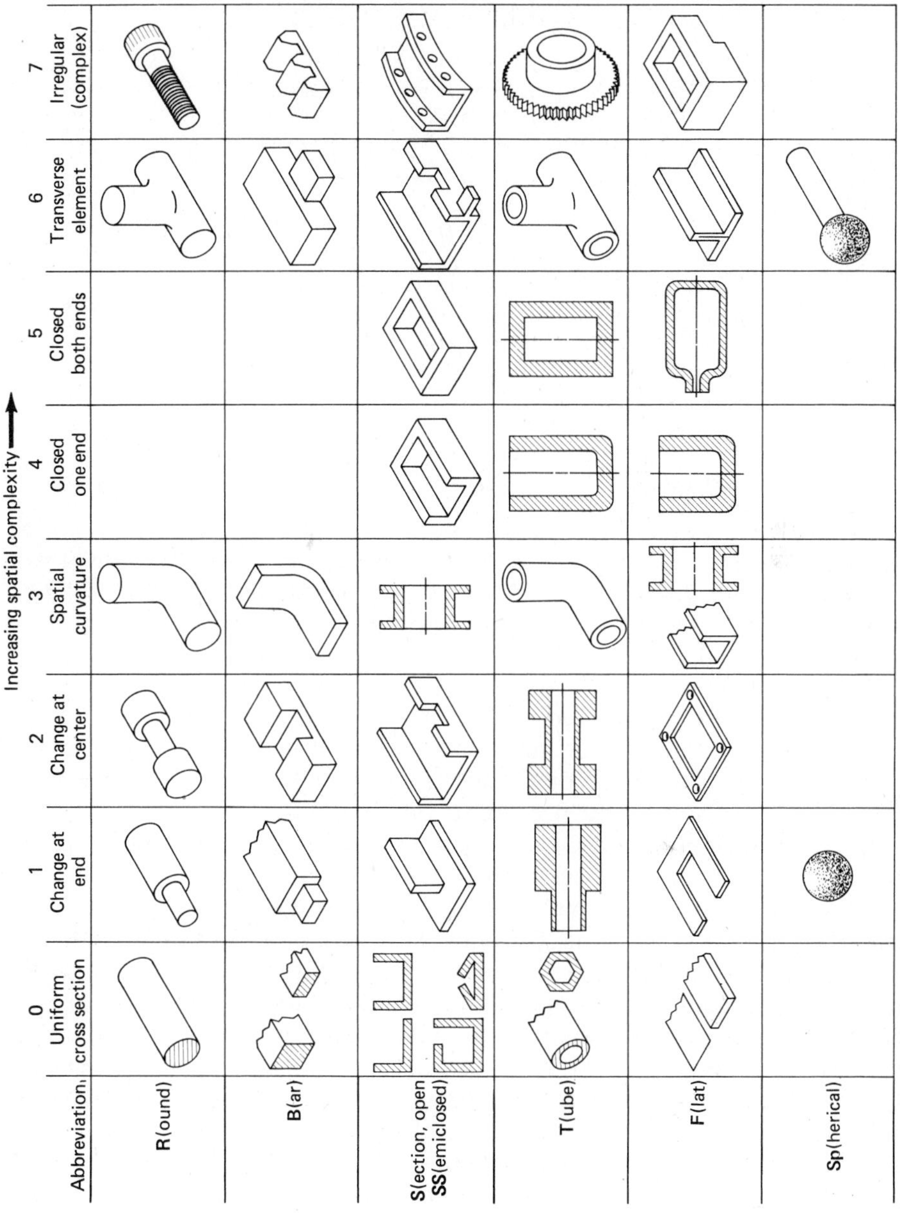

FIG. 11.9 Some basic workpiece shapes.

Abbreviation	0 Uniform cross section	1 Change at end	2 Change at center	3 Spatial curvature	4 Closed one end	5 Closed both ends	6 Transverse element	7 Irregular (complex)
R(ound)								
B(ar)								
S(ection, open) SS(emiclosed)								
T(ube)								
F(lat)								
Sp(herical)								

Increasing spatial complexity →

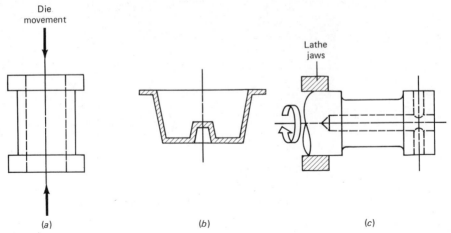

FIG. 11.10 Some examples of shape complexities: (*a*) undercut, (*b*) reentrant shape, and (*c*) cross-positioned hole.

pends on a great many factors, among them also the size, wall thickness, wall-thickness ratio, and slenderness (length-to-diameter ratio) of the part.

11.2.2 Size Limitations

The *maximum size producible* by any one technique is most often limited simply by the availability of large-size equipment. In some processes, however, there are limitations due to process conditions themselves. Thus, a casting mold may not stand up to the excessive solidification times imposed by very heavy walls, or a welding process may be limited to a maximum metal thickness if only single-pass welding is permissible.

More frequently, however, the limitation is on the *minimum producible size* or *wall thickness*. Thus, the wall thickness of a casting is limited by the fluidity of the metal, and that of a forging by the die pressures developed with increasing D/h ratios. The limits given in Tables 11.2 to 11.7 reflect both practical and fundamental limitations and as such are not inflexible. Thinner, smaller, and larger parts may well be made, but usually under special circumstances and at extra expense. With the development of technology, these limits are continually pushed outward.

Minimum attainable wall thicknesses are also a function of the extent (width) of the thinnest section, and more detailed guidance is given in Fig. 11.12.

11.3 MANUFACTURING COSTS

In choosing the optimum manufacturing process, no decision—not even a preliminary one—can be made without considering at least the major cost

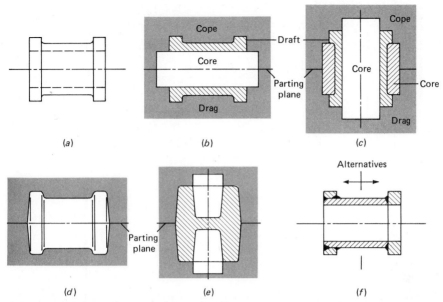

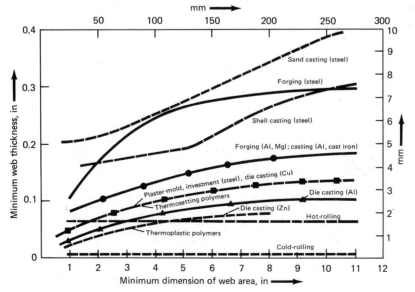

FIG. 11.11 Some possible methods of making a spool-shaped part: (a) machining from bar or tube, (b) casting in a horizontal position, (c) casting in a vertical position, (d) forging in a horizontal position, (e) forging in a vertical position, and (f) welding.

FIG. 11.12 Minimum web thicknesses normally obtained in various processes.

TABLE 11.2 GENERAL CHARACTERISTICS OF CASTING PROCESSES

	Casting process					
Characteristics	Sand	Shell	Plaster	Investment	Permanent mold	Die
Part:						
Material (casting)	All	All	Zn to Cu	All	Zn to cast iron	Zn to Cu
Porosity*	C–E	D–E	D–E	E	B–C	A–C
Shape†	All	All	All	All	Not T3, 5, F5 with solid core	Not T3, 5, F5
Size, kg	0.01–300,000	0.01–100	0.01–1000	0.01–10(100)	0.1–100	<0.01 to 50
Min. section, mm	3–6	2–4	1	1	2–4	0.5–1
Min. core dia., mm	4–6	3–6	10	0.5–1	4–6	3 (Zn: 0.8)
Surface detail*	C	B	A	A	B–C	A–B
Cost:						
Equipment*	C–E	C	C–E	C–E	B	A
Die (or pattern)*	C–E	B–C	C–E	B–C	B	A
Labor*	A–C	C	A–B	A–B	C	E
Finishing*	A–C	B–D	C–D	C–D	B–D	C–E
Production:						
Operator skill*	A–C	C	A–B	A–B	C	C–D
Lead time	Days	Weeks	Days	Hours–weeks	Weeks	Weeks–months
Rates (piece/h·mold)	1–20	5–50	1–10	1–1000	5–50	20–200
Min. quantity	~1–100	~100	~10	~10–1000	~1000	~100,000

* Comparative ratings, with A indicating the highest value, E the lowest.

† From Fig. 11.9.

TABLE 11.3 GENERAL CHARACTERISTICS OF BULK DEFORMATION PROCESSES

Characteristics	Deformation process						
	Hot forging		Hot extrusion	Cold forging, extrusion	Shape drawing	Shape rolling	Transverse rolling
	Open die	Impression					
Part:							
Material (wrought)	All	All	All	All	All	All	All
Shape†	R0-3; B; T1,2; F0; Sp6	R; B; S; T1,2,4; (T6,7); Sp	R; B; S; SS, T1,4; Sp	As hot	R0; B0; S0; T0	R0; B0; S0	R1-2,7; T1-2; Sp
Size, kg	0.1–200,000	0.01–100	1–500	0.001–50	10–1000	10–1000	0.001–10
Min. section, mm	5	3	1	(0.005) 1	0.1	0.5	1
Min. hole dia., mm	(10) 20	10	20	(1) 5	0.1	0.5	
Surface detail*	E	C	B-C	A-B	A	A-B	A-C
Cost:							
Equipment*	A-D	A-B	A-B	A-C	B-D	A-C	A-C
Die*	F	B-C	C-D	A-B	C-D	A-C	A-C
Labor*	A	B-D	B-C	C-E	C-E	C-E	C-E
Finishing*	A	B-C	C-D	D-E	E	E	D-E
Production:							
Operator skill*	A	B-C	C-E	C-E	D-E	B	B-C
Lead time	Hours	Weeks	Days–weeks	Weeks	Days	Weeks	Weeks–months
Rates, pieces/machine	1–50/h	10–300/h	10–100/h	100–10,000/h	10–2000 m/min	20–500 m/min	
Min. quantity	1	100–1000	1–10	1000–100,000	1000 m	50,000 m	1000–10,000

* Comparative ratings with A indicating the highest value, E the lowest.   † From Fig. 11.9.

TABLE 11.4 GENERAL CHARACTERISTICS OF SHEET METALWORKING PROCESSES INCLUDING BAR AND TUBE BENDING

Characteristics	Forming process					
	Blanking	Bending	Spinning	Stretching	Deep drawing	Rubber forming
Part:						
Material (wrought)	All	All	All	All	All	All
Shape†	F0-2; T7	R3; B3; S0,3,7; SS; T3; F3,6	T1,2,4,5; F4,5	F4; S7	T4; F4,7	As blanking, bending, deep drawing
Max. thickness, mm	>10	>100	>25	>2	>10	2
Min. hole dia.	½-1 thickness				<3	50 (for h = 1 mm Al)
Cost:						
Equipment*	B-D	C-E	B-D	B-C	A-C	A-C
Die*	B-D	B-E	B-D	A-C	A-B	C-D
Labor*	C-E	B-E	B-C	B-E	C-E	A-D
Finishing*	D-E	D-E	D-E	C-E	D-E	C-E
Production:						
Operator skill*	D-E	B-E	B-C	B-E	B-E	C-E
Lead time	Days	Hours-days	Days	Days-months	Weeks-months	Days
Rates, pieces/h	10^2-10^5	10-10^4	10-10^2	10-10^4	10-10^4	10-10^2
Min. quantity	10^2-10^4	1-10^4	1-10^2	10-10^5	10^3-10^5	10-10^2

* Comparative ratings, with A indicating the highest value, E the lowest.

† From Fig. 11.9.

TABLE 11.5 GENERAL CHARACTERISTICS OF POLYMER PROCESSES

| Characteristics | Manufacturing process | | | | | Thermoforming | |
| | Molding | | | Extrusion | Casting | | |
	Compression	Transfer	Injection			Vacuum	Pressure
Part:							
Material	Plastics, glass	Plastics	Plastics	Thermoplastics	Plastics, glass	Thermoplastics	Thermoplastics, glass
Preferred Shape†	Thermosets All but T3,5, 6 and F5	Thermosets All but T3,5, and F5	As transfer	As transfer	Thermosets All	T4; F4,7	T4,5; F4,5,7
Min. section, mm	(0.8) 1.5	(0.8) 1.5	0.4 thermoplastic, 1 thermoset	0.4	4	<1	<1
Cost:							
Equipment*	B–C	B–C	A–C	A–B	D–E	B–D	B–D
Tooling*	A–C	A–C	A–C	A–C	B–E	B–C	B–C
Labor*	C–E	C–E	D–E	D–E	A–C	B–E	B–E
Production:							
Operator's skill*	D–E	D–E	D–E	D–E	B–E	B–E	A–E
Lead time	Weeks	Weeks	Weeks	Weeks	Days	Days–weeks	Days–weeks
Cycle time (s)	20–600	10–300	10–60	10–60		10–60	(1) 10–60
Min. quantity	100–1000	100–1000	1000	10,000	1	10–1000	10–1000

* Comparative ratings with A indicating the highest value, E the lowest.

† From Fig. 11.9.

TABLE 11.6 GENERAL CHARACTERISTICS OF MACHINING PROCESSES

	Machining process						
Characteristics	Lathe turning	Automatic screw machine	Shaping, planing	Drilling	Milling	Grinding	Honing, lapping
Part:							
Material Preferred	All†	All† Free-machining	All†	All† Free-machining	All† Free-machining	All Hard	All Hard
Shape*	R0-2.7; T0-2,4,5:Sp	As turning	B; S0-2; F0	T0	B; S; SS; F0-4,7	As turning, shaping, milling	R0-2; T0-2, 4-7; F0-2: Sp
Min. section, mm	<1 dia.	<1 dia.	<2	0.1 (hole dia.)	<1	<0.5	<0.5
Surface configuration	Axially symmetrical	As turning	Straight generatrix	Cylindrical	Three-dimensional	All (mostly flat, axially symmetrical)	Flat, cylindrical, three-dimensional
Cost:							
Equipment‡	B-D	A-C	B-D	D-E	A-C	A-C	B-D
Tooling‡	D-E	A-D	D-E	D-E	A-D	B-D	A-E
Labor‡	A-C	D-E	B-D	B-D	A-B	A-E	B-D
Production:							
Operator skill‡	A-C	D-E	B-D	B-E	A-B	A-D	C-E
Setup time‡	C-D	A-C	C-D	C-E	A-C	B-D	C-E
Rates (pieces/h)	1-50	10-500	1-50	10-500	1-50	1-1000	10-1000
Min. quantity	1	500	1	1	1	1	1

* From Fig. 11.9.

† Except most ceramics.

‡ Comparative ratings with A indicating the highest value, E the lowest.

TABLE 11.7 GENERAL CHARACTERISTICS OF WELDING PROCESSES

Characteristics	Welding process					Electron beam	Oxyacetylene
	Arc welding						
	Shielded metal	Flux cored	Submerged	Gas-metal	Gas-W		
Part (assembly)							
Material	All but Zn	All steels	All steels	All but Zn	All but Zn	All but Zn	All but refractory metals
Preferred	Steels	Low-C steels	Low-C steels	Steels; non HT Al; Cu	All but Zn	All but Zn	Cast iron, steels
Thickness, min., mm	(1.5) 3	1.5	5	0.5	0.2	0.05	0.6
Single pass max.	8–10	3–6	40	5	5	75	10
Multiple pass max.	>25	>15	>200	>25	>6		>20
Unequal thickness	Difficult	Difficult	Very difficult	Difficult	Difficult	Easy	Difficult
Distortion*	A–B	A–C	A–B	B–C	B–C	C–E	B–D
Jigging needed	Minimum	Minimum	Full	Variable	Variable	Full	Minimum
Deslagging for multipass	Yes	Yes	Yes	No	No		No
Current: type	Alternating or direct	Direct (reverse polarity)	Alternating or direct	Direct (reverse polarity)	Alternating or direct (straight polarity)		
Volts	40 or 70↓	40 to 70↓	25–55	20–40 or 70↓	60–150	30–175 kV	
Amperes	30–800	30–800	300–2500	70–700	100–500	0.05–1	
Costs:							
Equipment*	D	B–D	B–C	B–C	B–C	A	D–E
Labor*	A	A–D	B–D	A–C	A–C	A–D	A
Finishing*	A–B	A–C	A–C	B–D	B–E	C–E	A
Production:							
Operator skill*	A	A–D	C–D	A–D	A–D	A–D	A
Welding rate, m/min	(1–6 kg/h)	0.02–1.5	0.1–5	0.2–15	0.2–1.5	0.2–2.5	(0.3–0.6 kg/h)
Operation	Manual	All	Automatic	All	All	All	Manual

* Comparative ratings, with A indicating the highest value, E the lowest.

elements. A complete discussion of *manufacturing cost analysis* is outside the scope of this book and only an outline of the approach can be given here.

Before any meaningful calculation can be made, it is essential that at least the major *production sequences* and their technical feasibility be established. This also allows definition of the *starting* and *finished shapes*.

11.3.1 Operating Costs (Direct Costs)

Direct costs can be clearly allocated to the product, and they are proportional to the number of units produced.

Material costs

The *net material cost* is the cost of purchasing the *raw material* (whether it be a casting, forging, rolled section or plate, powder, or any other starting material), less the value of the *scrap* produced. Thus the weight of the purchased starting material, W_0 (including the weight loss in cutting off, etc.), and of the finished part, W_f, is determined. If the unit price of the starting material is C_0 and that of the scrap is C_s (separated into heavy and light scrap if their resale value is widely different), the material cost per piece is

$$W_0 C_0 - (W_0 - W_f)\, C_s \tag{11.4}$$

In general, a process that generates less scrap, or generates it in a more valuable (usually heavier) form, is more economical.

Labor costs

With a possible process sequence and the appropriate equipment settled, the *number* and *skill of operating personnel* can also be predetermined. From experience, time studies, or a breakdown of operator functions into identifiable physical and mental action elements, the time required for completion of one piece can be calculated. Strictly speaking, the *net production time* is the time period during which the material is actually shaped or processed; thus in machining (Sec. 8.7.3) the net cutting time t_c is that during which material is actually removed. However, in other processes where the truly productive time is only a fraction of a longer but essential cycle—for example, in a press—the total cycle time is usually taken as the productive time. Whether the time required for moving the material from the floor to the machine and off again is regarded as productive or nonproductive is a matter of philosophy (a truer picture of productivity is obtained if it is classified as nonproductive). In the cost analysis, it usually appears in the total or floor-to-floor time, and is charged at direct labor rates, especially if the movement is performed by the machine-tool operator personally.

When *energy costs* are a significant fraction of the total cost, they too are allocated directly to production units, as are tools and dies if used only for that particular purpose.

11.3.2 Indirect Costs

As seen in the example of loading and unloading, the distinction between direct and indirect costs may be a fuzzy one. *Indirect costs* arise from the functions and services that contribute to the efficient performance of the actual production process. Traditionally, they include indirect labor (including material movement, repair and maintenance, cleaning services, etc.), supervision from foreman to plant superintendent, engineering (manufacturing and industrial engineering, quality control, laboratory, etc.), research and development, sales (and sometimes the entire management hierarchy of the company), lighting and heating (and sometimes all energy supplies and materials not directly used in production), office and sales expenses, etc. Some of these are somewhat flexible and vary with production, but the relationship is never as direct as with the operating costs. If properly controlled, indirect activities represent an indispensable part of the total production effort. In the ultimate development, a fully automated manufacturing process would generate no direct labor costs in the classical sense; yet, many of the indirect costs (including programming) obviously have to be regarded as production related. Similarly, development—and even long-range research if properly managed—is a vital part of the production process.

Indirect costs are also called *overhead* or, sometimes, *burden.* Unless properly controlled, indirect cost can indeed become a burden that finally nullifies the productivity gains in direct production. Therefore, breakdown of indirect costs into their elements is essential, and is one of the purposes of CAM, discussed in Sec. 10.2.1. All too often, the actual production process is carefully analyzed, improved, and made more productive, while the indirect cost sector is allowed to grow out of proportion, finally impairing the competitive position of a company or, for that matter, of national economy.

In manufacturing cost estimates the indirect cost may appear as a multiplying factor applied to direct labor costs, as a fixed cost per unit product, or a fixed cost per hour worked on a machine.

11.3.3 Fixed Costs

Fixed costs include the costs of equipment, buildings, and of total facilities in general, taking into account depreciation, interest, taxes, and insurance. For a given capital outlay, the fixed costs depend on the interest rates prevailing at the time of purchase and during the life of the equipment, the tax treatment accorded to investments, and the useful life of the production facility. In general, the life of equipment can be extended, but at the expense of rapidly rising maintenance costs and at the danger of obsolescence, and a replacement decision must sooner or later be made.

In manufacturing-cost estimates, fixed costs are allocated on the basis of anticipated equipment utilization. Thus, if a press was purchased on the premise that it will be used in a two-shift operation, the fixed cost per unit production doubles if only a one-shift operation can be sustained. In the simplest estimating procedure, the fixed cost appears as a machine-hour rate or burden.

11.4 ALTERNATIVE PROCESSES

In the ideal manufacturing situation the designer cooperates with the manufacturing engineer in assuring that the part should have features that make it eminently producible by a selected, optimum process. Indeed, this can happen in the largest mass-producing industries that do both their design and manufacturing in house. In addition to assembly plants, they operate their own manufacturing plants that make parts, often in competition with specialized, independent suppliers. Process selection then follows a formalized procedure.

In one approach, representatives of groups having knowledge of and interest in alternative processes get together in a *brainstorming session.* The basic rule is that no idea is regarded as too silly; no criticism is allowed until all ideas emerge, and only then does the process of whittling down alternatives begin. Those processes that stand concentrated criticism are then evaluated in detail and the optimum is finally chosen, often after detailed scrutiny by estimators.

The same function is performed, but in a much more organized form, by *value engineering* groups, headed by a value engineer who reports directly to higher management. The task of such groups is to review the functional requirements and ease of manufacturing of the part, and attach a value to functions and processes. The group is empowered to initiate design changes that make manufacturing (including assembly) easier. In addition, economic aspects are also fully weighed. It is not unusual that a part should be entirely redesigned, or several parts forming an assembly be made as a single unit, as a result of *value analysis.*

In many smaller companies the situation is quite different. The manufacturing engineer or technologist may have to rely on his own resources, without the benefit of interactions with specialists in other fields. There is then real danger of settling on the first obvious solution, conditioned by experience with a given process. As a minimum, the manufacturing engineer must consider alternatives in a "brainstorming session" with himself.

In many instances, a company performs only assembly operations, or possesses only machining facilities of its own. The raw material is purchased in the form of forgings, castings, sheet-metal stampings, and semifabricated products such as bars, tubes, sections, plate, and sheet. A vital function is then fulfilled by the *purchasing department* that translates a part design into semifabricated product requirements, and often also acts as an intermediary between the designer (or manufacturing specialist) and the supplier. Identification of alternative production methods and location of suppliers of appropriate capacity is essential. Without proper coordination between purchasing and manufacturing, the part is often "hacked out of the solid." While there are instances in which this approach is indeed the most economical, alternatives must always be explored.

First, one would assume that the designer has taken full advantage of the innumerable *standardized* and *semistandardized components* obtainable from specialized producers and thus benefiting from massproduction techniques. Screws, rivets, clips, springs, packings, seals, bearings, gears, etc., are available in almost any justifiable size and material, and should seldom if ever have to be custom made.

Second, geometrically similar parts can often be made identical without sacrificing functional performance, or they may be made into a family suitable for group technology treatment.

Third, in specifying materials, function as well as ease and cost of manufacturing must be considered. Machining costs alone can vary over several orders of magnitude (Table 11.8), but the overall economy cannot

TABLE 11.8 RELATIVE MACHINING COSTS FOR A GIVEN PART
Lathe turning, 60 min tool life

Material	Hardness	Machining cost
7075-T6 Al		10
1020 steel	111 BHN	25
410 stainless steel	163 BHN	40
310 stainless steel	168 BHN	55
Ti-6Al-4V		75
4340 steel	52 R_C	100
Inconel X		170
Inconel 700 (aged)	400 BHN	340

* From *Profile Milling Requirements for Hard Metals 1965–1970*, Report of the Ad Hoc Machine Tool Advisory Committee to the Department of the U.S. Air Force, May 1965.

FIG. 11.13 Spark-plug shell production: (a) hexagonal bar stock for machining, and (b) cold-headed slug for extrusion into (c) shell of (d) thin wall with 70 percent material saving. (Braun Engineering Company, Detroit.)

be determined without considering the alternatives in manufacturing approaches. Habits are difficult to change, but many parts now machined from steel could be cut much faster and cheaper from a fully heat-treated aluminum alloy, even though the starting material is slightly more expensive by volume. A sound decision is possible only if the true function of the part is known and no unnecessary constraints are imposed.

An example of the economies attainable by choosing the proper process is shown in Fig. 11.13. Spark plug shells used to be machined from hexagonal bar stock with a cycle time of 6 s. Today, the starting material is round wire (bar), sheared and upset (preformed) in a cold header at a rate of 250–400 per minute to form a slug which, after application of a phosphate/soap lubricant system, is cold extruded at a rate of 160 per minute in a single operation. With only minimum machining, at a rate of 60 per minute, the shell is finished, saving 70 percent of the material once used in machining.

11.5 SUMMARY

In the preceding chapters we concentrated on the unit processes of manufacturing and, to a much more limited extent, on bringing these processes together into a viable system. In the final analysis, the success of all this hinges on the product designer; while striving to satisfy functional requirements, the designer must be aware of manufacturing implications. This demands at least some familiarity with processes and process limitations as affected by materials, and a willingness to discuss the design with the manufacturing specialist. Some basic rules are always worth considering:

1 Specify the broadest tolerance and coarsest surface finish that still satisfy functional requirements.

2 Subject the design to value analysis, formally or informally. Start with the real functions, strip off all unnecessary constraints, and seek alternatives with the early involvement of people with knowledge of various manufacturing techniques (as opposed to the narrow specialists).

3 Remember that a seemingly minor change in shape, wall thickness, or radius can make the part suitable for manufacturing by a different, more economical technique.

4 Look at the relation of parts to each other; it may well happen that several parts are much easier to make as a single unit.

Manufacturing is a peculiar blend of art, science, and economics, with very broad social implications. It would be unreasonable to expect the designer or manufacturing engineer to consider formally all these elements whenever a detail decision is made. They must, nevertheless, guide the general approach to the profession; without sound scientific foundations, manufacturing remains but a collection of myriads of isolated rules; without practical sense, the best theory and research will fail to make impact; and without conscious attention to cost, the best product and process will lose out against competition. The Industrial Revolution was but a consequence of manufacturing growth; advancing technologies and the impact of their products may well bring comparable social changes within the next few decades.

Further Reading

DATA ON MATERIALS:
    CLAUSER, H. R. (ed.): *The Encyclopedia of Engineering Materials and Processes,* Reinhold, New York, 1963.

PARKER, E. R.: *Materials Data Book for Engineers and Scientists,* McGraw-Hill, New York, 1967.

SMITHELLS, C. J.: *Metals Reference Book,* 4th ed., Butterworths, London, 1967.

Materials Selector, special annual issue of *Materials Engineering,* Reinhold, Stamford, Conn.

Metals Handbook, 8th ed., American Society for Metals, Metals Park, Ohio, vol. 1, "Properties and Selection of Metals," 1961.

METHODS:

EARY, D. F., and G. E. JOHNSON: *Process Engineering for Manufacturing,* Prentice-Hall, Englewood Cliffs, N.J., 1962.

JELEN, F. C. (ed.): *Cost and Optimization Engineering,* McGraw-Hill, New York, 1970.

MILES, L. D.: *Techniques of Value Analysis and Engineering,* 2d ed., McGraw-Hill, New York, 1972.

MUDGE, A. E.: *Value Engineering,* McGraw-Hill, New York, 1971.

WAGE, H. W.: *Manufacturing Engineering,* McGraw-Hill, New York, 1963.

WILSON, F. W., and P. D. HARVEY (ed.): *Manufacturing Planning and Estimating Handbook,* American Society of Tool and Manufacturing Engineers, McGraw-Hill, New York, 1963.

Example of Process Selection

11.1　Once the dimensions, surface finish, and material of a workpiece are specified, process selection is limited in its scope. Assuming that no communication (at least no meaningful communication) with the designer is possible, the choices can be readily identified as shown on the example of a simple flange (see the part defined by solid lines in figure for Example 11.1).

If production alternatives are to be considered by one person, the danger of discarding profitable alternatives is best avoided by following the *process of elimination;* that is, one first considers all alternatives,

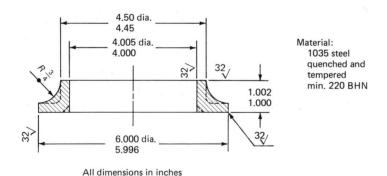

Material:
1035 steel
quenched and
tempered
min. 220 BHN

All dimensions in inches

(Broken lines show alternative design)

throws out those that are obviously impossible, and then concentrates on the feasible ones.

Casting. In the case of the flange, only casting processes capable of withstanding the casting temperatures of steel need be considered at all. This immediately excludes die casting (in anything but graphite molds) and plaster-mold casting (Table 11.2).

None of the casting processes can satisfy the tolerances and surface finish specified for the end faces and the internal surface; therefore, machining will be inevitable, and the lowest-cost process capable of producing the shape is acceptable (Fig. 11.7). Investment casting can be rejected because shape complexity does not justify the extra expense, but sand casting, shell molding or permanent-mold casting should all be investigated.

Further narrowing of the choices requires some economic analysis and, in all likelihood, shell molding will emerge as the most suitable process for the given dimensions (Fig. 11.12). With the aid of data from appropriate books or industry publications (see Further Reading for Chap. 3 and 11), the casting can be designed.

Deformation Processes. Of the deformation processes, tube rolling or extrusion could produce thick-walled tubes for further machining; these are, however, essentially semifabrication processes providing an input to machining. More directly, one should consider those deformation processes that yield a semifinished part.

The surface finish requirement is within attainable range for cold-deformation processes (Fig. 11.7) such as sheet-metal forming and cold forging-extrusion. Subsequent coining of a hot-worked part is also possible and a part finished exclusively by deformation techniques could be economical.

Hot forging is a very reasonable proposition, particularly on a hot upsetter which would upset a flange and then pierce out the bar, without creating scrap. The feasibility of cold forging can be determined only after making pressure and force calculations (Secs. 4.4.1 and 4.4.2). Piercing (or back-extrusion) is almost certainly uneconomical because of the inevitable scrap loss in the bore. Piercing of a smaller ring followed by ring rolling, however, may be economical in larger quantities.

Machining. The basic choice is between machining from a solid bar, a tube, or a cast or deformed part. The decisive factor is cost. The solid bar is by far the cheapest and can be machined on a screw-type multi-spindle automatic machine, but much scrap of low value is produced. The higher price of tubing may or may not be offset by reduced input weight and scrap losses. Cast or formed parts must be machined on a chucking automatic and could be turned around, although a much better solution would be machining of one face and the bore, followed by finishing the other face on a vertical-spindle, rotary table grinder.

Redesign. The problem would change entirely if discussions with the designer would reveal, say, that the functions of the part call simply for a minimum YS of 80,000 psi, and a minimum wall thickness of 0.25 in without need for the sharp edge (broken lines in the figure for Example 11.1).

A review of materials satisfying the minimum yield strength requirements brings to light as possible candidates: 1050 steel (cold worked), high-strength low-alloy steel (hot rolled in a controlled manner), nodular cast iron (quenched and tempered), malleable iron, and cast steel grades (quenched and tempered), as well as powder-metallurgy products.

Of the casting grades, nodular cast iron is likely to be the most economical, whereas 1050 and HSLA steels open the possibility of proceeding with cold or warm forward-extrusion and, since the sharp edge is no longer required, with sheet-metalworking techniques. Whether blanking of rings followed by flanging, or blanking of circles followed by drawing and punching out the base is more practical will be determined by the elongation capacity of the material (Sec. 5.3). Sheared edges will not satisfy surface finish and tolerance requirements and machining is still necessary, but it could now be grinding as well as turning.

The very simple shape of this part actually complicates the selection of an optimum process because the final choice will hinge on a rather detailed economic analysis. Unless the quantities are very large, one would concentrate on hot upsetting, machining from the solid or tube, powder metallurgy and perhaps also cold forging or drawing. Parts of more complex shape often limit the number of feasible processes (Tables 11.2–11.7) and simplify the search for the optimum.

Problems

11.1 Trace the coordinate system of Fig. 11.7 on a transparent paper, then construct a line corresponding to a tolerance of 20 R_a. From Fig. 11.7, judge whether a 16-μin R_a surface finish specified for a journal of 1.0000 $\pm$ 0.0004-in diameter can be produced (a) by grinding; (b) by cold drawing. (c) If the answer to (a) and/or (b) is yes, is the chosen finish reasonable? (Refer to the line constructed above.)

11.2 Collect five types of metal cans (at home and at the grocer's), selected from among containers for fizzy drinks, fruit juices, baby food, canned meat, and sardines. Make sure you have samples of two- and three-piece containers. Carefully section them and investigate their structure. Write an essay describing their method of manufacture as deduced from the evidence available.

11.3   What processes could you envisage for making (*a*) a high-pressure gas cylinder; (*b*) a CO_2 cartridge?   (*c*) What other processes could one consider if the part would be of shape F5 (Fig. 11.9) without the need to sustain internal pressures?

11.4   A component is found to fail in service by fatigue.   Laboratory examination reveals the presence of residual surface tensile stresses. Describe the methods that could be considered for changing the residual stresses to compressive, assuming that the original cause of surface tensile stresses cannot be eliminated.

11.5   Obtain two samples each of socket-head cap screws, and recessed- and slotted-head machine screws.   With the aid of a magnifying glass or, preferably, a stereomicroscope, investigate the heads for evidences of the method of manufacture.   Report your findings, including any defects discovered, in a professional manner.

11.6   Carry out the tasks described in Prob. 11.5, but collect six different self-tapping screws and include the threaded portion in your investigation.

11.7   Carry out the tasks of Prob. 11.6, but on six wood screws.

A

FLOW STRESS DETERMINATION

The discussion of the calculation of forces (Sec. 4.3.1) emphasized the need for using appropriate flow stress values even for an approximate estimate of forces and power requirements. If reliable data are not available, the flow stress must be determined on the actual workpiece material, under conditions relevant to the deformation process contemplated. In the following, the most useful methods are briefly outlined.

A.1 THE TENSION TEST

A round or flat specimen of standard dimensions (resembling the shape in Fig. 2.5a) is machined. The head is gripped securely but without imposing bending stresses on the specimen, and the tensile force is applied. Tensile testing machines usually record force vs. displacement of the moving head of the machine, resulting in a curve similar to that shown in broken lines in the figure for Prob. 2.5. It is much more preferable to attach a displacement transducer to the specimen and use an X-Y recorder to obtain a force vs. true elongation curve (the full line in the figure of Prob. 2.5).

To evaluate the recordings, the forces are read at, say, 5 points, between initial yielding and the onset of necking (where the curve begins to turn down), and the following data are calculated and tabulated:

1 Width of the starting specimen w_0

2 Thickness of the specimen h_0

3   Original gage length l_0

4   Volume of the specimen between gage-length marks $V = h_0 l_0 w_0$

5   New gage length at the selected points l_1

6   Engineering tensile strain $e_t = (l_1 - l_0)/l_0$

7   Natural strain $\epsilon = \ln (l_1/l_0)$

8   Force recorded at the selected points P_1

9   Instantaneous cross-sectional area $A_1 = V/l_1$

10 Engineering stress at the selected points $\sigma_{eng} = P_1/A_0$

11 True stress at these points $\sigma = P_1/A_1$

The results are then plotted as σ_{eng} vs. e_t or σ vs. e_t or σ vs. ϵ.

There is no point in calculating true stresses beyond the point of necking because the instantaneous minimum cross-sectional area needed is not revealed from length measurements.

The cross-sectional area after fracture A_f can be measured and used to calculate the reduction in area q.

Tensile tests are rather unsatisfactory in the hot-working temperature range because necking begins early, and the stress calculated from a constant volume will be misleadingly low. Furthermore, the required high strain rates call for very special machines. On the whole, it is better and easier to conduct compressive tests.

A.2 THE COMPRESSION TEST

Bulk deformation processes rely on essentially compressive deformation; therefore, properties are also best measured in a compression test. The main problem, then, is friction at the end faces of the specimen (Fig. 4.6). The measured compression force gives an average interface pressure p_a which is in excess of the true flow stress σ_f. To minimize the error, specimens of $(h_0/d_0) > 1$ should be tested with a very good lubricant. Alternatively, a ring-shaped workpiece can be compressed.

The ring-compression test has the advantage that dimensional changes of the ring reveal the magnitude of interface friction. Usually, rings of an OD:ID ratio of 2:1 are compressed between flat platens of smooth, random (e.g., lapped) finish; the height does not influence the shape change and can conveniently be taken as OD/3. A good lubricant is applied and either a number of rings are compressed to various reductions in height and the peak forces noted, or if a recorder is available, one (or, preferably, more) rings are compressed to various (up to 70 percent) re-

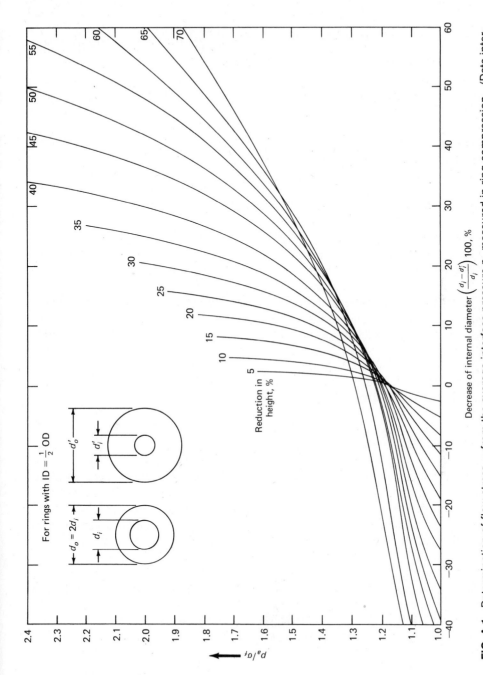

FIG. A.1 Determination of flow stress σ_f from the average interface pressure p_a measured in ring compression. (Data interpolated and plotted from G. Saul, A. T. Male, and V. DePierre: *A New Method for the Determination of Material Flow Stress Values under Metalworking Conditions, Technical Report AFML-TR-70-19*, Air Force Materials Laboratory, WPAFB, Ohio, 1970.)

ductions. The internal diameter of the deformed ring is measured; the change in internal diameter is a function of interface friction. With zero friction the diameter expands as though it would be part of a solid cylinder. With increasing friction, free radial flow is obstructed and the ring begins to flow both out and inward. From the force and the measured change in internal diameter, the flow stress σ_f can be obtained from Fig. A.1.

This method has several advantages. Presses and hammers that provide strain rates typical of real deformation processes are readily available; therefore, both cold- and hot-working flow-stress data can be determined. In testing at elevated temperatures, cooling can be prevented if the test temperatures are low enough to heat the workpiece inside a two-piece container made of a material that is at least 3 times as strong as the workpiece material at the testing temperature. Otherwise, the workpiece must be preheated separately, the anvils heated to a lower temperature, and the effect of cooling at the end faces limited by very brief contact times and lubricants (such as glasses) that also provide heat insulation.

Friction causes barreling of the ring just as of a cylinder, and an average internal diameter should be used for calculations. Preferably, however, a very good lubricant is used whenever possible, because the curves in Fig. A.1 will not give adequate compensation for friction when $p_a/\sigma_f > 1.3$.

A.3   HARDNESS TEST

As indicated by Eq. (4.3), the hardness equals 3 times the flow stress (in consistent units). Because of localized strain hardening, the value thus obtained is somewhat higher than the flow stress of the starting material, but this just provides some extra safety in calculations. The hardness test is the quickest way of determining the flow stress, but suffers from the difficulty that indentation proceeds at very low strain rates. Therefore the hardness value is typical of only very slow deformation processes, but for cold-working the method is reasonably satisfactory.

CONVERSION FACTORS

CONVERSION FACTORS FROM U.S. CONVENTIONAL UNITS TO METRIC (SI) UNITS

Quantity	Symbol	Name	Multiply by	Symbol	Name
	To convert from USCS units			**To obtain SI unit**	
Length	in	inch	*25.4	mm	millimeter
	ft	foot	*0.3048	m	meter
Area	in^2	square inch	*6.4516×10^{-4}	m^2	
Volume	in^3	cubic inch	1.639×10^{-5}	m^3	
	gallon	US gallon	3.785×10^{-3}	m^3	
Time	min	minute	*60	s	second
Velocity	fpm	ft/min	*5.08×10^{-3}	m/s	
Mass	lb	pound	0.4536	kg	kilogram
Acceleration (gravitational)	ft/sec^2 (32 ft/sec^2)		*0.3048 *(9.80665)	m/s^2 m/s^2	
Force	lbf (or lb)	pound force	4.448	N	newton
	tonf	ton force (2000 lb)	8.9	kN	
Stress (pressure)	lbf/in^2	psi	6.895×10^3	Pa	pascal (=N/m^2)
	kips (or kpsi)	1000 psi	6.895	MPa	(or N/mm^2)
Torque (work)	lbf·ft	foot-pound	1.356	N·m	newton-meter
Energy (work)	Btu	British thermal unit	1055	J	joule (=N·m)
	cal	gram calorie	*4.1868	J	
Power	hp	550 ft·lb/sec	745.7	W	watt (=J/s)
Viscosity	P	poise ($dyn·s/cm^2$)	0.1	Pa·s	(or $N·s/m^2$)
Temperature interval	F	Fahrenheit	0.5555	C or K	Celsius degree or kelvin
Temperature	t_F		$(t_F - 32)\,5/9$	t_C	degree Celsius
	t_C		$t_C + 273.15$	t_K or T	absolute degrees

Notes: Exact conversion factors are recorded with an asterisk.

The Celsius degree is often written °C to avoid confusion with C (coulomb)

Most frequently used multipliers:

	Prefix	Symbol
10^6	mega	M
10^3	kilo	k
10^{-3}	milli	m
10^{-6}	micro	μ

The International Committee of Weights and Measures (CIPM) modernized the metric system in 1960. The resulting SI units are now used worldwide in the literature; all industrialized nations have already committed themselves to conversion to the International System (SI).

For a detailed discussion see, for example, *ASME Orientation and Guide for Use of Metric Units*, 3d ed., American Society of Mechanical Engineers, New York, or *The International System of Units*, National Bureau of Standards SP330 (SD cat. no. C13.10:330/2), Government Printing Office, Washington, D.C.

SOLUTIONS TO SELECTED PROBLEMS

Chapter 2

2.1 $r = 1.51$
2.5   (c) $n = 0.125$
(f) Elastic deflection of tensile testing machine
(g) UTS $= 197$ MPa; $e_f = 17\%$

Chapter 3

3.1   (a) (b), (c), (f): 2% Si; (d), (e), (g): 12% Si
3.5   (a) $e = 0.0046$; (b) $\dot{e} = 7.6 \times 10^{-6}$/s
(c) $\sigma_f = 37$ MPa; (d) $\sigma = 580$ MPa
3.7   75 tonf/ring

Chapter 4

4.1 $P_c = 700$ kN
4.2   (a) $p_a = 233 \times 10^3$ psi; $P_a = 38$ tonf
(b) $p_a = 485 \times 10^3$ psi; $P_a = 80$ tonf
4.3   (a) $d_o < 40$ (say, 38 mm); $l_o = 84.5$ mm
(b) $p_1 = 1220$ MPa
(c) $p_1 = 440$ MPa; $p_i = 714$ MPa (or up to 1190 MPa)
4.6   (a) $\sigma_f = 76 \times 10^3$ psi; (b) $p_r = 183 \times 10^3$ psi; (c) yes
4.7   (a) $P_r = 25.7$ kN; (b) 1.85 kW
4.10   (a) $p_i = 1488$ MPa (possibly up to 2480 MPa)
4.13 $d_o = 0.375$ in
4.14   (a) $\sigma_{fm} = 305$ PMa; (b) $P_{dr} = 3.36$ kN; (c) no

Chapter 5

5.1 $P_s = 85 \times 10^3$ lbf
5.4 (a) $R_b = 0.06$ mm; (c) $P_b = 388$ kN
5.6 (a) 1180 lbf; (b) $P_d = 70 \times 10^3$ lbf; (c) 4.2 in
5.11 (a) $P_s = 1680$ kN; (b) $P_s = 48$ kN

Chapter 6

6.5 7940 safe, 7740 marginal, others fracture
6.6 0.417 and 0.265 mm

Chapter 7

7.2 (a) 8.2 g/cm³; (b) 92.13%; (c) 7.87%
7.3 $d = 29.16$ mm; $h = 48.6$ mm
7.4 (a) 27.1%; (b) 13.3%; (c) 8.68 × 17.34 × 130 mm; 19.54 cm³
7.6 (b) 942 tonf; (c) no

Chapter 8

8.2 $v_r = 2.54$ m/s
8.3 (c) 9.43 min
8.4 (a) $v = 63$ fpm; (b) 963 rpm; (c) $f = 0.025$ in/rev;
(d) 2.4 in/min; (e) 0.34 hp
8.6 (c) $t_c = 0.625$ min
8.11 (a) 125 passes; (b) $t_c = 2.43$ min; (d) 0.48 hp
8.12 Roughing 2.77 hr, finishing 5.6 hr

Chapter 9

9.3 (f) 0.103 kWh/20 spot welds
9.5 (b) 130 in/h

INDEX

Page numbers in *italic* indicate tables.